Lecture Notes in Mathematics 2079

For further volumes:
http://www.springer.com/series/304

Péter Major

On the Estimation of Multiple Random Integrals and *U*-Statistics

Springer

Péter Major
Alfréd Rényi Mathematical Institute
Hungarian Academy of Sciences
Budapest, Hungary

ISBN 978-3-642-37616-0 ISBN 978-3-642-37617-7 (eBook)
DOI 10.1007/978-3-642-37617-7
Springer Heidelberg New York Dordrecht London

Lecture Notes in Mathematics ISSN print edition: 0075-8434
ISSN electronic edition: 1617-9692

Library of Congress Control Number: 2013939618

Mathematics Subject Classification (2010): 60H05 (Stochastic Integral) 60G60 (Random fields)
60F10 (Large deviations)

Printed on acid-free paper

Springer is part of Springer Science+Business Media (www.springer.com)

Preface

This lecture note has a fairly long history. Its starting point was an attempt to solve some limit problems about the behaviour of non-linear functionals of a sequence of independent random variables. These problems could not be solved by means of classical probabilistic methods. I tried to solve them with the help of some sort of Taylor expansion. The idea was to represent the functional we are investigating as a sum with a leading term whose asymptotic behaviour can be well described by means of classical results of probability theory and with some error terms whose effect is negligible. This approach worked well, but to bound the error terms I needed some non-trivial estimates. The proof of these estimates was interesting in itself, it was a problem worth of a closer study on its own right. So I tried to work out the details and to present the most important and most interesting results I met during this research. This lecture note is the result of these efforts.

To solve the problems I met I had to give a good estimate on the tail distribution of the integral of a function of several variables with respect to the appropriate power of a normalized empirical distribution. Beside this I also had to consider a generalized version of this problem when the tail distribution of the supremum of such integrals has to be bounded. The difficulties in these problems concentrate around two points.

(a) We consider non-linear functionals of independent random variables, and we have to work out some techniques to deal with such problems.

(b) The idea behind several arguments is the observation that independent random variables behave in many respects almost as if they were Gaussian. But we have to understand how strong this similarity is, and when we can apply the techniques worked out for Gaussian random variables. Beside this we have to find methods to deal with our problems also in such cases when the techniques related to Gaussian and almost Gaussian random variables do not work.

To deal with problem (a) I have discussed the theory of multiple random integrals and their most important properties together with the properties of the so-called (degenerate) U-statistics. I considered the Wiener–Itô integrals which are multiple Gaussian type integrals and provide a useful tool to handle non-linear functionals of Gaussian sequences. I also proved some results about a good representation of the product of Wiener–Itô integrals or degenerate U-statistics as a sum of Wiener–Itô integrals or degenerate U-statistics. A comparison of these results indicates some similarity between the behaviour of Wiener–Itô integrals and degenerate U-statistics. I tried to present a fairly detailed discussion of Wiener–Itô integrals and degenerate U-statistics which contains their most important properties.

Problem (b) appeared in particular in the study of the supremum of a class of random integrals. It may be worth mentioning that there is a deep theory worked out mainly by Michel Talagrand which gives good estimates in such problems, at least in the case if only onefold integrals are considered. It turned out however that the results and methods of this theory are not appropriate to prove such estimates that I needed in this work. Roughly speaking, the problems I met have a different character than those investigated in Talagrand's theory. This point is discussed in more detail in the main text of this work, in particular in Chap. 18, which gives an overview of the problems investigated in this work together with their history. The problems get even harder if the supremum not only of onefold but also of multiple random integrals has to be estimated. Here some new methods are needed which we can find by refining some symmetrization arguments appearing in the theory of the so-called Vapnik–Červonenkis classes.

I have also considered an example in Chap. 2 which shows how to apply the estimates proved in this work in the study of some limit theorem problems in mathematical statistics. Actually this was the starting point of the research described in this work. I discussed only one example, but I consider it more than just an example. My goal was to explain a method that can help in solving some non-trivial limit problems and to show why the results of this lecture notes are useful in their investigation. I think that this approach works in a very general setting, but this is the task of future research. Let me also remark that to understand how this method works and how to apply it one does not have to learn the whole material of this lecture note. It is enough to understand the content of the results in Chap. 8 together with some results of Chap. 9 about the properties of U-statistics.

I had two kinds of readers in mind when writing this lecture note. The first kind of them would like to learn more about such problems in which relatively few independence is available, and as a consequence the methods of classical probability theory do not work in their study. They would like to acquire some results and methods useful in such cases, too. The second kind of readers would not like to go into the details of complicated, unpleasant arguments. They would restrict their attention to some useful methods which may help them in proving the limit theorem problems of probability theory they meet also in such cases when the standard methods do not work. This lecture note can be considered as an attempt to satisfy the wishes of both kinds of readers.

Acknowledgement

This research was supported by the Hungarian National Foundation for Science (OTKA) under grant number 91928.

Budapest, Hungary Péter Majjor
January 2013

Contents

Acronyms

$\Phi(u)$ — Standard normal distribution function.

$\mathscr{F}$ — It denotes generally a class of functions with some nice property. See e.g. page 22.

$S_n(f)$ — The normalized sum $\frac{1}{\sqrt{n}}\sum_{k=1}^{n} f(\xi_k)$ of independent identically distributed random variables with some test function f.

$\mu_n(A)$ — The value of the empirical distribution on the set A.

$J_n(f)$ — Onefold random integral with respect to a normalized empirical distribution.

$J_{n,k}(f)$ — k-fold random integral with respect to a normalized empirical distribution.

$\int'$ — The prime in the integral means that the diagonals are omitted from the domain of integration of a multiple integral.

$|S|$ — The cardinality of a (finite) set S.

$I_{n,k}(f)$ — U-statistic of order k with n sample points and kernel function f.

$I_{n,0}(c)$ — U-statistic of order zero, where c is a constant.

$\mathrm{Sym}\, f$ — Symmetrization of the function f.

μ_W — White noise with reference measure μ.

$Z_{\mu,k}(f)$ — k-fold Wiener–Itô integral with respect of a white noise with reference measure μ.

$P_j f$ — The projection of the function f defined in the Euclidean space R^k to the subspace consisting of the functions not depending on the j-th coordinate.

$Q_j f$ — The projection orthogonal to the projection P_j in the space of functions on R^k.

$f_V(x_{j_1}, \dots, x_{j_{|V|}})$ — The canonical function depending on the arguments indexed by the set V which appears in the Hoeffding decomposition of the U-statistic $I_{n,k}(f)$.

$\mathcal{H}_{\mu,k}$	The class of functions which can be chosen as the kernel function of a k-fold Wiener–Itô integral with respect to a white noise with reference measure μ.
$\Gamma(k,l)$	The class of diagrams in the diagram formula for the product of a k-fold and an l-fold Wiener–Itô integral.
$F_\gamma(f,g)$	The kernel function of the Wiener–Itô integral corresponding to the diagram γ in the diagram formula for the product of two Wiener–Itô integrals.
	The kernel function $F_\gamma(f_1,f_2)$ corresponding to the coloured diagram γ in the diagram formula for the product of two degenerate U-statistics appears at page 125.
$\Gamma(k_1,\dots,k_m)$	The class of diagrams in the diagram formula for the product of Wiener–Itô integrals of order $k_1, k_2, \dots k_m$.
	The same notation is applied for the class of coloured diagrams in the diagram formula for the product of degenerate U-statistics.
$F_\gamma(f_1,\dots,f_m)$	The kernel function of the Wiener–Itô integral in the general form of the diagram formula corresponding to the diagram γ.
	The same notation is applied for the kernel function corresponding to a coloured diagram γ in the diagram formula for the product of degenerate U-statistics.
$\bar\Gamma(k_1,\dots,k_m)$	The class of closed diagrams in the diagram formula.
	The same notation for the class of closed coloured diagrams.
$H_k(u)$	The k-th Hermite polynomial with leading coefficient 1.
$\mathrm{Exp}\,(\mathcal{H}_\mu)$	The Fock space.
$O(\gamma)$ and $C(\gamma)$	The open and closed chains of a coloured diagram γ.
$O_2(\gamma)$	The set of open chains of length 2 in a coloured diagram with two rows.
$W(\gamma)$	An appropriate function of a coloured diagram γ appearing in the diagram formula for the product of degenerate U-statistics. It is defined in the case of the product of two degenerate U-statictics at page 126, in the general case at page 132.
$\bar I_{n,k}(f)$	Decoupled U-statistic of order k with n sample points.
$\bar I^{\varepsilon}_{n,k}(f)$	Randomized decoupled U-statistic of order k with n sample points.
$H_{n,k}(f)$	A random variable appearing in the definition of good tail behaviour for a class of integrals of decoupled U-statistics in Chap. 15.
$H_{n,k}(f\mid G,V_1,V_2)$	A random variable playing central role in the proofs of Chaps. 16 and 17. It depends of a function of k variables, a diagram G and two subsets V_1 and V_2 of the set $\{1,\dots,k\}$.

$I_{n,k}(f(\ell))$	Generalized U-statistics.
$\bar{I}_{n,k}(f(\ell))$	Generalized decoupled U-statistics.
$\tilde{I}_{n,k}(f)$ and $\tilde{I}^{\varepsilon}_{n,k}(f)$	Some linear combinations of decoupled U-statistics and randomized decoupled U-statistics applied in the symmetrization argument of Chapter 15.
$\mathcal{G}$	A class of diagram defined in Chapter 16. applied in the proof of the main result.

Chapter 1
Introduction

First I briefly describe the main subject of this work.

Fix a positive integer n, consider n independent and identically distributed random variables $\xi_1, \dots, \xi_n$ on a measurable space $(X, \mathcal{X})$ with some distribution μ and take their empirical distribution μ_n together with its normalization $\sqrt{n}(\mu_n - \mu)$. Beside this, take a function $f(x_1, \dots, x_k)$ of k variables on the k-fold product $(X^k, \mathcal{X}^k)$ of the space $(X, \mathcal{X})$, introduce the k-th power of the normalized empirical measure $\sqrt{n}(\mu_n - \mu)$ on $(X^k, \mathcal{X}^k)$ and define the integral of the function f with respect to this signed product measure. This integral is a random variable, and we want to give a good estimate on its tail distribution. More precisely, we take the integrals not on the whole space, the diagonals $x_s = x_{s'}$, $1 \le s, s' \le k$, $s \ne s'$, of the space X^k are omitted from the domain of integration. Such a modification of the integral seems to be natural.

We shall also be interested in the following generalized version of the above problem. Let us have a nice class of functions $\mathcal{F}$ of k variables on the product space $(X^k, \mathcal{X}^k)$, and consider the integrals of all functions in this class with respect to the k-fold direct product of our normalized empirical measure. Give a good estimate on the tail distribution of the supremum of these integrals.

One may ask why the above problems deserve a closer study. I found them important, because they may help in solving some essential problems in probability theory and mathematical statistics. I met such problems when I tried to adapt the method of proof about the Gaussian limit behaviour of the maximum likelihood estimate to some similar but more difficult questions. In the original problem the asymptotic behaviour of the solution of the so-called maximum likelihood equation has to be investigated. The study of this problem is hard in its original form. But by applying an appropriate Taylor expansion of the function that appears in this equation and throwing out its higher order terms we get an approximation whose behaviour can be well understood. So to describe the limit behaviour of the maximum likelihood estimate it suffices to show that this approximation causes only a negligible error.

P. Major, *On the Estimation of Multiple Random Integrals and U-Statistics*, Lecture Notes in Mathematics 2079, DOI 10.1007/978-3-642-37617-7_1,

One would try to apply a similar method in the study of more difficult questions. I met some non-parametric maximum likelihood problems, for instance the description of the limit behaviour of the so-called Kaplan–Meyer product limit estimate when such an approach could be applied. But in these problems it was harder to show that the simplifying approximation causes only a negligible error. In this case the solution of the above mentioned problems was needed. In the non-parametric maximum likelihood estimate problems I met, the estimation of multiple (random) integrals played a role similar to the estimation of the coefficients in the Taylor expansion in the study of maximum likelihood estimates. Although I could apply this approach only in some special cases, I believe that it works in very general situations. But it demands some further work to show this.

The above formulated problems about random integrals are interesting and non-trivial even in the special case $k = 1$. Their solution leads to some interesting and non-trivial generalization of the fundamental theorem of the mathematical statistics about the difference of the empirical and real distribution of a large sample.

These problems have a natural counterpart about the behaviour of so-called U-statistics, which is a fairly popular subject in probability theory. The investigation of multiple random integrals and U-statistics are closely related, and it turned out to be useful to consider them simultaneously.

Let us try to get some feeling about what kind of results can be expected in these problems. For a large sample size n the normalized empirical measure $\sqrt{n}(\mu_n - \mu)$ behaves similarly to a Gaussian random measure. This suggests that in the problems we are interested in similar results should hold as in the analogous problems about multiple Gaussian integrals. The behaviour of multiple Gaussian integrals, called Wiener–Itô integrals in the literature, is fairly well understood, and it suggests that the tail distribution of a k-fold random integral with respect to a normalized empirical measure should satisfy such estimates as the tail distribution of the k-th power of a Gaussian random variable with expectation zero and appropriate variance. Beside this, if we consider the supremum of multiple random integrals of a class of functions with respect to a normalized empirical measure or with respect to a Gaussian random measure, then we expect that under not too restrictive conditions this supremum is not much larger than the "worst" random integral with the largest variance taking part in this supremum. We may also hope that the methods of the theory of multiple Gaussian integrals can be adapted to the investigation of our problems.

The above presented heuristic considerations supply a fairly good description of the situation, but they do not take into account a very essential difference between the behaviour of multiple Gaussian integrals and multiple integrals with respect to a normalized empirical measure. If the variance of a multiple integral with respect to a normalized empirical measure is very small, what turns out to be equivalent to a very small L_2-norm of the function we are integrating, then the behaviour of this integral is different from that of a multiple Gaussian integral with the same kernel function. In this case the effect of some irregularities of the normalized empirical distribution turns out to be non-negligible, and no good Gaussian approximation holds any longer. This case must be better understood, and some new methods have to be worked out to handle it. The hardest problems discussed in this work are related to this phenomenon.

The precise formulation of the results will be given in the main part of the work. Beside their proofs I also tried to explain the main ideas behind them and the notions introduced in their investigation. This work contains some new results, and also the proof of some already rather classical theorems is presented. The results about Gaussian random variables and their non-linear functionals, in particular multiple integrals with respect to a Gaussian field, have a most important role in the study of the present work. Hence they are discussed in detail together with some of their counterparts about multiple random integrals with respect to a normalized empirical measure and some results about U-statistics.

The proofs apply results from different parts of the probability theory. Papers investigating similar results refer to works dealing with quite different subjects, and this makes their reading rather hard. To overcome this difficulty I tried to work out the details and to present a self-contained discussion even at the price of a longer text. Thus I wrote down (in the main text or in the Appendix) the proof of many interesting and basic results, like results about Vapnik–Červonenkis classes, about U-statistics and their decomposition to sums of so-called degenerate U-statistics, about so-called decoupled U-statistics and their relation to ordinary U-statistics, the diagram formula about the product of Wiener–Itô integrals, their counterpart about the product of degenerate U-statistics, etc. I tried to give such an exposition where different parts of the problem are explained independently of each other, and they can be understood in themselves.

As all the topics treated in the individual chapters relate to each other it seemed natural to me to tell the history of how the various results were reached in one last chapter. This last chapter, Chap. 18, just before the Appendix, also contains the complete reference list. I tried to give satisfactory referencing to all essential problems discussed, concentrate on explaining the main ideas behind the proofs and indicate where they were published. I did not attempt to provide an exhaustive literature list for fear that more would be less. As a consequence the reference list reflects my subjective preferences, my way of thinking.

Chapter 2
Motivation of the Investigation: Discussion of Some Problems

In this chapter I try to show by means of an example why the solution of the problems mentioned in the introduction may be useful in the study of some important problems of probability theory. I try to give a good picture about the main ideas, but I do not work out all details. Actually the elaboration of some details omitted from this discussion would demand hard work. But as the present chapter is quite independent of the rest of the work, these omissions cause no problem in understanding the subsequent part.

I start with a short discussion of the maximum likelihood estimate in the simplest case. The following problem is considered. Let us have a class of density functions $f(x,\vartheta)$ on the real line depending on a parameter $\vartheta \in R^1$, and observe a sequence of independent random variables $\xi_1(\omega),\dots,\xi_n(\omega)$ with a density function $f(x,\vartheta_0)$, where ϑ_0 is an unknown parameter we want to estimate with the help of the above sequence of random variables.

The maximum likelihood method suggests the following approach. Choose that value $\hat\vartheta_n = \hat\vartheta_n(\xi_1,\dots,\xi_n)$ as the estimate of the parameter ϑ_0 where the density function of the random vector $(\xi_1,\dots,\xi_n)$, i.e. the product

$$\prod_{k=1}^{n} f(\xi_k,\vartheta) = \exp\left\{\sum_{k=1}^{n} \log f(\xi_k,\vartheta)\right\}$$

takes its maximum. This point can be found as the solution of the so-called maximum likelihood equation

$$\sum_{k=1}^{n} \frac{\partial}{\partial\vartheta} \log f(\xi_k,\vartheta) = 0. \tag{2.1}$$

We are interested in the asymptotic behaviour of the random variable $\hat\vartheta_n - \vartheta_0$, where $\hat\vartheta_n$ is the (appropriate) solution of (2.1).

P. Major, *On the Estimation of Multiple Random Integrals and U-Statistics*, Lecture Notes in Mathematics 2079, DOI 10.1007/978-3-642-37617-7_2,

The direct study of this equation is rather hard, but a Taylor expansion of the expression at the left-hand side of (2.1) around the (unknown) point ϑ_0 yields a good and simple approximation of $\hat{\vartheta}_n$, and it enables us to describe the asymptotic behaviour of $\hat{\vartheta}_n - \vartheta_0$.

This Taylor expansion yields that

$$\begin{aligned}
&\sum_{k=1}^{n} \frac{\partial}{\partial \vartheta} \log f(\xi_k, \hat{\vartheta}_n) \\
&= \sum_{k=1}^{n} \frac{\frac{\partial}{\partial \vartheta} f(\xi_k, \vartheta_0)}{f(\xi_k, \vartheta_0)} + (\hat{\vartheta}_n - \vartheta_0) \left(\sum_{k=1}^{n} \left(\frac{\frac{\partial^2}{\partial \vartheta^2} f(\xi_k, \vartheta_0)}{f(\xi_k, \vartheta_0)} - \frac{\left(\frac{\partial}{\partial \vartheta} f(\xi_k, \vartheta_0)\right)^2}{f^2(\xi_k, \bar{\vartheta}_0)} \right) \right) \\
&\quad + O\left(n(\hat{\vartheta}_n - \vartheta_0)^2 \right) \\
&= \sum_{k=1}^{n} \left(\eta_k + \zeta_k (\hat{\vartheta}_n - \vartheta_0) \right) + O\left(n(\hat{\vartheta}_n - \vartheta_0)^2 \right), \qquad (2.2)
\end{aligned}$$

where

$$\eta_k = \frac{\frac{\partial}{\partial \vartheta} f(\xi_k, \vartheta_0)}{f(\xi_k, \vartheta_0)} \quad \text{and} \quad \zeta_k = \frac{\frac{\partial^2}{\partial \vartheta^2} f(\xi_k, \vartheta_0)}{f(\xi_k, \vartheta_0)} - \frac{\left(\frac{\partial}{\partial \vartheta} f(\xi_k, \vartheta_0)\right)^2}{f^2(\xi_k, \bar{\vartheta}_0)}$$

for $k = 1, \dots, n$. We want to understand the asymptotic behaviour of the (random) expression on the right-hand side of (2.2). The relation

$$E\eta_k = \int \frac{\frac{\partial}{\partial \vartheta} f(x, \vartheta_0)}{f(x, \vartheta_0)} f(x, \vartheta_0)\, dx = \frac{\partial}{\partial \vartheta} \int f(x, \vartheta_0)\, dx = 0$$

holds, since $\int f(x, \vartheta)\, dx = 1$ for all ϑ, and a differentiation of this relation gives the last identity. Similarly, $E\eta_k^2 = -E\zeta_k = \int \frac{\left(\frac{\partial}{\partial \vartheta} f(x,\vartheta_0)\right)^2}{f(x,\vartheta_0)}\, dx > 0$, $k = 1, \dots, n$. Hence by the central limit theorem $\chi_n = \frac{1}{\sqrt{n}} \sum\limits_{k=1}^{n} \eta_k$ is asymptotically normal with expectation zero and variance $I^2 = \int \frac{\left(\frac{\partial}{\partial \vartheta} f(x,\vartheta_0)\right)^2}{f(x,\vartheta_0)}\, dx > 0$. In the statistics literature this number I is called the Fisher information. By the laws of large numbers $\frac{1}{n} \sum\limits_{k=1}^{n} \zeta_k \sim -I^2$.

Thus relation (2.2) suggests the approximation of the maximum-likelihood estimate $\hat{\vartheta}_n$ by the random variable $\tilde{\vartheta}_n$ given by the identity $\tilde{\vartheta}_n - \vartheta_0 = -\frac{\sum\limits_{k=1}^{n} \eta_k}{\sum\limits_{k=1}^{n} \zeta_k}$, and the previous calculations imply that $\sqrt{n}(\tilde{\vartheta}_n - \vartheta_0)$ is asymptotically normal with expectation zero and variance $\frac{1}{I^2}$. The random variable $\tilde{\vartheta}_n$ is not a solution of (2.1), the value of the expression at the left-hand side is of order $O(n(\tilde{\vartheta}_n - \vartheta_0)^2) = O(1)$ in this point. On the other hand, some calculations show that the derivative of the function at the left-hand side is large in this point, it is greater than const. n with some const. > 0. This implies that the maximum-likelihood equation has a solution $\hat{\vartheta}_n$ such that $\hat{\vartheta}_n - \tilde{\vartheta}_n = O\left(\frac{1}{n}\right)$. Hence $\sqrt{n}(\hat{\vartheta}_n - \vartheta_0)$ and $\sqrt{n}(\tilde{\vartheta}_n - \vartheta_0)$ have the same asymptotic limit behaviour.

The previous method can be summarized in the following way: Take a simpler linearized version of the expression we want to estimate by means of an appropriate Taylor expansion, describe the limit distribution of this linearized version and show that the linearization causes only a negligible error.

We want to show that such a method also works in more difficult situations. But in some cases it is harder to show that the error committed by a replacement of the original expression by a simpler linearized version is negligible, and to show this the solution of the problems mentioned in the introduction is needed. The discussion of the following problem, called the Kaplan–Meyer method for the estimation of the empirical distribution function with the help of censored data shows such an example.

The following problem is considered. Let (X_i, Z_i), $i = 1, \dots, n$, be a sequence of independent, identically distributed random vectors such that the components X_i and Z_i are also independent with some unknown, continuous distribution functions $F(x)$ and $G(x)$. We want to estimate the distribution function F of the random variables X_i, but we cannot observe the variables X_i, only the random variables $Y_i = \min(X_i, Z_i)$ and $\delta_i = I(X_i \le Z_i)$. In other words, we want to solve the following problem. There are certain objects whose lifetime X_i are independent and F distributed. But we cannot observe this lifetime X_i, because after a time Z_i the observation must be stopped. We also know whether the real lifetime X_i or the censoring variable Z_i was observed. We make n independent experiments and want to estimate with their help the distribution function F.

Kaplan and Meyer, on the basis of some maximum-likelihood estimation type considerations, proposed the following so-called product limit estimator $S_n(u)$ to estimate the unknown survival function $S(u) = 1 - F(u)$:

$$1 - F_n(u) = S_n(u) = \begin{cases} \prod\limits_{i=1}^{n} \left(\frac{N(Y_i)}{N(Y_i)+1}\right)^{I(Y_i \le u, \delta_i = 1)} & \text{if } u \le \max(Y_1, \dots, Y_n) \\ 0 \text{ if } u \ge \max(Y_1, \dots, Y_n), \text{ and } \delta_n = 1, \\ \text{undefined if } u \ge \max(Y_1, \dots, Y_n), \text{ and } \delta_n = 0, \end{cases} \tag{2.3}$$

where

$$N(t) = \#\{Y_i,\ Y_i > t,\ 1 \le i \le n\} = \sum_{i=1}^{n} I(Y_i > t).$$

We want to show that the above estimate (2.3) is really good. For this goal we shall approximate the random variables $S_n(u)$ by some appropriate random variables. To do this first we introduce some notations.

Put

$$H(u) = P(Y_i \le u) = 1 - \bar{H}(u),$$
$$\tilde{H}(u) = P(Y_i \le u,\ \delta_i = 1), \quad \tilde{\tilde{H}}(u) = P(Y_i \le u,\ \delta_i = 0) \tag{2.4}$$

and

$$H_n(u) = \frac{1}{n}\sum_{i=1}^{n} I(Y_i \le u)$$
$$\tilde{H}_n(u) = \frac{1}{n}\sum_{i=1}^{n} I(Y_i \le u,\ \delta_i = 1), \quad \tilde{\tilde{H}}_n(u) = \frac{1}{n}\sum_{i=1}^{n} I(Y_i \le u,\ \delta_i = 0). \tag{2.5}$$

Clearly $H(u) = \tilde{H}(u) + \tilde{\tilde{H}}(u)$ and $H_n(u) = \tilde{H}_n(u) + \tilde{\tilde{H}}_n(u)$. We shall estimate $F_n(u) - F(u)$ for $u \in (-\infty, T]$ if

$$1 - H(T) > \delta \quad \text{with some fixed } \delta > 0. \tag{2.6}$$

Condition (2.6) implies that there are more than $\frac{\delta}{2}n$ sample points Y_j larger than T with probability almost 1. The complementary event has only an exponentially small probability. This observation helps to show in the subsequent calculations that some events have negligibly small probability.

We introduce the so-called cumulative hazard function and its empirical version

$$\Lambda(u) = -\log(1 - F(u)), \quad \Lambda_n(u) = -\log(1 - F_n(u)). \tag{2.7}$$

Since $F_n(u) - F(u) = \exp(-\Lambda(u))\,(1 - \exp(\Lambda(u) - \Lambda_n(u)))$ a simple Taylor expansion yields

$$F_n(u) - F(u) = (1 - F(u))\,(\Lambda_n(u) - \Lambda(u)) + R_1(u), \tag{2.8}$$

and it is easy to see that $R_1(u) = O\left((\Lambda(u) - \Lambda_n(u))^2\right)$. It follows from the subsequent estimations that $\Lambda(u) - \Lambda_n(u) = O(n^{-1/2})$, thus $R_1(u) = O(\frac{1}{n})$. Hence it is enough to investigate the term $\Lambda_n(u)$. We shall show that $\Lambda_n(u)$ has an expansion with $\Lambda(u)$ as the main term plus $n^{-1/2}$ times a term which is a linear functional of an appropriate normalized empirical distribution function plus an error term of order $O(n^{-1})$.

From (2.3) it is obvious that

$$\Lambda_n(u) = -\sum_{i=1}^{n} I(Y_i \le u,\ \delta_i = 1)\log\left(1 - \frac{1}{1 + N(Y_i)}\right).$$

It is not difficult to get rid of the unpleasant logarithmic function in this formula by means of the relation $-\log(1-x) = x + O(x^2)$ for small x. It yields that

$$\Lambda_n(u) = \sum_{i=1}^{n} \frac{I(Y_i \le u,\, \delta_i = 1)}{N(Y_i)} + R_2(u) = \tilde{\Lambda}_n(u) + R_2(u) \tag{2.9}$$

with an error term $R_2(u)$ such that $nR_2(u)$ is smaller than a constant with probability almost one. (The probability of the exceptional set is exponentially small.)

The expression $\tilde{\Lambda}_n(u)$ is still inappropriate for our purposes. Since the denominators $N(Y_i) = \sum\limits_{j=1}^{n} I(Y_j > Y_i)$ are dependent for different indices i we cannot see directly the limit behaviour of $\tilde{\Lambda}_n(u)$.

We try to approximate $\tilde{\Lambda}_n(u)$ by a simpler expression. A natural approach would be to approximate the terms $N(Y_i)$ in it by their conditional expectation $(n-1)\bar{H}(Y_i) = (n-1)(1-H(Y_i)) = E(N(Y_i)|Y_i)$ with respect to the σ-algebra generated by the random variable Y_i. This is a too rough "first order" approximation, but the following "second order approximation" will be sufficient for our goals. Put

$$N(Y_i) = \sum_{j=1}^{n} I(Y_j > Y_i) = n\bar{H}(Y_i)\left(1 + \frac{\sum\limits_{j=1}^{n} I(Y_j > Y_i) - n\bar{H}(Y_i)}{n\bar{H}(Y_i)}\right)$$

and express the terms $\frac{1}{N(Y_i)}$ in the sum defining $\tilde{\Lambda}_n$, (with $\tilde{\Lambda}_n$ introduced in (2.9)) by means of the relation $\frac{1}{1+z} = \sum\limits_{k=0}^{\infty} (-1)^k z^k = 1 - z + \varepsilon(z)$ with the choice $z = \frac{\sum_{j=1}^{n} I(Y_j > Y_i) - n\bar{H}(Y_i)}{n\bar{H}(Y_i)}$. As $|\varepsilon(z)| < 2z^2$ for $|z| < \frac{1}{2}$ we get that

$$\begin{aligned}
\tilde{\Lambda}_n(u) &= \sum_{i=1}^{n} \frac{I(Y_i \le u,\, \delta_i = 1)}{n\bar{H}(Y_i)}\left(1 + \sum_{k=1}^{\infty}\left(-\frac{\sum\limits_{j=1}^{n} I(Y_j > Y_i) - n\bar{H}(Y_i)}{n\bar{H}(Y_i)}\right)^k\right) \\
&= \sum_{i=1}^{n} \frac{I(Y_i \le u,\, \delta_i = 1)}{n\bar{H}(Y_i)}\left(1 - \frac{\sum\limits_{j=1}^{n} I(Y_j > Y_i) - n\bar{H}(Y_i)}{n\bar{H}(Y_i)}\right) + R_3(u) \\
&= 2A(u) - B(u) + R_3(u),
\end{aligned} \tag{2.10}$$

where

$$A(u) = A(n,u) = \sum_{i=1}^{n} \frac{I(Y_i \le u,\, \delta_i = 1)}{n\bar{H}(Y_i)}$$

and

$$B(u) = B(n,u) = \sum_{i=1}^{n}\sum_{j=1}^{n} \frac{I(Y_i \le u,\, \delta_i = 1)I(Y_j > Y_i)}{n^2\bar{H}^2(Y_i)}.$$

It can be proved by means of standard methods that $nR_3(u)$ is exponentially small. Thus relations (2.9) and (2.10) yield that

$$\Lambda_n(u) = 2A(u) - B(u) + \text{negligible error.} \tag{2.11}$$

This means that to solve our problem the asymptotic behaviour of the random variables $A(u)$ and $B(u)$ has to be given. We can get a better insight to this problem by rewriting the sum $A(u)$ as an integral and the double sum $B(u)$ as a twofold integral with respect to empirical measures. Then these integrals can be rewritten as sums of random integrals with respect to normalized empirical measures and deterministic measures. Such an approach yields a representation of $\Lambda_n(u)$ in the form of a sum whose terms can be well understood.

Let us write

$$A(u) = \int_{-\infty}^{+\infty} \frac{I(y \le u)}{1 - H(y)}\, d\tilde{H}_n(y),$$

$$B(u) = \int_{-\infty}^{+\infty}\int_{-\infty}^{+\infty} \frac{I(y \le u)I(x > y)}{(1 - H(y))^2}\, dH_n(x) d\tilde{H}_n(y).$$

We rewrite the terms $A(u)$ and $B(u)$ in a form better for our purposes. We express these terms as a sum of integrals with respect to $dH(u)$, $d\tilde{H}(u)$ and the normalized empirical processes $d\sqrt{n}(H_n(x) - H(x))$ and $d\sqrt{n}(\tilde{H}_n(y) - \tilde{H}(y))$. For this goal observe that

$$\begin{aligned} H_n(x)\tilde{H}_n(y) = {} & H(x)\tilde{H}(y) + H(x)(\tilde{H}_n(y) - \tilde{H}(y)) + (H_n(x) - H(x))\tilde{H}(y) \\ & + (H_n(x) - H(x))(\tilde{H}_n(y) - \tilde{H}(y)). \end{aligned}$$

Hence we can write that $B(u) = B_1(u) + B_2(u) + B_3(u) + B_4(u)$, where

$$B_1(u) = \int_{-\infty}^{u}\int_{-\infty}^{+\infty} \frac{I(x > y)}{(1 - H(y))^2}\, dH(x)\, d\tilde{H}(y),$$

$$B_2(u) = \frac{1}{\sqrt{n}}\int_{-\infty}^{u}\int_{-\infty}^{+\infty} \frac{I(x > y)}{(1 - H(y))^2}\, dH(x)\, d\left(\sqrt{n}(\tilde{H}_n(y) - \tilde{H}(y))\right),$$

$$B_3(u) = \frac{1}{\sqrt{n}} \int_{-\infty}^{u} \int_{-\infty}^{+\infty} \frac{I(x > y)}{(1 - H(y))^2} \, d\left(\sqrt{n}\,(H_n(x) - H(x))\right) d\tilde{H}(y),$$

$$B_4(u) = \frac{1}{n} \int_{-\infty}^{u} \int_{-\infty}^{+\infty} \frac{I(x > y)}{(1 - H(y))^2} \, d\left(\sqrt{n}\,(H_n(x) - H(x))\right)$$

$$d\left(\sqrt{n}(\tilde{H}_n(y) - \tilde{H}(y))\right).$$

In the above decomposition of $B(u)$ the term B_1 is a deterministic function, B_2, B_3 are linear functionals of normalized empirical processes and B_4 is a nonlinear functional of normalized empirical processes. The deterministic term $B_1(u)$ can be calculated explicitly. Indeed,

$$B_1(u) = \int_{-\infty}^{u} \int_{-\infty}^{+\infty} \frac{I(x > y)}{(1 - H(y))^2} \, dH(x) d\tilde{H}(y) = \int_{-\infty}^{u} \frac{d\tilde{H}(y)}{1 - H(y)}.$$

Then the relations $\tilde{H}(u) = \int_{-\infty}^{u} (1 - G(t)) \, dF(t)$ and $1 - H = (1 - F)(1 - G)$ imply that

$$B_1(u) = \int_{-\infty}^{u} \frac{dF(y)}{1 - F(y)} = -\log(1 - F(u)) = \Lambda(u). \qquad (2.12)$$

Observe that

$$A(u) = \int_{-\infty}^{u} \frac{d\tilde{H}_n(y)}{1 - H(y)}$$

$$= \int_{-\infty}^{u} \frac{d\tilde{H}(y)}{1 - H(y)} + \frac{1}{\sqrt{n}} \int_{-\infty}^{u} \frac{d\left(\sqrt{n}(\tilde{H}_n(y) - \tilde{H}(y))\right)}{1 - H(y)}$$

$$= B_1(u) + B_2(u). \qquad (2.13)$$

From relations (2.11)–(2.13) it follows that

$$\Lambda_n(u) - \Lambda(u) = B_2(u) - B_3(u) - B_4(u) + \text{negligible error}. \qquad (2.14)$$

Integration of B_2 and B_3 with respect to the variable x and then integration by parts in the expression B_2 yields that

$$B_2(u) = \frac{1}{\sqrt{n}} \int_{-\infty}^{u} \frac{d\left(\sqrt{n}(\tilde{H}_n(y) - \tilde{H}(y))\right)}{1 - H(y)}$$

$$= \frac{\sqrt{n}\,(\tilde{H}_n(u) - \tilde{H}(u))}{\sqrt{n}(1 - H(u))} - \frac{1}{\sqrt{n}} \int_{-\infty}^{u} \frac{\sqrt{n}(\tilde{H}_n(y) - \tilde{H}(y))}{(1 - H(y))^2} \, dH(y),$$

$$B_3(u) = \frac{1}{\sqrt{n}} \int_{-\infty}^{u} \frac{\sqrt{n}\,(H(y) - H_n(y))}{(1 - H(y))^2} \, d\tilde{H}(y).$$

With the help of the above expressions for B_2 and B_3 (2.14) can be rewritten as

$$\sqrt{n}\,(\Lambda_n(u)-\Lambda(u)) = \frac{\sqrt{n}\,(\tilde{H}_n(u)-\tilde{H}(u))}{1-H(u)} - \int_{-\infty}^{u} \frac{\sqrt{n}(\tilde{H}_n(y)-\tilde{H}(y))}{(1-H(y))^2}\,dH(y)$$
$$+\int_{-\infty}^{u} \frac{\sqrt{n}\,(H_n(y)-H(y))}{(1-H(y))^2}\,d\tilde{H}(y)$$
$$-\sqrt{n}\,B_4(u) + \text{negligible error.} \tag{2.15}$$

Formula (2.15) (together with formula (2.8)) almost agrees with the statement we wanted to prove. Here the random variable $\sqrt{n}\,(\Lambda_n(u)-\Lambda(u))$ is expressed as a sum of linear functionals of normalized empirical distributions plus some negligible error terms plus the error term $\sqrt{n}\,B_4(u)$. So to get a complete proof it is enough to show that $\sqrt{n}\,B_4(u)$ also yields a negligible error. But $n\,B_4(u)$ is a double integral of a bounded function (here we apply again formula (2.6)) with respect to a normalized empirical distribution. Hence to bound this term we need a good estimate of multiple stochastic integrals (with multiplicity 2), and this is just the problem formulated in the introduction. The estimate we need here follows from Theorem 8.1 of the present work. Let us remark that the problem discussed here corresponds to the estimation of the coefficient of the second term in the Taylor expansion considered in the study of the maximum likelihood estimation. One may worry a little bit how to bound $n\,B_4(u)$ with the help of estimations of double stochastic integrals, since in the definition of $B_4(u)$ integration is taken with respect to different normalized empirical processes in the two coordinates. But this is a not too difficult technical problem. It can be simply overcome for instance by rewriting the integral as a double integral with respect to the empirical process $\left(\sqrt{n}\,(H_n(x)-H(x)),\,\sqrt{n}\,(\tilde{H}_n(y)-\tilde{H}(y))\right)$ in the space R^2.

By working out the details of the above calculation we get that the linear functional $B_2(u)-B_3(u)$ of normalized empirical processes yields a good estimate on the expression $\sqrt{n}(\Lambda_n(u)-\Lambda(u))$ for a fixed parameter u. But we want to prove somewhat more, we want to get an estimate uniform in the parameter u, i.e. to show that even the random variable $\sup\limits_{u\le T}\left|\sqrt{n}(\Lambda_n(u)-\Lambda(u))-B_2(u)+B_3(u)\right|$ is small. This can be done by making estimates uniform in the parameter u in all steps of the above calculation. There appears only one difficulty when trying to carry out this program. Namely, we need an estimate on $\sup\limits_{u\le T}|n\,B_4(u)|$, i.e. we have to bound the supremum of multiple random integrals with respect to a normalized random measure for a nice class of kernel functions. This can be done, but at this point the second problem mentioned in the introduction appears. This difficulty can be overcome by means of Theorem 8.2 of this work.

Thus the limit behaviour of the Kaplan–Meyer estimate can be described by means of an appropriate expansion. The steps of the calculation leading to such an expansion are fairly standard, the only hard part is the solution of the problems mentioned in the introduction. It can be expected that such a method also works in a much more general situation.

I finish this chapter with a remark of Richard Gill he made in a personal conversation after my talk on this subject at a conference. While he accepted my proof he missed an argument in it about the maximum likelihood character of the Kaplan–Meyer estimate. This was a completely justified remark, since if we do not restrict our attention to this problem, but try to generalize it to general non-parametric maximum likelihood estimates, then we have to understand how the maximum likelihood character of the estimate can be exploited. I believe that this can be done, but only with the help of some further studies.

Chapter 3
Some Estimates About Sums of Independent Random Variables

We shall need a good bound on the tail distribution of sums of independent random variables bounded by a constant with probability one. Later only the results about sums of independent and identically distributed variables will be interesting for us. But since they can be generalized without any effort to sums of not necessarily identically distributed random variables the condition about identical distribution of the summands will be dropped. We are interested in the question when these estimates give such a good bound as the central limit theorem suggests, and what can be told otherwise.

More explicitly, the following problem will be considered: Let $X_1, \dots, X_n$ be independent random variables, $EX_j = 0$, $\operatorname{Var} X_j = \sigma_j^2$, $1 \le j \le n$, and take the random sum $S_n = \sum_{j=1}^{n} X_j$ and its variance $\operatorname{Var} S_n = V_n^2 = \sum_{j=1}^{n} \sigma_j^2$. We want to get a good bound on the probability $P(S_n > uV_n)$. The central limit theorem suggests that under general conditions an upper bound of the order $1 - \Phi(u)$ should hold for this probability, where $\Phi(u)$ denotes the standard normal distribution function. Since the standard normal distribution function satisfies the inequality $\left(\frac{1}{u} - \frac{1}{u^3}\right)\frac{e^{-u^2/2}}{\sqrt{2\pi}} < 1 - \Phi(u) < \frac{1}{u}\frac{e^{-u^2/2}}{\sqrt{2\pi}}$ for all $u > 0$ it is natural to ask when the probability $P(S_n > uV_n)$ is comparable with the value $e^{-u^2/2}$. More generally, we shall call an upper bound of the form $P(S_n > uV_n) \le e^{-Cu^2}$ with some constant $C > 0$ a Gaussian type estimate.

First I formulate Bernstein's inequality which tells for which values u the probability $P(S_n > uV_n)$ has a Gaussian type estimate. It supplies such an estimate if $u \le \text{const.}\, V_n$. On the other hand, for $u \ge \text{const.}\, V_n$ it yields a much weaker bound. I shall formulate another result, called Bennett's inequality, which is a slight improvement of Bernstein's inequality. It helps us to tell what can be expected if Bernstein's inequality does not provide a Gaussian type estimate. I shall also present an example which shows that Bennett's inequality is in some sense sharp. The main difficulties we meet in this work are closely related to the weakness of the estimates

P. Major, *On the Estimation of Multiple Random Integrals and U-Statistics*, Lecture Notes in Mathematics 2079, DOI 10.1007/978-3-642-37617-7_3,

we have for the probability $P(S_n > uV_n)$ if it does not satisfy a Gaussian type estimate. As we shall see this happens if $u \gg \text{const.}\, V_n$.

In the usual formulation of Bernstein's inequality a real number M is introduced, and it is assumed that the terms in the sum we investigate are bounded by this number. But since the problem can be simply reduced to the case $M = 1$ I shall consider only this special case.

Theorem 3.1 (Bernstein's inequality). *Let $X_1, \dots, X_n$ be independent random variables, $P(|X_j| \le 1) = 1$, $EX_j = 0$, $1 \le j \le n$. Put $\sigma_j^2 = EX_j^2$, $1 \le j \le n$, $S_n = \sum\limits_{j=1}^{n} X_j$ and $V_n^2 = \operatorname{Var} S_n = \sum\limits_{j=1}^{n} \sigma_j^2$. Then*

$$P\left(S_n > uV_n\right) \le \exp\left\{-\frac{u^2}{2\left(1+\frac{1}{3}\frac{u}{V_n}\right)}\right\} \quad \textit{for all } u > 0. \tag{3.1}$$

Proof of Theorem 3.1. Let us give a good bound on the exponential moments Ee^{tS_n} for appropriate parameters $t > 0$. Since $EX_j = 0$ and $E|X_j^{k+2}| \le \sigma_j^2$ for $k \ge 0$ we can write $Ee^{tX_j} = \sum\limits_{k=0}^{\infty} \frac{t^k}{k!} EX_j^k \le 1 + \frac{t^2\sigma_j^2}{2}\left(1 + \sum\limits_{k=1}^{\infty} \frac{2t^k}{(k+2)!}\right) \le 1 + \frac{t^2\sigma_j^2}{2}\left(1 + \sum\limits_{k=1}^{\infty} 3^{-k}t^k\right) = 1 + \frac{t^2\sigma_j^2}{2}\frac{1}{1-\frac{t}{3}} \le \exp\left\{\frac{t^2\sigma_j^2}{2}\frac{1}{1-\frac{t}{3}}\right\}$ if $0 \le t < 3$. Hence

$$Ee^{tS_n} = \prod_{j=1}^{n} Ee^{tX_j} \le \exp\left\{\frac{t^2V_n^2}{2}\frac{1}{1-\frac{t}{3}}\right\} \qquad \text{for } 0 \le t < 3.$$

The above relation implies that

$$P\left(S_n > uV_n\right) = P(e^{tS_n} > e^{tuV_n}) \le Ee^{tS_n}e^{-tuV_n} \le \exp\left\{\frac{t^2V_n^2}{2}\frac{1}{1-\frac{t}{3}} - tuV_n\right\}$$

if $0 \le t < 3$. Choose the number t in this inequality as the solution of the equation $t^2V_n^2\frac{1}{1-\frac{t}{3}} = tuV_n$, i.e. put $t = \frac{u}{V_n+\frac{u}{3}}$. Then $0 \le t < 3$, and we get that $P(S_n > uV_n) \le e^{-tuV_n/2} = \exp\left\{-\frac{u^2}{2\left(1+\frac{1}{3}\frac{u}{V_n}\right)}\right\}$. □

If the random variables $X_1, \dots, X_n$ satisfy the conditions of Bernstein's inequality, then also the random variables $-X_1, \dots, -X_n$ satisfy them. By applying the above result in both cases we get that $P(|S_n| > uV_n) \le 2\exp\left\{-\frac{u^2}{2\left(1+\frac{1}{3}\frac{u}{V_n}\right)}\right\}$ under the conditions of Bernstein's inequality.

By Bernstein's inequality for all $\varepsilon > 0$ there is some number $\alpha(\varepsilon) > 0$ such that in the case $\frac{u}{V_n} < \alpha(\varepsilon)$ the inequality $P(S_n > uV_n) \le e^{-(1-\varepsilon)u^2/2}$ holds. Beside this, for all fixed numbers $A > 0$ there is some constant $C = C(A) > 0$ such that if $\frac{u}{V_n} < A$, then $P(S_n > uV_n) \le e^{-Cu^2}$. This can be interpreted as a Gaussian type estimate for the probability $P(S_n > uV_n)$ if $u \le \text{const.}\, V_n$.

On the other hand, if $\frac{u}{V_n}$ is very large, then Bernstein's inequality yields a much worse estimate. The question arises whether in this case Bernstein's inequality can be replaced by a better, more useful result. Next I present Theorem 3.2, the so-called Bennett's inequality which provides a slight improvement of Bernstein's inequality. But if $\frac{u}{V_n}$ is very large, then also Bennett's inequality provides a much weaker estimate on the probability $P(S_n > uV_n)$ than the bound suggested by a Gaussian comparison. On the other hand, I shall present an example that shows that (without imposing some additional conditions) no real improvement of this estimate is possible.

Theorem 3.2 (Bennett's inequality). *Let $X_1, \dots, X_n$ be independent random variables, $P(|X_j| \le 1) = 1$, $EX_j = 0$, $1 \le j \le n$. Put $\sigma_j^2 = EX_j^2$, $1 \le j \le n$, $S_n = \sum_{j=1}^n X_j$ and $V_n^2 = \operatorname{Var} S_n = \sum_{j=1}^n \sigma_j^2$. Then*

$$P(S_n > u) \le \exp\left\{-V_n^2\left[\left(1+\frac{u}{V_n^2}\right)\log\left(1+\frac{u}{V_n^2}\right)-\frac{u}{V_n^2}\right]\right\} \quad \text{for all } u > 0. \tag{3.2}$$

As a consequence, for all $\varepsilon > 0$ there exists some $B = B(\varepsilon) > 0$ such that

$$P(S_n > u) \le \exp\left\{-(1-\varepsilon)u\log\frac{u}{V_n^2}\right\} \quad \text{if } u > BV_n^2, \tag{3.3}$$

and there exists some positive constant $K > 0$ such that

$$P(S_n > u) \le \exp\left\{-Ku\log\frac{u}{V_n^2}\right\} \quad \text{if } u > 2V_n^2. \tag{3.4}$$

Proof of Theorem 3.2. We have

$$Ee^{tX_j} = \sum_{k=0}^{\infty}\frac{t^k}{k!}EX_j^k \le 1+\sigma_j^2\sum_{k=2}^{\infty}\frac{t^k}{k!} = 1+\sigma_j^2\left(e^t-1-t\right) \le e^{\sigma_j^2(e^t-1-t)},$$
$$1 \le j \le n,$$

and $Ee^{tS_n} \le e^{V_n^2(e^t-1-t)}$ for all $t \ge 0$. Hence $P(S_n > u) \le e^{-tu}Ee^{tS_n} \le e^{-tu+V_n^2(e^t-1-t)}$ for all $t \ge 0$. We get relation (3.2) from this inequality with the choice $t = \log\left(1+\frac{u}{V_n^2}\right)$. (This is the place of minimum of the function $-tu + V_n^2(e^t-1-t)$ for fixed u in the parameter t.)

Relation (3.2) and the observation $\lim\limits_{v\to\infty}\frac{(v+1)\log(v+1)-v}{v\log v}=1$ with the choice $v=\frac{u}{V_n^2}$ imply formula (3.3). Because of relation (3.3) to prove formula (3.4) it is enough to check it for $2\le\frac{u}{V_n^2}\le B$ with some sufficiently large constant $B>0$. In this case relation (3.4) follows directly from formula (3.2). This can be seen for instance by observing that the expression $\frac{V_n^2\left[\left(1+\frac{u}{V_n^2}\right)\log\left(1+\frac{u}{V_n^2}\right)-\frac{u}{V_n^2}\right]}{u\log\frac{u}{V_n^2}}$ is a continuous and positive function of the variable $\frac{u}{V_n^2}$ in the interval $2\le\frac{u}{V_n^2}\le B$, hence its minimum in this interval is strictly positive. □

Let us make a short comparison between Bernstein's and Bennett's inequalities. Both results yield an estimate on the probability $P(S_n>u)$, and their proofs are very similar. They are based on an estimate of the moment generating functions $R_j(t)=Ee^{tX_j}$ of the summands X_j, but Bennett's inequality yields a better estimate. It may be worth mentioning that the estimate given for $R_j(t)=Ee^{tX_j}$ in the proof of Bennett's inequality agrees with the moment generating function $Ee^{t(Y_j-EY_j)}$ of the normalization Y_j-EY_j of a Poissonian random variable Y_j with parameter $\operatorname{Var}X_j$. As a consequence, we get, by using the standard method of estimating tail-distributions by means of the moment generating functions such an estimate for the probability $P(S_n>u)$ which is comparable with the probability $P(T_n-ET_n>u)$, where T_n is a Poissonian random variable with parameter $V_n=\operatorname{Var}S_n$. We can say that Bernstein's inequality yields a Gaussian and Bennett's inequality a Poissonian type estimate for the sums of independent, bounded random variables.

Remark. Bennett's inequality yields a sharper estimate for the probability $P(S_n>u)$ than Bernstein's inequality for all numbers $u>0$. To prove this it is enough to show that for all $0\le t<3$ the inequality $Ee^{tS_n}\le e^{V_n^2(e^t-1-t)}$ appearing in the proof of Bennett's inequality is a sharper estimate than the corresponding inequality $Ee^{tS_n}\le\exp\left\{\frac{t^2V_n^2}{2}\frac{1}{1-\frac{t}{3}}\right\}$ appearing in the proof of Bernstein's inequality. (Recall, how we estimate the probability $P(S_n>u)$ in these proofs with the help of the exponential moment Ee^{tS_n}.) But to prove this it is enough to check that $e^t-1-t\le\frac{t^2}{2}\frac{1}{1-\frac{t}{3}}$ for all $0\le t<3$. This inequality clearly holds, since $e^t-1-t=\sum\limits_{k=2}^{\infty}\frac{t^k}{k!}$, and $\frac{t^2}{2}\frac{1}{1-\frac{t}{3}}=\sum\limits_{k=2}^{\infty}\frac{1}{2}(\frac{1}{3})^{k-2}t^k$.

Next I present Example 3.3 which shows that Bennett's inequality yields a sharp estimate also in the case $u\gg V_n^2$ when Bernstein's inequality yields a weak bound. But Bennett's inequality provides only a small improvement which has only a limited importance. This may be the reason why Bernstein's inequality which yields a more transparent estimate is more popular.

Example 3.3 (Sums of independent random variables with bad tail distribution for large values). *Let us fix some positive integer n, real numbers u and σ^2 such* *that* $0<\sigma^2\le\frac18$, $n>4u\ge 6$ *and* $u>4n\sigma^2$. *Let* $\bar\sigma^2$ *be that solution of the equation* $x^2-x+\sigma^2=0$ *which is smaller than* $\frac12$. *Take a sequence of independent and identically distributed random variables* $\bar X_1,\dots,\bar X_n$ *such that* $P(\bar X_j=1)=\bar\sigma^2$, $P(\bar X_j=0)=1-\bar\sigma^2$ *for all* $1\le j\le n$. *Put* $X_j=\bar X_j-E\bar X_j=X_j-\bar\sigma^2$, $1\le j\le n$, $S_n=\sum\limits_{j=1}^n X_j$ *and* $V_n^2=n\sigma^2$. *Then* $P(|X_1|\le 1)=1$, $EX_1=0$, $\operatorname{Var} X_1=\sigma^2$, *hence* $ES_n=0$, *and* $\operatorname{Var} S_n=V_n^2$. *Beside this*

$$P(S_n\ge u)>\exp\left\{-Bu\log\frac{u}{V_n^2}\right\}$$

with some appropriate constant $B>0$ *not depending on* n, σ *and* u.

Proof of Example 3.3. Simple calculation shows that $EX_j=0$, $\operatorname{Var} X_j=\bar\sigma^2-\bar\sigma^4=\sigma^2$, $P(|X_j|\le 1)=0$, and also the inequality $\sigma^2\le\bar\sigma^2\le\frac32\sigma^2$ holds. To see the upper bound in the last inequality observe that $\bar\sigma^2\le\frac13$, i.e. $1-\bar\sigma^2\ge\frac23$, hence $\sigma^2=\bar\sigma^2(1-\bar\sigma^2)\ge\frac23\bar\sigma^2$. In the proof of the inequality of Example 3.3, we can restrict our attention to the case when u is an integer, because in the general case we can apply the inequality with $\bar u=[u]+1$ instead of u, where $[u]$ denotes the integer part of u, and since $u\le\bar u\le 2u$, the application of the result in this case supplies the desired inequality with a possibly worse constant $B>0$.

Put $\bar S_n=\sum\limits_{j=1}^n\bar X_j$. We can write $P(S_n\ge u)=P(\bar S_n\ge u+n\bar\sigma^2)\ge P(\bar S_n\ge 2u)\ge P(\bar S_n=2u)=\binom{n}{2u}\bar\sigma^{4u}(1-\bar\sigma^2)^{(n-2u)}\ge\left(\frac{n\bar\sigma^2}{2u}\right)^{2u}(1-\bar\sigma^2)^{(n-2u)}$, since $u\ge n\bar\sigma^2$, and $n\ge 2u$. On the other hand $(1-\bar\sigma^2)^{(n-2u)}\ge e^{-2\bar\sigma^2(n-2u)}\ge e^{-2n\bar\sigma^2}\ge e^{-u}$, hence

$$\begin{aligned}P(S_n\ge u)&\ge\exp\left\{-2u\log\left(\frac{u}{n\bar\sigma^2}\right)-2u\log 2-u\right\}\\&=\exp\left\{-2u\log\left(\frac{u}{n\sigma^2}\right)-2u\log\frac{\bar\sigma^2}{\sigma^2}-2u\log 2-u\right\}\\&\ge\exp\left\{-100u\log\left(\frac{u}{V_n^2}\right)\right\}.\end{aligned}$$

Example 3.3 is proved. □

In the case $u>4V_n^2$ Bernstein's inequality yields the estimate $P(S_n>u)\le e^{-\alpha u}$ with some universal constant $\alpha>0$, and the above example shows that at most an additional logarithmic factor $K\log\frac{u}{V_n^2}$ can be expected in the exponent of the upper bound in an improvement of this estimate. Bennett's inequality shows that such an improvement is really possible.

I finish this chapter with another estimate due to Hoeffding which will be later useful in some symmetrization arguments.

Theorem 3.4 (Hoeffding's inequality). *Let $\varepsilon_1,\dots,\varepsilon_n$ be independent random variables, $P(\varepsilon_j = 1) = P(\varepsilon_j = -1) = \frac{1}{2}$, $1 \le j \le n$, and let $a_1,\dots,a_n$ be arbitrary real numbers. Put $V = \sum\limits_{j=1}^{n} a_j\varepsilon_j$. Then*

$$P(V > u) \le \exp\left\{-\frac{u^2}{2\sum_{j=1}^{n} a_j^2}\right\} \quad \textit{for all } u > 0. \tag{3.5}$$

Remark 1. Clearly $EV = 0$ and $\operatorname{Var} V = \sum\limits_{j=1}^{n} a_j^2$, hence Hoeffding's inequality yields such an estimate for $P(V > u)$ which the central limit theorem suggests. This estimate holds for all real numbers $a_1,\dots,a_n$ and $u > 0$.

Remark 2. The Rademacher functions $r_k(x)$, $k = 1,2,\dots$, defined by the formulas $r_k(x) = 1$ if $(2j-1)2^{-k} \le x < 2j2^{-k}$ and $r_k(x) = -1$ if $2(j-1)2^{-k} \le x < (2j-1)2^{-k}$, $1 \le j \le 2^{k-1}$, for all $k = 1,2,\dots$, can be considered as random variables on the probability space $\Omega = [0,1]$ with the Borel σ-algebra and the Lebesgue measure as probability measure on the interval $[0,1]$. They are independent random variables with the same distribution as the random variables $\varepsilon_1,\dots,\varepsilon_n$ considered in Theorem 3.4. Therefore results about such sequences of random variables whose distributions agree with those in Theorem 3.4 are also called sometimes results about Rademacher functions in the literature. At some points we will also apply this terminology.

Proof of Theorem 3.4. Let us give a good bound on the exponential moment Ee^{tV} for all $t > 0$. The identity $Ee^{tV} = \prod\limits_{j=1}^{n} Ee^{ta_j\varepsilon_j} = \prod\limits_{j=1}^{n} \frac{(e^{a_jt}+e^{-a_jt})}{2}$ holds, and $\frac{(e^{a_jt}+e^{-a_jt})}{2} = \sum\limits_{k=0}^{\infty} \frac{a_j^{2k}}{(2k)!}t^{2k} \le \sum\limits_{k=0}^{\infty} \frac{(a_jt)^{2k}}{2^k k!} = e^{a_j^2t^2/2}$, since $(2k)! \ge 2^k k!$ for all $k \ge 0$. This implies that $Ee^{tV} \le \exp\left\{\frac{t^2}{2}\sum\limits_{j=1}^{n} a_j^2\right\}$. Hence $P(V > u) \le \exp\left\{-tu + \frac{t^2}{2}\sum\limits_{j=1}^{n} a_j^2\right\}$, and we get relation (3.5) with the choice $t = u\left(\sum\limits_{j=1}^{n} a_j^2\right)^{-1}$. □

Chapter 4
On the Supremum of a Nice Class of Partial Sums

This chapter contains an estimate about the supremum of a nice class of normalized sums of independent and identically distributed random variables together with an analogous result about the supremum of an appropriate class of onefold random integrals with respect to a normalized empirical distribution. The second result deals with a one-variate version of the problem about the estimation of multiple integrals with respect to a normalized empirical distribution. This problem was mentioned in the introduction. Some natural questions related to these results will be also discussed. It will be examined how restrictive their conditions are. In particular, we are interested in the question how the condition about the countable cardinality of the class of random variables can be weakened. A natural Gaussian counterpart of the supremum problems about random onefold integrals will be also considered. Most proofs will be postponed to later chapters.

To formulate these results first a notion will be introduced that plays a most important role in the sequel.

Definition of L_p-dense classes of functions. *Let a measurable space $(Y,\mathcal{Y})$ be given together with a class $\mathcal{G}$ of $\mathcal{Y}$ measurable real valued functions on this space. The class of functions $\mathcal{G}$ is called an L_p-dense class of functions, $1\le p<\infty$, with parameter D and exponent L if for all numbers $0<\varepsilon\le 1$ and probability measures ν on the space $(Y,\mathcal{Y})$ there exists a finite ε-dense subset $\mathcal{G}_{\varepsilon,\nu}=\{g_1,\dots,g_m\}\subset\mathcal{G}$ in the space $L_p(Y,\mathcal{Y},\nu)$ with $m\le D\varepsilon^{-L}$ elements, i.e. there exists such a set $\mathcal{G}_{\varepsilon,\nu}\subset\mathcal{G}$ with $m\le D\varepsilon^{-L}$ elements for which $\inf\limits_{g_j\in\mathcal{G}_{\varepsilon,\nu}}\int|g-g_j|^p\,d\nu<\varepsilon^p$ for all functions $g\in\mathcal{G}$. (Here the set $\mathcal{G}_{\varepsilon,\nu}$ may depend on the measure ν, but its cardinality is bounded by a number depending only on ε.)*

In most results of this work the above defined L_p-dense classes will be considered only for the parameter $p=2$. But at some points it will be useful to work also with L_p-dense classes with a different parameter p. Hence to avoid some repetitions I introduced the above definition for a general parameter p. When working with L_p-dense classes we shall consider only such classes of functions $\mathcal{G}$

P. Major, *On the Estimation of Multiple Random Integrals and U-Statistics*, Lecture Notes in Mathematics 2079, DOI 10.1007/978-3-642-37617-7_4,

whose elements are functions with bounded absolute value. Hence all integrals appearing in the definition of L_p-dense classes of functions are finite.

The following estimate will be proved.

Theorem 4.1 (Estimate on the supremum of a class of partial sums). *Let us consider a sequence of independent and identically distributed random variables* $\xi_1, \dots, \xi_n$, $n \geq 2$, *with values in a measurable space* $(X, \mathcal{X})$ *and with some distribution* μ. *Beside this, let a countable and* L_2*-dense class of functions* $\mathcal{F}$ *with some parameter* $D \geq 1$ *and exponent* $L \geq 1$ *be given on the space* $(X, \mathcal{X})$ *which satisfies the conditions*

$$\|f\|_\infty = \sup_{x \in X} |f(x)| \leq 1, \qquad \text{for all } f \in \mathcal{F} \tag{4.1}$$

$$\|f\|_2^2 = \int f^2(x)\mu(dx) \leq \sigma^2 \qquad \text{for all } f \in \mathcal{F} \tag{4.2}$$

with some constant $0 < \sigma \leq 1$, *and*

$$\int f(x)\mu(dx) = 0 \quad \text{for all } f \in \mathcal{F}. \tag{4.3}$$

Define the normalized partial sums $S_n(f) = \frac{1}{\sqrt{n}} \sum_{k=1}^{n} f(\xi_k)$ *for all* $f \in \mathcal{F}$.

There exist some universal constants $C > 0$, $\alpha > 0$ *and* $M > 0$ *such that the supremum of the normalized random sums* $S_n(f)$, $f \in \mathcal{F}$, *satisfies the inequality*

$$P\left(\sup_{f \in \mathcal{F}} |S_n(f)| \geq u\right) \leq C \exp\left\{-\alpha \left(\frac{u}{\sigma}\right)^2\right\} \quad \text{for those numbers } u$$
$$\text{for which } \sqrt{n}\sigma^2 \geq u \geq M\sigma(L^{3/4} \log^{1/2} \tfrac{2}{\sigma} + (\log D)^{3/4}), \tag{4.4}$$

where the numbers D *and* L *in formula (4.4) agree with the parameter and exponent of the* L_2*-dense class* $\mathcal{F}$.

Remark. Here and also in the subsequent part of this work we consider random variables which take their values in a general measurable space $(X, \mathcal{X})$. The only restriction we impose on these spaces is that all sets consisting of one point are measurable, i.e. $\{x\} \in \mathcal{X}$ for all $x \in X$.

The condition $\sqrt{n}\sigma^2 \geq u \geq M\sigma(L^{3/4} \log^{1/2} \frac{2}{\sigma} + D^{3/4})$ for the numbers u for which inequality (4.4) holds is natural. I discuss this after the formulation of Theorem 4.2 which can be considered as the Gaussian counterpart of Theorem 4.1. I also formulate a result in Example 4.3 which can be considered as part of this discussion.

The condition about the countable cardinality of $\mathcal{F}$ can be weakened with the help of the notion of countable approximability introduced below. For the sake of later applications I define it in a more general form than needed in this chapter. In the subsequent part of this work I shall assume that the probability measure I work with is complete, i.e. for all such pairs of sets A and B in the probability space $(\Omega, \mathcal{A}, P)$ for which $A \in \mathcal{A}$, $P(A) = 0$ and $B \subset A$ we have $B \in \mathcal{A}$ and $P(B) = 0$.

Definition of countably approximable classes of random variables. *Let us have a class of random variables $U(f)$, $f \in \mathcal{F}$, indexed by a class of functions $f \in \mathcal{F}$ on a measurable space $(Y, \mathcal{Y})$. This class of random variables is called countably approximable if there is a countable subset $\mathcal{F}' \subset \mathcal{F}$ such that for all numbers $u > 0$ the sets $A(u) = \{\omega\colon \sup\limits_{f \in \mathcal{F}} |U(f)(\omega)| \ge u\}$ and $B(u) = \{\omega\colon \sup\limits_{f \in \mathcal{F}'} |U(f)(\omega)| \ge u\}$ satisfy the identity $P(A(u) \setminus B(u)) = 0$.*

Clearly, $B(u) \subset A(u)$. In the above definition it was demanded that for all $u > 0$ the set $B(u)$ should be almost as large as $A(u)$. The following corollary of Theorem 4.1 holds.

Corollary of Theorem 4.1. *Let a class of functions $\mathcal{F}$ satisfy the conditions of Theorem 4.1 with the only exception that instead of the condition about the countable cardinality of $\mathcal{F}$ it is assumed that the class of random variables $S_n(f)$, $f \in \mathcal{F}$, is countably approximable. Then the random variables $S_n(f)$, $f \in \mathcal{F}$, satisfy relation (4.4).*

This corollary can be simply proved, only Theorem 4.1 has to be applied for the class $\mathcal{F}'$. To do this it has to be checked that if $\mathcal{F}$ is an L_2-dense class with some parameter D and exponent L, and $\mathcal{F}' \subset \mathcal{F}$, then $\mathcal{F}'$ is also an L_2-dense class with the same exponent L, only with a possibly different parameter D'.

To prove this statement let us choose for all numbers $0 < \varepsilon \le 1$ and probability measures ν on $(Y, \mathcal{Y})$ some functions $f_1, \dots, f_m \in \mathcal{F}$ with $m \le D\left(\frac{\varepsilon}{2}\right)^{-L}$ elements, such that the sets $\mathcal{D}_j = \left\{f\colon \int |f - f_j|^2 \, d\nu \le \left(\frac{\varepsilon}{2}\right)^2\right\}$ satisfy the relation $\bigcup\limits_{j=1}^{m} \mathcal{D}_j = Y$. For all sets $\mathcal{D}_j$ for which $\mathcal{D}_j \cap \mathcal{F}'$ is non-empty choose a function $f'_j \in \mathcal{D}_j \cap \mathcal{F}'$. In such a way we get a collection of functions f'_j from the class $\mathcal{F}'$ containing at most $2^L D\varepsilon^{-L}$ elements which satisfies the condition imposed for L_2-dense classes with exponent L and parameter $2^L D$ for this number ε and measure ν.

Next I formulate in Theorem 4.1′ a result about the supremum of the integral of a class of functions with respect to a normalized empirical distribution. It can be considered as a simple version of Theorem 4.1. I formulated this result, because Theorems 4.1 and 4.1′ are special cases of their multivariate counterparts about the supremum of so-called U-statistics and multiple integrals with respect to a normalized empirical distribution discussed in Chap. 8. These results are also closely related, but the explanation of their relation demands some work.

Given a sequence of independent μ distributed random variables $\xi_1,\dots,\xi_n$ taking values in $(X,\mathcal{X})$ let us introduce their empirical distribution on $(X,\mathcal{X})$ as

$$\mu_n(A)(\omega)=\frac{1}{n}\#\{j\colon 1\le j\le n,\ \xi_j(\omega)\in A\}\quad \text{for all } A\in\mathcal{X}, \tag{4.5}$$

and define for all measurable and μ integrable functions f the (random) integral

$$J_n(f)=J_{n,1}(f)=\sqrt{n}\int f(x)(\mu_n(dx)-\mu(dx)). \tag{4.6}$$

Clearly

$$J_n(f)=\frac{1}{\sqrt{n}}\sum_{j=1}^{n}(f(\xi_j)-Ef(\xi_j))=S_n(\hat f)$$

with $\hat f(x)=f(x)-\int f(x)\mu(dx)$. It is not difficult to see that $\sup\limits_{x\in X}|\hat f(x)|\le 2$ if $\sup\limits_{x\in X}|f(x)|\le 1$, $\int \hat f(x)\mu(dx)=0$, $\int \hat f^2(x)\mu(dx)\le \int f^2(x)\mu(dx)$, and if $\mathcal{F}$ is an L_2-dense class of functions with parameter D and exponent L, then the class of functions $\bar{\mathcal{F}}$ consisting of the functions $\bar f(x)=\frac{1}{2}\left(f(x)-\int f(x)\mu(dx)\right)$, $f\in\mathcal{F}$, is an L_2-dense class of functions with parameter D and exponent L. Indeed, since $\int(\bar f-\bar g)^2\,d\nu\le\frac{1}{2}\int(f-g)^2\,d\nu+\frac{1}{2}\int(f-g)^2\,d\mu=\int(f-g)^2\frac{d\mu+d\nu}{2}$, hence $\{\bar f_1,\dots,\bar f_m\}$ is an ε-dense set of $\bar{\mathcal{F}}$ in the $L_2(\nu)$-norm if $\{f_1,\dots,f_m\}$ is an ε-dense set of $\mathcal{F}$ in the $L_2(\frac{\mu+\nu}{2})$-norm. Hence Theorem 4.1 implies the following result.

Theorem 4.1′ (Estimate on the supremum of random integrals with respect to a normalized empirical distribution). *Let us have a sequence of independent and identically distributed random variables $\xi_1,\dots,\xi_n$, $n\ge 2$, with distribution μ on a measurable space $(X,\mathcal{X})$ together with some class of functions $\mathcal{F}$ on this space which satisfies the conditions of Theorem 4.1 with the possible exception of condition (4.3). The estimate (4.4) remains valid if the random sums $S_n(f)$ are replaced in it by the random integrals $J_n(f)$ defined in (4.6). Moreover, similarly to the corollary of Theorem 4.1, the condition about the countable cardinality of the set $\mathcal{F}$ can be replaced by the condition that the class of random variables $J_n(f)$, $f\in\mathcal{F}$, is countably approximable.*

All finite dimensional distributions of the set of random variables $S_n(f)$, $f\in\mathcal{F}$, considered in Theorem 4.1 converge to those of a Gaussian random field $Z(f)$, $f\in\mathcal{F}$, with expectation $EZ(f)=0$ and correlation $EZ(f)Z(g)=\int f(x)g(x)\mu(dx)$, $f,g\in\mathcal{F}$ as $n\to\infty$. Here, and in the subsequent part of the paper a collection of random variables indexed by some set of parameters will be called a Gaussian random field if for all finite subsets of these parameters the random variables indexed by this finite set are jointly Gaussian. We shall also define so-called linear Gaussian random fields. They consist of jointly Gaussian random variables $Z(f)$, $f \in \mathcal{G}$, indexed by the elements of a linear space $f \in \mathcal{G}$ which satisfy the relation $Z(af+bg) = aZ(f) + bZ(g)$ with probability 1 for all real numbers a and b and $f, g \in \mathcal{G}$. (Let us observe that a set of Gaussian random variables $Z(f)$, indexed by the elements of a linear space $f \in \mathcal{G}$ such that $EZ(f) = 0$, and $EZ(f)Z(g) = \int f(x)g(x)\mu(dx)$ for all $f, g \in \mathcal{F}$ is a linear Gaussian random field. This can be seen by checking the identity $E[Z(af+bg)-(aZ(f)+bZ(g))]^2 = 0$ for all real numbers a and b and $f, g \in \mathcal{G}$ in this case.)

Let us consider a linear Gaussian random field $Z(f)$, $f \in \mathcal{G}$, where the set of indices $\mathcal{G} = \mathcal{G}_\mu$ consists of the functions f square integrable with respect to a σ-finite measure μ, and take an appropriate restriction of this field to some parameter set $\mathcal{F} \subset \mathcal{G}$. In the next Theorem 4.2 I present a natural Gaussian counterpart of Theorem 4.1 by means of an appropriate choice of $\mathcal{F}$. Let me also remark that in Chap. 10 the multiple Wiener–Itô integrals of functions of k variables with respect to a white noise will be defined for all $k \ge 1$. In the special case $k = 1$ the Wiener–Itô integrals for an appropriate class of functions $f \in \mathcal{F}$ yield a model for which Theorem 4.2 is applicable. Before formulating this result let us introduce the following definition which is a version of the definition of L_p-dense functions.

Definition of L_p-dense classes of functions with respect to a measure μ. *Let a measurable space $(X, \mathcal{X})$ be given together with a measure μ on the σ-algebra $\mathcal{X}$ and a set $\mathcal{F}$ of $\mathcal{X}$ measurable real valued functions on this space. The set of functions $\mathcal{F}$ is called an L_p-dense class of functions, $1 \le p < \infty$, with respect to the measure μ with parameter D and exponent L if for all numbers $0 < \varepsilon \le 1$ there exists a finite ε-dense subset $\mathcal{F}_\varepsilon = \{f_1, \dots, f_m\} \subset \mathcal{F}$ in the space $L_p(X, \mathcal{X}, \mu)$ with $m \le D\varepsilon^{-L}$ elements, i.e. such a set $\mathcal{F}_\varepsilon \subset \mathcal{F}$ with $m \le D\varepsilon^{-L}$ elements for which $\inf\limits_{f_j \in \mathcal{F}_\varepsilon} \int |f - f_j|^p \, d\mu < \varepsilon^p$ for all functions $f \in \mathcal{F}$.*

Theorem 4.2 (Estimate on the supremum of a class of Gaussian random variables). *Let a probability measure μ be given on a measurable space $(X, \mathcal{X})$ together with a linear Gaussian random field $Z(f)$, $f \in \mathcal{G}$, such that $EZ(f) = 0$, $EZ(f)Z(g) = \int f(x)g(x)\mu(dx)$, $f, g \in \mathcal{G}$, where $\mathcal{G}$ is the space of square integrable functions with respect to this measure μ. Let $\mathcal{F} \subset \mathcal{G}$ be a countable and L_2-dense class of functions with respect to the measure μ with some exponent $L \ge 1$ and parameter $D \ge 1$ which also satisfies condition (4.2) with some $0 < \sigma \le 1$.*

Then there exist some universal constants $C > 0$ and $M > 0$ (for instance $C = 4$ and $M = 16$ is a good choice) such that the inequality

$$P\left(\sup_{f \in \mathcal{F}} |Z(f)| \ge u\right) \le C(D+1)\exp\left\{-\frac{1}{256}\left(\frac{u}{\sigma}\right)^2\right\} \quad \text{if } u \ge ML^{1/2}\sigma\log^{1/2}\frac{2}{\sigma} \tag{4.7}$$

holds with the parameter D and exponent L introduced in this theorem.

Remark. In formulas (4.4) of Theorem 4.1 and in (4.7) of Theorem 4.2 we had a slightly different lower bound on the numbers u for which these results give an estimate on the probability that the supremum of certain random variables is larger then u. Nevertheless in the most interesting cases when the exponent L and the parameter D of the L_2-dense class of functions we consider in these theorems are separated both from zero and infinity these bounds behave similarly. In such cases they have the magnitude const. $\sigma \log^{1/2} \frac{2}{\sigma}$. In (4.7) the lower bound on the number u did not depend on the parameter D, since the dependence on this parameter appeared in the coefficient at the right-hand side of the inequality in this relation. The formula providing a lower bound on the number u had a coefficient $L^{3/4}$ in (4.4) and not a coefficient $L^{1/2}$ as in (4.7). This is a weak bound if L is very large, and it could be improved. But we did not work on this problem, because we were mainly interested in a good bound in the case when the exponent L is separated from infinity.

The exponent at the right-hand side of inequality (4.7) does not contain the best possible universal constant. One could choose the coefficient $\frac{1-\varepsilon}{2}$ with arbitrary small $\varepsilon > 0$ instead of the coefficient $\frac{1}{256}$ in the exponent at the right-hand side of (4.7) if the universal constants $C > 0$ and $M > 0$ are chosen sufficiently large in this inequality. Actually, later in Theorem 8.6 such an estimate will be proved which can be considered as the multivariate generalization of Theorem 4.2 with the expression $-\frac{(1-\varepsilon)u^2}{2\sigma^2}$ in the exponent.

The condition about the countable cardinality of the set $\mathcal{F}$ in Theorem 4.2 could be weakened similarly to Theorem 4.1. But I omit the discussion of this question, since Theorem 4.2 was only introduced for the sake of a comparison between the Gaussian and non-Gaussian case. An essential difference between Theorems 4.1 and 4.2 is that the class of functions $\mathcal{F}$ considered in Theorem 4.1 had to be L_2-dense, while in Theorem 4.2 a weaker version of this property was needed. In Theorem 4.2 it was demanded that there exists a finite subset of $\mathcal{F}$ of relatively small cardinality which is dense in the $L_2(\mu)$ norm. In the L_2-density property imposed in Theorem 4.1 a similar property was demanded for all probability measures ν. The appearance of such a condition may be unexpected. It is not clear why we demand this property for such probability measures ν which have nothing to do with our problem. But as we shall see, the proof of Theorem 4.1 contains a conditioning argument where a lot of new conditional measures appear, and the L_2-density property is needed to work with all of them. One would also like to know some results that enable us to check when this condition holds. In the next chapter a notion popular in probability theory, the notion of Vapnik–Červonenkis classes will be introduced, and it will be shown that a Vapnik–Červonenkis class of functions bounded by 1 is L_2-dense.

Another difference between Theorems 4.1 and 4.2 is that the conditions of formula (4.4) contain the upper bound $\sqrt{n}\sigma^2 > u$, and no similar condition was imposed in formula (4.7). The appearance of this condition in Theorem 4.1 can be explained by comparing this result with those of Chap. 3. As we have seen, we do not loose much information if we restrict our attention to the case

$u \le \text{const.}\, V_n^2 = \text{const.}\, n\sigma^2$ in Bernstein's inequality (if sums of independent and identically distributed random variables are considered). Theorem 4.1 gives an almost as good estimate for the supremum of normalized partial sums under appropriate conditions for the class $\mathcal{F}$ of functions we consider in this theorem as Bernstein's inequality yields for the normalized partial sums of independent and identically distributed random variables with variance bounded by σ^2. But we could prove the estimate of Theorem 4.1 only under the condition $\sqrt{n}\sigma^2 > u$. (Actually we could slightly improve this result. We could impose the condition $B\sqrt{n}\sigma^2 > u$ with an arbitrary constant $B > 0$ in (4.4) if the remaining constants are appropriately chosen in dependence of B in this formula.) It has also a natural reason why condition (4.1) about the supremum of the functions $f \in \mathcal{F}$ appeared in Theorems 4.1 and 4.1′, and no such condition was needed in Theorem 4.2.

The lower bounds for the level u were imposed in formulas (4.4) and (4.7) because of a similar reason. To understand why such a condition is needed in formula (4.7) let us consider the following example.

Take a Wiener process $W(t)$, $0 \le t \le 1$, define for all $0 \le s < t \le 1$ the functions $f_{s,t}(\cdot)$ on the interval $[0,1]$ as $f_{s,t}(u) = 1$ if $s \le u \le t$, $f_{s,t}(u) = 0$ if $0 \le u < s$ or $t < u \le 1$, and introduce for all $\sigma > 0$ the following class of functions $\mathcal{F}_\sigma$. $\mathcal{F}_\sigma = \{f_{s,t}\colon 0 \le s < t \le 1,\ t - s \le \sigma^2,\ s \text{ and } t \text{ are rational numbers.}\}$. The integral $Z(f) = \int_0^1 f(x)W(dx)$ can be defined for all square integrable functions f on the interval $[0,1]$, and this yields a linear Gaussian random field on the space of square integrable functions. In the special case $f = f_{s,t}$ we have $Z(f_{s,t}) = \int f_{s,t}(u)W(du) = W(t) - W(s)$. It is not difficult to see that the Gaussian random field $Z(f)$, $f \in \mathcal{F}_\sigma$, satisfies the conditions of Theorem 4.2 with the number σ in formula (4.2). It is natural to expect that $P\left(\sup\limits_{f \in \mathcal{F}_\sigma} Z(f) > u\right) \le e^{-\text{const.}\,(u/\sigma)^2}$. However, this relation does not hold if $u = u(\sigma) < 2(1-\varepsilon)\sigma \log^{1/2}\frac{1}{\sigma}$ with some $\varepsilon > 0$. In such cases $P\left(\sup\limits_{f \in \mathcal{F}_\sigma} Z(f) > u\right) \to 1$, as $\sigma \to 0$. This can be proved relatively simply with the help of the estimate $P(Z(f_{s,t}) > u(\sigma)) \ge \text{const.}\, \sigma^{2(1-\varepsilon)^2}$ if $|t - s| = \sigma^2$ and the independence of the random integrals $Z(f_{s,t})$ if the functions $f_{s,t}$ are indexed by such pairs (s,t) for which the intervals (s,t) are disjoint. This means that in this example formula (4.7) holds only under the condition $u \ge M\sigma \log^{1/2}\frac{1}{\sigma}$ with $M = 2$.

There is a classical result about the modulus of continuity of Wiener processes, and actually this result helped us to find the previous example. It is also worth mentioning that there are some concentration inequalities, see Ledoux [31] and Talagrand [56], which state that under very general conditions the distribution of the supremum of a class of partial sums of independent random variables or of the elements of a Gaussian random field is strongly concentrated around the expected value of this supremum. (Talagrand's result in this direction is also formulated in Theorem 18.1 of this lecture note.) These results imply that the problems discussed in Theorems 4.1 and 4.2 can be reduced to a good estimate of the expected value

$E \sup\limits_{f\in\mathcal{F}} |S_n(f)|$ and $E \sup\limits_{f\in\mathcal{F}} |Z(f)|$ of the supremum considered in these results. However, the estimation of the expected value of these suprema is not much simpler than the original problem.

Theorem 4.2 implies that under its conditions

$$E \sup_{f\in\mathcal{F}} |Z(f)| \le \text{const.}\,\sigma \log^{1/2}\frac{2}{\sigma}$$

with an appropriate multiplying constant depending on the parameter D and exponent L of the class of functions $\mathcal{F}$. In the case of Theorem 4.1 a similar estimate holds, but under more restrictive conditions. We also have to impose that $\sqrt{n}\sigma^2 \ge \text{const.}\,\sigma \log^{1/2}\frac{2}{\sigma}$ with a sufficiently large constant. This condition is needed to guarantee that the set of numbers u satisfying condition (4.4) is not empty. If this condition is violated, then Theorem 4.1 supplies a weaker estimate which we get by replacing σ by an appropriate $\bar\sigma > \sigma$, and by applying Theorem 4.1 with this number $\bar\sigma$.

One may ask whether the above estimate on the expected value of the supremum of normalized partial sums holds without the condition $\sqrt{n}\sigma^2 \ge \text{const.}\,\sigma \log^{1/2}\frac{2}{\sigma}$. We show an example which gives a negative answer to this question. Since here we discuss a rather particular problem which is outside of our main interest in this work I give a rather sketchy explanation of this example. I present this example together with a Poissonian counterpart of it which may help to explain its background.

Example 4.3 (Supremum of partial sums with bad tail behaviour). *Let $\xi_1,\dots,\xi_n$ be a sequence of independent random variables with uniform distribution in the interval* $[0,1]$. *Choose a sequence of real numbers, ε_n, $n = 3,4,\dots$, such that $\varepsilon_n \to 0$ as $n\to\infty$, and $\frac12 \ge \varepsilon_n \ge n^{-\delta}$ with a sufficiently small number $\delta > 0$. Put $\sigma_n = \varepsilon_n\sqrt{\frac{\log n}{n}}$, and define the set of functions $\bar f_{j,n}(\cdot)$ and $f_{j,n}(\cdot)$ on the interval* $[0,1]$ *by the formulas $\bar f_{j,n}(x) = 1$ if $(j-1)\sigma_n^2 \le x < j\sigma_n^2$, $\bar f_{j,n}(x) = 0$ otherwise, and $f_{j,n}(x) = \bar f_{j,n}(x) - \sigma_n^2$, $n = 3,4,\dots$, $1 \le j \le \frac{1}{\sigma_n^2}$. Put $\mathcal{F}_n = \{f_{j,n}(\cdot)\colon 1 \le j \le \frac{1}{\sigma_n^2}\}$, $S_n(f) = \frac{1}{\sqrt n}\sum\limits_{k=1}^{n} f(\xi_k)$ for $f \in \mathcal{F}_n$ and $u_n = \frac{A}{\log\frac{1}{\varepsilon_n}}\frac{\log n}{\sqrt n}$ with a sufficiently small $A > 0$. Then*

$$\lim_{n\to\infty} P\left(\sup_{f\in\mathcal{F}_n} S_n(f) > u_n\right) = 1.$$

This example has the following Poissonian counterpart.

Example 4.3′ (A Poissonian counterpart of Example 4.3). *Let $\bar P_n(x)$ be a Poisson process on the interval* $[0,1]$ *with parameter n and $P_n(x) = \frac{1}{\sqrt n}[\bar P_n(x) - nx]$, $0 \le x \le 1$. Consider the same sequences of numbers ε_n, σ_n and u_n as in*

Example 4.3, and define the random variables $Z_{n,j} = P_n(j\sigma_n^2) - P_n((j-1)\sigma_n^2)$ *for all* $n = 3, 4, \dots$ *and* $1 \le j \le \frac{1}{\sigma_n^2}$. *Then*

$$\lim_{n\to\infty} P\left(\sup_{1\le j\le \frac{1}{\sigma_n}} Z_{n,j} > u_n\right) = 1.$$

The classes of functions $\mathcal{F}_n$ in Example 4.3 are L_2-dense classes of functions with some exponent L and parameter D not depending on the parameter n and the choice of the numbers σ_n. It can be seen that even the class of function $\mathcal{F} = \{f_{s,t}\colon f_{s,t}(x) = 1, \text{ if } s \le x < t,\ f_{s,t}(x) = 0 \text{ otherwise.}\}$ consisting of functions defined on the interval $[0,1]$ is an L_2-dense class with some exponent L and parameter D. This follows from the results discussed in the later part of this work (mainly Theorem 5.2), but it can be proved directly that this statement holds e.g. with $L = 1$ and $D = 8$. The classes of functions $\mathcal{F}_n$ also satisfy conditions (4.1)–(4.3) of Theorem 4.1 with $\sigma^2 = \bar\sigma_n^2 = \sigma_n^2 - \sigma_n^4$, $\lim\limits_{n\to\infty}\frac{\bar\sigma_n}{\sigma_n} = 1$, and the number u_n satisfies the second condition $u_n \ge M\bar\sigma_n(L^{3/4}\log^{1/2}\frac{2}{\bar\sigma_n} + (\log D)^{3/4})$ in (4.4) for sufficiently large n. But it does not satisfy the first condition $\sqrt{n}\bar\sigma_n^2 \ge u_n$ of (4.4), and as a consequence Theorem 4.1 cannot be applied in this case. On the other hand, some calculation shows that $u_n \ge (\frac{2}{1+4\delta})^{1/2}\frac{A}{\varepsilon_n\log\frac{1}{\varepsilon_n}}\sigma_n\log^{1/2}\frac{2}{\sigma_n}$. Hence $\liminf\limits_{n\to\infty}\varepsilon_n\log\frac{1}{\varepsilon_n}\cdot\frac{1}{\bar\sigma_n\log^{1/2}\frac{2}{\bar\sigma_n}}E\sup\limits_{f\in\mathcal{F}_n}S_n(f) > 0$ in this case. As $\varepsilon_n\log\frac{1}{\varepsilon_n}\to 0$ as $n\to\infty$, this means that the expected value of the supremum of the random sums considered in Example 4.3 does not satisfy the estimate $\limsup\limits_{n\to\infty}\frac{1}{\bar\sigma_n\log^{1/2}\frac{2}{\bar\sigma_n}}E\sup\limits_{f\in\mathcal{F}_n}S_n(f) < \infty$ suggested by Theorem 4.1. Observe that $\sqrt{n}\bar\sigma_n^2 \sim \text{const.}\,\varepsilon_n\bar\sigma_n\log^{1/2}\frac{2}{\bar\sigma_n}$ in this case, since $\sqrt{n}\bar\sigma_n^2 \sim \varepsilon_n^2\frac{\log n}{\sqrt{n}}$, and $\bar\sigma_n\log^{1/2}\frac{2}{\bar\sigma_n}\sim \text{const.}\,\varepsilon_n\frac{\log n}{\sqrt{n}}$.

The proof of Examples 4.3 and 4.3′. First we prove the statement of Example 4.3′. For a fixed index n the number of random variables $Z_{n,j}$ equals $\frac{1}{\sigma_n^2} \ge \frac{1}{\varepsilon_n^2}\frac{n}{\log n} \ge \frac{n}{\log n}$, and they are independent. Hence it is enough to show that $P(Z_{n,j} > u_n) \ge n^{-1/2}$ if first $A > 0$ and then $\delta > 0$ (appearing in the condition $\varepsilon_n > n^{-\delta}$) are chosen sufficiently small, and $n \ge n_0$ with some threshold index $n_0 = n_0(A,\delta)$.

Put $\bar u_n = [\sqrt{n}u_n + n\sigma_n^2] + 1$, where $[\cdot]$ denotes integer part. Then $P(Z_{n,j} > u_n) \ge P(\bar P_n(\sigma_n^2) \ge \bar u_n) \ge P(\bar P_n(\sigma_n^2) = \bar u_n) = \frac{(n\sigma_n^2)^{\bar u_n}}{\bar u_n!}e^{-n\sigma_n^2} \ge \left(\frac{n\sigma_n^2}{\bar u_n}\right)^{\bar u_n}e^{-n\sigma_n^2}$. Some calculation shows that $\bar u_n \le \frac{A\log n}{\log\frac{1}{\varepsilon_n}} + \varepsilon_n^2\log n + 1 \le \frac{2A\log n}{\log\frac{1}{\varepsilon_n}}$, $\frac{n\sigma_n^2}{\bar u_n} \ge \frac{\varepsilon_n^2\log\frac{1}{\varepsilon_n}}{2A}$, and $\log\frac{n\sigma_n^2}{\bar u_n} \ge -2\log\frac{1}{\varepsilon_n}$ if the constants $A > 0$, $\delta > 0$ and threshold index n_0 are appropriately chosen. Hence $P(Z_{n,j} > u_n) \ge e^{-2\bar u_n\log(1/\varepsilon_n) - n\sigma_n^2} \ge e^{-2A\log n - \varepsilon_n^2\log n} \ge \frac{1}{\sqrt{n}}$ if $A_0 > 0$ is small enough.

The statement of Example 4.3 can be deduced from Example 4.3′ by applying Poissonian approximation. Let us apply the result of Example 4.3′ for a Poisson process $\bar{P}_{n/2}$ with parameter $\frac{n}{2}$ and with such a number $\bar{\varepsilon}_{n/2}$ with which the value of $\sigma_{n/2}$ equals the previously defined σ_n. Then $\bar{\varepsilon}_{n/2} \sim \frac{\varepsilon_n}{\sqrt{2}}$, and the number of sample points of $\bar{P}_{n/2}$ is less than n with probability almost 1. Attaching additional sample points to get exactly n sample points we can get the result of Example 4.3. I omit the details. □

In formulas (4.4) and (4.7) we formulated such a condition for the validity of Theorem 4.1 and Theorem 4.2 which contains a large multiplying constant $ML^{3/4}$ and $ML^{1/2}$ of $\sigma \log^{1/2} \frac{2}{\sigma}$ in the lower bound for the number u if we deal with such an L_2-dense class of functions $\mathcal{F}$ which has a large exponent L. At a heuristic level it is clear that in such a case a large multiplying constant appears. On the other hand, I did not try to find the best possible coefficients in the lower bound in relations (4.4) and (4.7).

In Theorem 4.1 (and in its version, Theorem 4.1′) it was demanded that the class of functions $\mathcal{F}$ should be countable. Later this condition was replaced by a weaker one about countable approximability. By restricting our attention to countable or countably approximable classes we could avoid some unpleasant measure theoretical problems which would have arisen if we had worked with the supremum of non-countably many random variables which may be non-measurable. There are some papers where possibly non-measurable models are also considered with the help of some rather deep results of the analysis and measure theory. Here I chose a different approach. I proved a simple result in the following Lemma 4.4 which enables us to show that in many interesting problems we can restrict our attention to countably approximable classes of random variables. In Chap. 18, in the discussion of the content of Chap. 4 I write more about the relation of this approach to the results of other works.

Lemma 4.4. *Let a class of random variables $U(f)$, $f \in \mathcal{F}$, indexed by some set $\mathcal{F}$ of functions be given on a space $(Y, \mathcal{Y})$. If there exists a countable subset $\mathcal{F}' \subset \mathcal{F}$ of the set $\mathcal{F}$ such that the sets $A(u) = \{\omega\colon \sup\limits_{f \in \mathcal{F}} |U(f)(\omega)| \ge u\}$ and $B(u) = \{\omega\colon \sup\limits_{f \in \mathcal{F}'} |U(f)(\omega)| \ge u\}$ introduced for all $u > 0$ in the definition of countable approximability satisfy the relation $A(u) \subset B(u - \varepsilon)$ for all $u > \varepsilon > 0$, then the class of random variables $U(f)$, $f \in \mathcal{F}$, is countably approximable.*

The above property holds if for all $f \in \mathcal{F}$, $\varepsilon > 0$ and $\omega \in \Omega$ there exists a function $\bar{f} = \bar{f}(f, \varepsilon, \omega) \in \mathcal{F}'$ such that $|U(\bar{f})(\omega)| \ge |U(f)(\omega)| - \varepsilon$.

Proof of Lemma 4.4. If $A(u) \subset B(u - \varepsilon)$ for all $\varepsilon > 0$, then $P^*(A(U) \setminus B(u)) \le \lim\limits_{\varepsilon \to 0} P(B(u - \varepsilon) \setminus B(u)) = 0$, where $P^*(X)$ denotes the outer measure of a not necessarily measurable set $X \subset \Omega$, since $\bigcap\limits_{\varepsilon \to 0} B(u - \varepsilon) = B(u)$, and this is what we had to prove. If $\omega \in A(u)$, then for all $\varepsilon > 0$ there exists some $f = f(\omega) \in \mathcal{F}$ such that $|U(f)(\omega)| > u - \frac{\varepsilon}{2}$. If there exists some $\bar{f} = \bar{f}(f, \frac{\varepsilon}{2}, \omega)$, $\bar{f} \in \mathcal{F}'$ such that $|U(\bar{f})(\omega)| \ge |Uf(\omega)| - \frac{\varepsilon}{2}$, then $|U(\bar{f})(\omega)| > u - \varepsilon$, and $\omega \in B(u - \varepsilon)$. This means that $A(u) \subset B(u - \varepsilon)$. □

The question about countable approximability also appears in the case of multiple random integrals with respect to a normalized empirical measure. To avoid some repetition we prove a result which also covers such cases. For this goal first we introduce the notion of multiple integrals with respect to a normalized empirical distribution.

Given a measurable function $f(x_1,\dots,x_k)$ on the k-fold product space $(X^k,\mathcal{X}^k)$ and a sequence of independent random variables $\xi_1,\dots,\xi_n$ with some distribution μ on the space $(X,\mathcal{X})$ we define the integral $J_{n,k}(f)$ of the function f with respect to the k-fold product of the normalized version of the empirical distribution μ_n introduced in (4.5) by the formula

$$J_{n,k}(f) = \frac{n^{k/2}}{k!}\int' f(x_1,\dots,x_k)(\mu_n(dx_1)-\mu(dx_1))\dots(\mu_n(dx_k)-\mu(dx_k)),$$

$$\text{where the prime in } \int' \text{ means that the diagonals } x_j = x_l,$$

$$1 \le j < l \le k, \text{ are omitted from the domain of integration.} \tag{4.8}$$

In the case $k \ge 2$ it will be assumed that the probability measure μ has no atoms.

Lemma 4.4 enables us to prove that certain classes of random integrals $J_{n,k}(f)$, $f \in \mathcal{F}$, defined with the help of some set of functions $f \in \mathcal{F}$ of k variables are countably approximable. I present an example for a class of such random integrals. I restrict my attention in this work to this case, because this seems to be the most important case in possible statistical applications. The result I formulate says roughly speaking that if we take the (multiple) integral of a function restricted to all possible rectangles (with respect to a normalized empirical distribution), then the class of these integrals is countably approximable. Hence the results of this lecture note is applicable for them.

Let us consider the case when $X = R^s$, the s-dimensional Euclidean space with some $s \ge 1$. For two vectors $u = (u^{(1)},\dots,u^{(s)}) \in R^s$, $v = (v^{(1)},\dots,v^{(s)}) \in R^s$ such that $u < v$, i.e. $u^{(j)} < v^{(j)}$ for all $1 \le j \le s$ let $B(u,v)$ denote the s-dimensional rectangle $B(u,v) = \{z\colon u < z < v\}$. Let us fix some function $f(x_1,\dots,x_k)$ of k variables such that $\sup|f(x_1,\dots,x_k)| \le 1$, on the space $(X^k,\mathcal{X}^k) = (R^{ks},\mathcal{B}^{ks})$, where $\mathcal{B}^t$ denotes the Borel σ-algebra on the Euclidean space R^t, together with some probability measure μ on $(R^s,\mathcal{B}^s)$. For all pairs of vectors $(u_1,\dots,u_k)$, $(v_1,\dots,v_k)$ such that $u_j, v_j \in R^s$ and $u_j \le v_j$, $1 \le j \le k$, let us define the function $f_{u_1,\dots,u_k,v_1,\dots,v_k}$ which equals the function f on the rectangle $(u_1,v_1)\times\dots\times(u_k,v_k)$, and it is zero outside of this rectangle. Let us call a class of functions $\mathcal{F}$ consisting of functions of the form $f_{u_1,\dots,u_k,v_1,\dots,v_k}$ closed if it has the following property. If $f_{u_1,\dots,u_k,v_1,\dots,v_k} \in \mathcal{F}$ for some vectors $(u_1,\dots,u_k)$ and $(v_1,\dots,v_k)$, and $u_j \le \bar u_j < \bar v_j \le v_j$, $1 \le j \le k$, then $f_{\bar u_1,\dots,\bar u_k,\bar v_1,\dots,\bar v_k} \in \mathcal{F}$. In Lemma 4.5 a closed class $\mathcal{F}$ of functions will be considered, and it will be proved that the random integrals of the functions from this class of functions $\mathcal{F}$ introduced in formula (4.8) constitute a countably approximable class.

Lemma 4.5. *Let a function f on the Euclidean space R^{ks} satisfy the condition $|f| \le 1$ in all points, and let us consider a closed class $\mathcal{F}$ of functions of the form $f_{u_1,\dots,u_k,v_1,\dots,v_k} \in (R^{sk}, \mathcal{B}^{sk})$, $u_j, v_j \in R^s$, $u_j \le v_j$, $1 \le j \le k$, introduced in the previous paragraph with the help of this function f. Let us take n independent and identically distributed random variables $\xi_1, \dots, \xi_n$ with some distribution μ and values in the space $(R^s, \mathcal{B}^s)$. Let μ_n denote the empirical distribution of this sequence. Then the class of random integrals $J_{n,k}(f_{u_1,\dots,u_k,v_1,\dots,v_k})$ defined in formula (4.8) with functions $f_{u_1,\dots,u_k,v_1,\dots,v_k} \in \mathcal{F}$ is countably approximable.*

Proof of Lemma 4.5. We shall prove that the definition of countable approximability is satisfied in this model if the class of functions $\mathcal{F}'$ consists of those functions $f_{u_1,\dots,u_k,v_1,\dots,v_k}$, $u_j \le v_j$, $1 \le j \le k$, for which all coordinates of the vectors u_j and v_j are rational numbers.

Given some function $f_{u_1,\dots,u_k,v_1,\dots,v_k}$, a real number $0 < \varepsilon < 1$ and $\omega \in \Omega$ let us choose a function $f_{\bar u_1,\dots,\bar u_k,\bar v_1,\dots,\bar v_k} \in \mathcal{F}'$ determined with some vectors $\bar u_j = \bar u_j(\varepsilon, \omega)$, $\bar v_j = \bar v_j(\varepsilon, \omega)$ $1 \le j \le k$, with rational coordinates $u_j \le \bar u_j < \bar v_j \le v_j$ in such a way that the sets $K_j = B(u_j, v_j) \setminus B(\bar u_j, \bar v_j)$ satisfy the relations $\mu(K_j) \le \varepsilon 2^{-2k+1} n^{-k/2}$, and $\xi_l(\omega) \notin K_j$ for all $j = 1, \dots, k$ and $l = 1, \dots, n$. Let us show that

$$|J_{n,k}(f_{\bar u_1,\dots,\bar u_k,\bar v_1,\dots,\bar v_k})(\omega) - J_{n,k}(f_{u_1,\dots,u_k,v_1,\dots,v_k})(\omega)| \le \varepsilon. \tag{4.9}$$

Lemma 4.4 (with the choice $U(f) = J_{n,k}(f)$) and relation (4.9) imply Lemma 4.5.

Relation (4.9) holds, since the difference of integrals at its left-hand side can be written as the sum of the $2^k - 1$ integrals of the function f with respect to the k-fold product of the measure $\sqrt{n}(\mu_n - \mu)$ on the domains $D_1 \times \dots \times D_k$ with the omission of the diagonals $x_j = x_{\bar j}$, $1 \le j, \bar j \le k$, $j \ne \bar j$, where D_j is either the set K_j or $B(u_j, v_j)$ and $D_j = K_j$ for at least one index j. It is enough to show that the absolute value of all these integrals is less than $\varepsilon 2^{-k}$. This follows from the observations that $|f(x_1, \dots, x_k)| \le 1$, $\sqrt{n}(\mu_n - \mu)(K_j) = -\sqrt{n}\mu(K_j)$, $\mu(K_j) \le \varepsilon 2^{-2k+1} n^{-k/2}$, and the total variation of the signed measure $\sqrt{n}(\mu_n - \mu)$ (restricted to the set $B(u_j, v_j)$) is less than $2\sqrt{n}$. □

In Lemma 4.5 we have shown with the help of Lemma 4.4 about an important class of functions that it is countably approximable. There are other interesting classes of functions whose countable approximability can be proved with the help of Lemma 4.4. But here we shall not discuss this problem.

Let us discuss the relation of the results in this chapter to an important result in probability theory, to the so-called fundamental theorem of the mathematical statistics. In that result a sequence of independent random variables $\xi_1(\omega), \dots, \xi_n(\omega)$ is taken with some distribution function $F(x)$, the empirical distribution function $F_n(x) = F_n(x, \omega) = \frac{1}{n}\#\{j\colon 1 \le j \le n,\ \xi_j(\omega) < x\}$ is introduced, and the difference $F_n(x) - F(x)$ is considered. This result states that $\sup_x |F_n(x) - F(x)|$ tends to zero with probability one.

Observe that $\sup_x |F_n(x) - F(x)| = n^{-1/2} \sup_{f \in \mathcal{F}} |J_n(f)|$, where $\mathcal{F}$ consists of the functions $f_x(\cdot)$, $x \in R^1$, defined by the relation $f_x(u) = 1$ if $u < x$, and $f_x(u) = 0$ if $u \geq x$. Theorem 4.1′ yields an estimate for the probabilities $P\left(\sup\limits_{f\in\mathcal{F}} |J_n(f)| > u\right)$. We have seen that the above class of functions $\mathcal{F}$ is countably approximable. The results of the next chapter imply that this class of functions is also L_2-dense. Let me remark that actually it is not difficult to check this property directly. Hence we can apply Theorem 4.1′ to the above defined class of functions with $\sigma = 1$, and it yields that $P\left(n^{-1/2} \sup\limits_{f\in\mathcal{F}} |J_n(f)| > u\right) \leq e^{-Cnu^2}$ if $1 \geq u \geq \bar{C}n^{-1/2}$ with some universal constants $C > 0$ and $\bar{C} > 0$. (The condition $1 \geq u$ can actually be dropped.) The application of this estimate for the numbers $\varepsilon > 0$ together with the Borel–Cantelli lemma imply the fundamental theorem of the mathematical statistics.

In short, the results of this chapter yield more information about the closeness the empirical distribution function F_n and distribution function F than the fundamental theorem of the mathematical statistics. Moreover, since these results can also be applied for other classes of functions, they yield useful information about the closeness of the probability measure μ to the empirical distribution μ_n.

Chapter 5
Vapnik–Červonenkis Classes and L_2-Dense Classes of Functions

In this chapter the most important notions and results will be presented about Vapnik–Červonenkis classes, and it will be explained how they help to show in some important cases that certain classes of functions are L_2-dense. The classes of L_2-dense classes played an important role in the previous chapter. The results of this chapter may help to find interesting classes of functions with this property. Some of the results of this chapter will be proved in Appendix A.

First I recall the definition of the following notion.

Definition of Vapnik–Červonenkis classes of sets and functions. *Let a set X be given, and let us select a class $\mathcal{D}$ of subsets of this set X. We call $\mathcal{D}$ a Vapnik–Červonenkis class if there exist two real numbers B and K such that for all positive integers n and subsets $S(n) = \{x_1, \dots, x_n\} \subset X$ of cardinality n of the set X the collection of sets of the form $S(n) \cap D$, $D \in \mathcal{D}$, contains no more than Bn^K subsets of $S(n)$. We shall call B the parameter and K the exponent of this Vapnik–Červonenkis class.*

A class of real valued functions $\mathcal{F}$ on a space $(Y, \mathcal{Y})$ is called a Vapnik–Červonenkis class if the collection of graphs of these functions is a Vapnik–Červonenkis class, i.e. if the sets $A(f) = \{(y,t)\colon y \in Y,\ \min(0, f(y)) \le t \le \max(0, f(y))\}$, $f \in \mathcal{F}$, constitute a Vapnik–Červonenkis class of subsets of the product space $X = Y \times R^1$.

The following result which was first proved by Sauer plays a fundamental role in the theory of Vapnik–Červonenkis classes. This result provides a relatively simple condition for a class $\mathcal{D}$ of subsets of a set X to be a Vapnik–Červonenkis class. Its proof is given in Appendix A. Before its formulation I introduce some terminology which is often applied in the literature.

Definition of shattering of a set. *Let a set S and a class $\mathcal{E}$ of subsets of S be given. A finite set $F \subset S$ is called shattered by the class $\mathcal{E}$ if all its subsets $H \subset F$ can be written in the form $H = E \cap F$ with some element $E \in \mathcal{E}$ of the class of sets of $\mathcal{E}$.*

P. Major, *On the Estimation of Multiple Random Integrals and U-Statistics*, Lecture Notes in Mathematics 2079, DOI 10.1007/978-3-642-37617-7_5,

Theorem 5.1 (Sauer's lemma). *Let a finite set $S = S(n)$ consisting of n elements be given together with a class $\mathcal{E}$ of subsets of S. If $\mathcal{E}$ shatters no subset of S of cardinality k, then $\mathcal{E}$ contains at most $\binom{n}{0} + \binom{n}{1} + \cdots + \binom{n}{k-1}$ subsets of S.*

The estimate of Sauer's lemma is sharp. Indeed, if $\mathcal{E}$ contains all subsets of S of cardinality less than or equal to $k-1$, then it shatters no subset of a set F of cardinality k (a set F of cardinality k cannot be written in the form $E \cap F$, $E \in \mathcal{E}$), and $\mathcal{E}$ contains $\binom{n}{0} + \binom{n}{1} + \cdots + \binom{n}{k-1}$ subsets of S. Sauer's lemma states, that this is an extreme case. Any class of subsets $\mathcal{E}$ of S with cardinality greater than $\binom{n}{0} + \binom{n}{1} + \cdots + \binom{n}{k-1}$ shatters at least one subset of S with cardinality k.

Let us have a set X and a class of subsets $\mathcal{D}$ of it. One may be interested in when $\mathcal{D}$ is a Vapnik–Červonenkis class. Sauer's lemma gives a useful condition for it. Namely, it implies that if there exists a positive integer k such that the class $\mathcal{D}$ shatters no subset of X of cardinality k, then $\mathcal{D}$ is a Vapnik–Červonenkis class. Indeed, let us take some number $n \ge k$, fix an arbitrary set $S(n) = \{x_1, \dots, x_n\} \subset X$ of cardinality n, and introduce the class of subsets $\mathcal{E} = \mathcal{E}(S(n)) = \{S(n) \cap D \colon D \subset \mathcal{D}\}$. If $\mathcal{D}$ shatters no subset of X of cardinality k, then $\mathcal{E}$ shatters no subset of $S(n)$ of cardinality k. Hence by Sauer's lemma the class $\mathcal{E}$ contains at most $\binom{n}{0} + \binom{n}{1} + \cdots + \binom{n}{k-1}$ elements. Let me remark that it is also proved that $\binom{n}{0} + \binom{n}{1} + \cdots + \binom{n}{k-1} \le 1.5\frac{n^{k-1}}{(k-1)!}$ if $n \ge k+1$. This estimate gives a bound on the parameter and exponent of a Vapnik–Červonenkis class which satisfies the above condition.

Moreover, Theorem 5.1 also has the following consequence. Take an (infinite) set X and a class of its subsets $\mathcal{D}$. There are two possibilities. Either there is some set $S(n) \subset X$ of cardinality n for all integers n such that $\mathcal{E}(S(n))$ contains all subsets of $S(n)$, i.e. $\mathcal{D}$ shatters this set, or $\sup\limits_{S\colon S\subset X, |S|=n} |\mathcal{E}(S)|$ tends to infinity at most in a polynomial order as $n \to \infty$, where $|S|$ and $|\mathcal{E}(S)|$ denote the cardinality of S and $\mathcal{E}(S)$.

To understand why Sauer's lemma plays an important role in the theory of Vapnik–Červonenkis classes let us formulate the following consequence of the above considerations.

Corollary of Sauer's lemma. *Let a set X be given together with a class $\mathcal{D}$ of subsets of this set X. This class of sets $\mathcal{D}$ is a Vapnik–Červonenkis class if there exists a positive integer k such that $\mathcal{D}$ shatters no subset $F \subset X$ of cardinality k. In other words if each set $F = \{x_1, \dots, x_k\} \subset X$ of cardinality k has a subset $G \subset F$ which cannot be written in the form $G = D \cap F$ with some $D \in \mathcal{D}$, then $\mathcal{D}$ is a Vapnik–Červonenkis class.*

The following Theorem 5.2, an important result of Richard Dudley, states that a Vapnik–Červonenkis class of functions bounded by 1 is an L_1-dense class of functions.

Theorem 5.2 (A relation between the L_1-dense class and Vapnik–Červonenkis class property). *Let $f(y)$, $f \in \mathcal{F}$, be a Vapnik–Červonenkis class of real valued* *functions on some measurable space* $(Y, \mathscr{Y})$ *such that* $\sup_{y\in Y} |f(y)| \le 1$ *for all* $f \in \mathscr{F}$. *Then* $\mathscr{F}$ *is an* L_1*-dense class of functions on* $(Y, \mathscr{Y})$*. More explicitly, if* $\mathscr{F}$ *is a Vapnik–Červonenkis class with parameter* $B \ge 1$ *and exponent* $K > 0$*, then it is an* L_1*-dense class with exponent* $L = 2K$ *and parameter* $D = CB^2(4K)^{2K}$ *with some universal constant* $C > 0$.

Proof of Theorem 5.2. Let us fix some probability measure ν on $(Y, \mathscr{Y})$ and a real number $0 < \varepsilon \le 1$. We are going to show that any finite set $\mathscr{D}(\varepsilon, \nu) = \{f_1, \dots, f_M\} \subset \mathscr{F}$ such that $\int |f_j - f_k|\, d\nu \ge \varepsilon$ if $j \ne k$, $f_j, f_k \in \mathscr{D}(\varepsilon, \nu)$ has cardinality $M \le D\varepsilon^{-L}$ with some $D > 0$ and $L > 0$. This implies that $\mathscr{F}$ is an L_1-dense class with parameter D and exponent L. Indeed, let us take a maximal subset $\bar{\mathscr{D}}(\varepsilon, \nu) = \{f_1, \dots, f_M\} \subset \mathscr{F}$ such that the $L_1(\nu)$ distance of any two functions in this subset is at least ε. Maximality means in this context that no function $f_{M+1} \in \mathscr{F}$ can be attached to $\bar{\mathscr{D}}(\varepsilon, \nu)$ without violating this condition. Thus the inequality $M \le D\varepsilon^{-L}$ means that $\bar{\mathscr{D}}(\varepsilon, \nu)$ is an ε-dense subset of $\mathscr{F}$ in the space $L_1(Y, \mathscr{Y}, \nu)$ with no more than $D\varepsilon^{-L}$ elements.

In the estimation of the cardinality M of a set $\mathscr{D}(\varepsilon, \nu) = \{f_1, \dots, f_M\} \subset \mathscr{F}$ with the property $\int |f_j - f_k|\, d\nu \ge \varepsilon$ if $j \ne k$ we exploit the Vapnik–Červonenkis class property of $\mathscr{F}$ in the following way. Let us choose relatively few $p = p(M, \varepsilon)$ points (y_l, t_l), $y_l \in Y$, $-1 \le t_l \le 1$, $1 \le l \le p$, in the space $Y \times [-1, 1]$ in such a way that the set $S_0(p) = \{(y_l, t_l),\ 1 \le l \le p\}$ and graphs $A(f_j) = \{(y, t)\colon y \in Y,\ \min(0, f_j(y)) \le t \le \max(0, f_j(y))\}$, $f_j \in \mathscr{D}(\varepsilon, \nu) \subset \mathscr{F}$ have the property that all sets $A(f_j) \cap S_0(p)$, $1 \le j \le M$, are different. Then the Vapnik–Červonenkis class property of $\mathscr{F}$ implies that $M \le Bp^K$. Hence if there exists a set $S_0(p)$ with the above property and with a relatively small number p, then this yields a useful estimate on M. Such a set $S_0(p)$ will be given by means of the following random construction.

Let us choose the p points (y_l, t_l), $1 \le l \le p$, of the (random) set $S_0(p)$ independently of each other in such a way that the coordinate y_l is chosen with distribution ν on $(Y, \mathscr{Y})$ and the coordinate t_l with uniform distribution on the interval $[-1, 1]$ independently of y_l. (The number p will be chosen later.) Let us fix some indices $1 \le j, k \le M$, and estimate from above the probability that the sets $A(f_j) \cap S_0(p)$ and $A(f_k) \cap S_0(p)$ agree, where $A(f)$ denotes the graph of the function f. Consider the symmetric difference $A(f_j)\Delta A(f_k)$ of the sets $A(f_j)$ and $A(f_k)$. The sets $A(f_j) \cap S_0(p)$ and $A(f_k) \cap S_0(p)$ agree if and only if $(y_l, t_l) \notin A(f_j)\Delta A(f_k)$ for all $(y_l, t_l) \in S_0(p)$. Let us observe that for a fixed l the estimate $P((y_l, t_l) \in A(f_j)\Delta A(f_k)) = \frac{1}{2}(\nu \times \lambda)(A(f_j)\Delta A(f_k)) = \frac{1}{2}\int |f_j - f_k|\, d\nu \ge \frac{\varepsilon}{2}$ holds, where λ denotes the Lebesgue measure. This implies that the probability that the (random) sets $A(f_j) \cap S_0(p)$ and $A(f_k) \cap S_0(p)$ agree can be bounded from above by $\left(1 - \frac{\varepsilon}{2}\right)^p \le e^{-p\varepsilon/2}$. Hence the probability that all sets $A(f_j) \cap S_0(p)$ are different is greater than $1 - \binom{M}{2} e^{-p\varepsilon/2} \ge 1 - \frac{M^2}{2} e^{-p\varepsilon/2}$. Choose p such that $\frac{7}{4} e^{p\varepsilon/2} > e^{(p+1)\varepsilon/2} > M^2 \ge e^{p\varepsilon/2}$. (We may assume that $M > 1$, in which case there is such a number $p \ge 1$. We may really assume that $M > 1$, since we want to give an upper bound on M. Moreover, the estimate we shall give on it, satisfies this inequality.) Then the above probability is greater than $\frac{1}{8}$, and there exists some set $S_0(p)$ with the desired property.

The inequalities $M \le Bp^K$ and $M^2 \ge e^{p\varepsilon/2}$ imply that $M \ge M^{p\varepsilon/4} \ge e^{\varepsilon M^{1/K}/4B^{1/K}}$, i.e. $\frac{\log M^{1/K}}{M^{1/K}} \ge \frac{\varepsilon}{4KB^{1/K}}$. As $\frac{\log M^{1/K}}{M^{1/K}} \le CM^{-1/2K}$ for $M \ge 1$ with some universal constant $C > 0$, this estimate implies that Theorem 5.2 holds with the exponent L and parameter D given in its formulation. □

Let us observe that if $\mathcal{F}$ is an L_1-dense class of functions on a measure space $(Y, \mathcal{Y})$ with some exponent L and parameter D, and also the inequality $\sup_{y \in Y} |f(y)| \le 1$ holds for all $f \in \mathcal{F}$, then $\mathcal{F}$ is an L_2-dense class of functions with exponent $2L$ and parameter $D2^L$. Indeed, if we fix some probability measure ν on $(Y, \mathcal{Y})$ together with a number $0 < \varepsilon \le 1$, and $\mathcal{D}(\varepsilon, \nu) = \{f_1, \dots, f_M\}$ is an $\frac{\varepsilon^2}{2}$-dense set of $\mathcal{F}$ in the space $L_1(Y, \mathcal{Y}, \nu)$, $M \le 2^L D\varepsilon^{-2L}$, then for all function $f \in \mathcal{F}$ some function $f_j \in \mathcal{D}(\varepsilon, \nu)$ can be chosen in such a way that $\int (f - f_j)^2 \, d\nu \le 2 \int |f - f_j| \, d\nu \le \varepsilon^2$. This implies that $\mathcal{F}$ is an L_2-dense class with the given exponent and parameter.

It is not easy to check whether a collection of subsets $\mathcal{D}$ of a set X is a Vapnik–Červonenkis class even with the help of Theorem 5.1. Therefore the following Theorem 5.3 which enables us to construct many non-trivial Vapnik–Červonenkis classes is of special interest. Its proof is given in Appendix A.

Theorem 5.3 (A way to construct Vapnik–Červonenkis classes). *Let us consider a k-dimensional subspace $\mathcal{G}_k$ of the linear space of real valued functions defined on a set X, and define the level-set $A(g) = \{x \colon x \in X, g(x) \ge 0\}$ for all functions $g \in \mathcal{G}_k$. Take the class of subsets $\mathcal{D} = \{A(g) \colon g \in \mathcal{G}_k\}$ of the set X consisting of the above introduced level sets. No subset $S = S(k+1) \subset X$ of cardinality $k+1$ is shattered by $\mathcal{D}$. Hence by Theorem 5.1 $\mathcal{D}$ is a Vapnik–Červonenkis class of subsets of X.*

Theorem 5.3 enables us to construct interesting Vapnik–Červonenkis classes. Thus for instance the class of all half-spaces in a Euclidean space, the class of all ellipses in the plane, or more generally the level sets of k-order algebraic functions of p variables with a fixed number k constitute a Vapnik–Červonenkis class in the p-dimensional Euclidean space R^p. It can be proved that if $\mathcal{C}$ and $\mathcal{D}$ are Vapnik–Červonenkis classes of subsets of a set S, then also their intersection $\mathcal{C} \cap \mathcal{D} = \{C \cap D \colon C \in \mathcal{C}, D \in \mathcal{D}\}$, their union $\mathcal{C} \cup \mathcal{D} = \{C \cup D \colon C \in \mathcal{C}, D \in \mathcal{D}\}$ and complementary sets $\mathcal{C}^c = \{S \setminus C \colon C \in \mathcal{C}\}$ are Vapnik–Červonenkis classes. These results are less important for us, and their proofs will be omitted. We are interested in Vapnik–Červonenkis classes not for their own sake. We are going to find L_2-dense classes of functions, and Vapnik–Červonenkis classes help us in this. Indeed, Theorem 5.2 implies that if $\mathcal{D}$ is a Vapnik–Červonenkis class of subsets of a set S, then their indicator functions constitute a Vapnik–Červonenkis class of functions, and as a consequence an L_1-dense, hence also an L_2-dense class of functions. Then the results of Lemma 5.4 formulated below enable us to construct new L_2-dense classes of functions.

Lemma 5.4 (Some useful properties of L_2-dense classes). *Let $\mathcal{G}$ be an L_2-dense class of functions on some space $(Y,\mathcal{Y})$ whose absolute values are bounded by one, and let f be a function on $(Y,\mathcal{Y})$ also with absolute value bounded by one. Then $f\cdot\mathcal{G}=\{f\cdot g\colon g\in\mathcal{G}\}$ is also an L_2-dense class of functions. Let $\mathcal{G}_1$ and $\mathcal{G}_2$ be two L_2-dense classes of functions on some space $(Y,\mathcal{Y})$ whose absolute values are bounded by one. Then the classes of functions $\mathcal{G}_1+\mathcal{G}_2=\{g_1+g_2\colon g_1\in\mathcal{G}_1, g_2\in\mathcal{G}_2\}$, $\mathcal{G}_1\cdot\mathcal{G}_2=\{g_1g_2\colon g_1\in\mathcal{G}_1,\ g_2\in\mathcal{G}_2\}$, $\min(\mathcal{G}_1,\mathcal{G}_2)=\{\min(g_1,g_2)\colon g_1\in\mathcal{G}_1,\ g_2\in\mathcal{G}_2\}$, $\max(\mathcal{G}_1,\mathcal{G}_2)=\{\max(g_1,g_2)\colon g_1\in\mathcal{G}_1,\ g_2\in\mathcal{G}_2\}$ are also L_2-dense. If $\mathcal{G}$ is an L_2-dense class of functions, and $\mathcal{G}'\subset\mathcal{G}$, then $\mathcal{G}'$ is also an L_2-dense class.*

The proof of Lemma 5.4. is rather straightforward. One has to observe for instance that if $g_1,\bar g_1\in\mathcal{G}_1$, $g_2,\bar g_2\in\mathcal{G}_2$ then $|\min(g_1,g_2)-\min(\bar g_1,\bar g_2)|\le|g_1-\bar g_1)|+|g_2-\bar g_2|$, hence if $g_{1,1},\dots,g_{1,M_1}$ is an $\frac{\varepsilon}{2}$-dense subset of $\mathcal{G}_1$ and $g_{2,1},\dots,g_{2,M_2}$ is an $\frac{\varepsilon}{2}$-dense subset of $\mathcal{G}_2$ in the space $L_2(Y,\mathcal{Y},\nu)$ with some probability measure ν, then the functions $\min(g_{1,j},g_{2,k})$, $1\le j\le M_1$, $1\le k\le M_2$ constitute an ε-dense subset of $\min(\mathcal{G}_1,\mathcal{G}_2)$ in $L_2(Y,\mathcal{Y},\nu)$. The last statement of Lemma 5.4 was proved after the Corollary of Theorem 4.1. The details are left to the reader. □

The above result enable us to construct some L_2-dense class of functions. We give an example for it in the following Example 5.5 which is a consequence of Theorem 5.2 and Lemma 5.4.

Example 5.5. *Take m measurable functions $f_j(x)$, $1\le j\le m$, on a measurable space $(X,\mathcal{X})$ which have the property $\sup\limits_{x\in X}|f_j(x)|\le 1$ for all $1\le j\le m$. Let $\mathcal{D}$ be a Vapnik–Červonenkis class consisting of measurable subsets of the set X. Define for all pairs (f_j,D), f_j, $1\le j\le m$, and $D\in\mathcal{D}$ the function $f_{j,D}(\cdot)$ as $f_{j,D}(x)=f_j(x)$ if $x\in D$, and $f_{j,D}(x)=0$ if $x\notin D$, i.e. $f_{j,D}(\cdot)$ is the restriction of the function $f_j(\cdot)$ to the set D. Then the set of functions $\mathcal{F}=\{f_{j,D}\colon 1\le j\le m$, $D\in\mathcal{D}\}$ is L_2-dense.*

Beside this, Theorem 5.3 helps us to construct Vapnik–Červonenkis classes of sets. Let me also remark that it follows from the result of this chapter that the random variables considered in Lemma 4.5 are not only countably approximable, but the class of functions $f_{u_1,\dots,u_k,v_1,\dots,v_k}$ appearing in their definition is L_2-dense.

Chapter 6
The Proof of Theorems 4.1 and 4.2 on the Supremum of Random Sums

In this chapter we prove Theorem 4.2, an estimate about the tail distribution of the supremum of an appropriate class of Gaussian random variables with the help of a method, called the chaining argument. We also investigate the proof of Theorem 4.1 which can be considered as a version of Theorem 4.2 about the supremum of partial sums of independent and identically distributed random variables. The chaining argument is not a strong enough method to prove Theorem 4.1, but it enables us to prove a weakened form of it formulated in Proposition 6.1. This result turned out to be useful in the proof of Theorem 4.1. It enables us to reduce the proof of Theorem 4.1 to a simpler statement formulated in Proposition 6.2. In this chapter we prove Proposition 6.1, formulate Proposition 6.2, and reduce the proof of Theorem 4.1 with the help of Proposition 6.1 to this result. The proof of Proposition 6.2 which demands different arguments is postponed to the next chapter. Before presenting the proofs I briefly describe the chaining argument.

Let us consider a countable class of functions $\mathcal{F}$ on a probability space $(X, \mathcal{X}, \mu)$ which is L_2-dense with respect to the probability measure μ. Let us have either a class of Gaussian random variables $Z(f)$ with zero expectation such that $EZ(f)Z(g) = \int f(x)g(x)\mu(dx)$, $f, g \in \mathcal{F}$, or a set of normalized partial sums $S_n(f) = \frac{1}{\sqrt{n}}\sum\limits_{j=1}^{n} f(\xi_j)$, $f \in \mathcal{F}$, where $\xi_1, \dots, \xi_n$ is a sequence of independent μ distributed random variables with values in the space $(X, \mathcal{X})$, and assume that $Ef(\xi_j) = 0$ for all $f \in \mathcal{F}$. We want to get a good estimate on the probability $P\left(\sup\limits_{f\in\mathcal{F}} Z(f) > u\right)$ or $P\left(\sup\limits_{f\in\mathcal{F}} S_n(f) > u\right)$ if the class of functions $\mathcal{F}$ has some nice properties. The chaining argument suggests to prove such an estimate in the following way.

Let us try to find an appropriate sequence of subset $\mathcal{F}_1 \subset \mathcal{F}_2 \subset \cdots \subset \mathcal{F}$ such that $\bigcup\limits_{N=1}^{\infty} \mathcal{F}_N = \mathcal{F}$, $\mathcal{F}_N$ is such a set of functions from $\mathcal{F}$ with relatively few elements for which $\inf\limits_{f\in\mathcal{F}_N} \int (f-\bar{f})^2\,d\mu \le \delta_N$ with an appropriately chosen number

P. Major, *On the Estimation of Multiple Random Integrals and U-Statistics*, Lecture Notes in Mathematics 2079, DOI 10.1007/978-3-642-37617-7_6,

δ_N for all functions $\bar{f} \in \mathcal{F}$, and let us give a good estimate on the probability $P\left(\sup\limits_{f \in \mathcal{F}_N} Z(f) > u_N\right)$ or $P\left(\sup\limits_{f \in \mathcal{F}_N} S_n(f) > u_N\right)$ for all $N = 1, 2, \dots$ with an appropriately chosen monotone increasing sequence u_N such that $\lim\limits_{N \to \infty} u_N = u$.

We can get a relatively good estimate under appropriate conditions for the class of functions $\mathcal{F}$ by choosing the classes of functions $\mathcal{F}_N$ and numbers δ_N and u_N in an appropriate way. We try to bound the difference of the probabilities

$$P\left(\sup_{f \in \mathcal{F}_{N+1}} Z(f) > u_{N+1}\right) - P\left(\sup_{f \in \mathcal{F}_N} Z(f) > u_N\right)$$

or of the analogous difference if $Z(f)$ is replaced by $S_n(f)$. For the sake of completeness define this difference also in the case $N = 1$ with the choice $\mathcal{F}_0 = \emptyset$, when the second probability in this difference equals zero.

The above mentioned difference of probabilities can be estimated in a natural way by taking for all functions $f_{j_{N+1}} \in \mathcal{F}_{N+1}$ a function $f_{j_N} \in \mathcal{F}_N$ which is close to it, more explicitly $\int (f_{j_{N+1}} - f_{j_N})^2 \, d\mu \le \delta_N^2$, and calculating the probability that the difference of the random variables corresponding to these two functions is greater than $u_{N+1} - u_N$. We can estimate these probabilities with the help of some results which give a relatively good bound on the tail distribution of $Z(g)$ or $S_n(g)$ if $\int g^2 \, d\mu$ is small. The sum of all such probabilities gives an upper bound for the above considered difference of probabilities. Then we get an estimate for the probability $P\left(\sup\limits_{f \in \mathcal{F}_N} Z(f) > u_N\right)$ for all $N = 1, 2, \dots$, by summing up the above estimate, and we get a bound on the probability we are interested in by taking the limit $N \to \infty$. This method is called the chaining argument. It got this name, because we estimate the contribution of a random variable corresponding to a function $f_{j_{N+1}} \in \mathcal{F}_{N+1}$ to the bound of the probability we investigate by taking the random variable corresponding to a function $f_{j_N} \in \mathcal{F}_N$ close to it, then we choose another random variable corresponding to a function $f_{j_{N-1}} \in \mathcal{F}_{N-1}$ close to this function, and by continuing this procedure we take a chain of subsequent functions and the random variables corresponding to them.

First we show how this method supplies the proof of Theorem 4.2. Then we turn to the investigation of Theorem 4.1. In the study of this problem the above method does not work well, because if two functions are very close to each other in the $L_2(\mu)$-norm, then the Bernstein inequality (or an improvement of it) supplies a much weaker estimate for the difference of the partial sums corresponding to these two functions than the bound suggested by the central limit theorem. On the other hand, we shall prove a weaker version of Theorem 4.1 in Proposition 6.1 with the help of the chaining argument. This result will be also useful for us.

Proof of Theorem 4.2. Let us list the elements of $\mathcal{F}$ as $\{f_0, f_1, \dots\} = \mathcal{F}$, and choose for all $p = 0, 1, 2, \dots$ a set of functions $\mathcal{F}_p = \{f_{a(1,p)}, \dots, f_{a(m_p,p)}\} \subset \mathcal{F}$

with $m_p \le (D+1)\,2^{2pL}\sigma^{-L}$ elements in such a way that $\inf\limits_{1\le j\le m_p}\int(f-f_{a(j,p)})^2\,d\mu \le 2^{-4p}\sigma^2$ for all $f\in\mathscr{F}$, and let the set $\mathscr{F}_p$ contain also the function f_p. (We imposed the condition $f_p\in\mathscr{F}_p$ to guarantee that the relation $f\in\mathscr{F}_p$ holds with some index p for all $f\in\mathscr{F}$. We could do this by slightly enlarging the upper bound we can give for the number m_p by replacing the factor D by $D+1$ in it.) For all indices $a(j,p)$ of the functions in $\mathscr{F}_p$, $p=1,2,\dots$, define a predecessor $a(j',p-1)$ from the indices of the set of functions $\mathscr{F}_{p-1}$ in such a way that the functions $f_{a(j,p)}$ and $f_{a(j',p-1))}$ satisfy the relation $\int(f_{a(j,p)}-f_{a(j',p-1)})^2\,d\mu\le 2^{-4(p-1)}\sigma^2$. With the help of the behaviour of the standard normal distribution function we can write the estimates

$$\begin{aligned}P(A(j,p)) &= P\left(|Z(f_{a(j,p)})-Z(f_{a(j',p-1)})|\ge 2^{-(1+p)}u\right)\\ &\le 2\exp\left\{-\frac{2^{-2(p+1)}u^2}{2\cdot 2^{-4(p-1)}\sigma^2}\right\}=2\exp\left\{-\frac{2^{2p}u^2}{128\sigma^2}\right\}\\ &\qquad 1\le j\le m_p,\ p=1,2,\dots,\end{aligned}$$

and

$$P(B(j))=P\left(|Z(f_{a(j,0)})|\ge\frac{u}{2}\right)\le\exp\left\{-\frac{u^2}{8\sigma^2}\right\},\qquad 1\le j\le m_0.$$

The above estimates together with the relation $\bigcup\limits_{p=0}^{\infty}\mathscr{F}_p=\mathscr{F}$ which implies that $\{|Z(f)|\ge u\}\subset\bigcup\limits_{p=1}^{\infty}\bigcup\limits_{j=1}^{m_p}A(j,p)\cup\bigcup\limits_{s=1}^{m_0}B(s)$ for all $f\in\mathscr{F}$ yield that

$$\begin{aligned}&P\left(\sup_{f\in\mathscr{F}}|Z(f)|\ge u\right)\\ &\quad\le P\left(\bigcup_{p=1}^{\infty}\bigcup_{j=1}^{m_p}A(j,p)\cup\bigcup_{s=1}^{m_0}B(s)\right)\\ &\quad\le\sum_{p=1}^{\infty}\sum_{j=1}^{m_p}P(A(j,p))+\sum_{s=1}^{m_0}P(B(s))\\ &\quad\le\sum_{p=1}^{\infty}2(D+1)2^{2pL}\sigma^{-L}\exp\left\{-\frac{2^{2p}u^2}{128\sigma^2}\right\}+2(D+1)\sigma^{-L}\exp\left\{-\frac{u^2}{8\sigma^2}\right\}.\end{aligned}$$

If $u \ge ML^{1/2}\sigma \log^{1/2}\frac{2}{\sigma}$ with $M \ge 16$ (and $L \ge 1$ and $0 < \sigma \le 1$), then

$$2^{2pL}\sigma^{-L}\exp\left\{-\frac{2^{2p}u^2}{256\sigma^2}\right\} \le 2^{2pL}\sigma^{-L}\left(\frac{\sigma}{2}\right)^{2^{2p}M^2L/256} \le 2^{-pL} \le 2^{-p}$$

for all $p = 0, 1 \dots$, hence the previous inequality implies that

$$P\left(\sup_{f\in\mathcal{F}}|Z(f)| \ge u\right) \le 2(D+1)\sum_{p=0}^{\infty} 2^{-p}\exp\left\{-\frac{2^{2p}u^2}{256\sigma^2}\right\}$$
$$= 4(D+1)\exp\left\{-\frac{u^2}{256\sigma^2}\right\}.$$

Theorem 4.2 is proved. □

With an appropriate choice of the bound of the integrals in the definition of the sets $\mathcal{F}_p$ in the proof of Theorem 4.2 and some additional calculation it can be proved that the coefficient $\frac{1}{256}$ in the exponent of the right-hand side (4.7) can be replaced by $\frac{1-\varepsilon}{2}$ with arbitrary small $\varepsilon > 0$ if the remaining (universal) constants in this estimate are chosen sufficiently large.

The proof of Theorem 4.2 was based on a sufficiently good estimate on the probabilities $P(|Z(f) - Z(g)| > u)$ for pairs of functions $f, g \in \mathcal{F}$ and numbers $u > 0$. In the case of Theorem 4.1 only a weaker bound can be given for the corresponding probabilities. There is no good estimate on the tail distribution of the difference $S_n(f) - S_n(g)$ if its variance is small. As a consequence, the chaining argument supplies only a weaker result in this case. This result, where the tail distribution of the supremum of the normalized random sums $S_n(f)$ is estimated on a relatively dense subset of the class of functions $f \in \mathcal{F}$ in the $L_2(\mu)$ norm will be given in Proposition 6.1. Another result will be formulated in Proposition 6.2 whose proof is postponed to the next chapter. It will be shown that Theorem 4.1 follows from Propositions 6.1 and 6.2.

Before the formulation of Proposition 6.1 I recall an estimate which is a simple consequence of Bernstein's inequality. If $S_n(f) = \frac{1}{\sqrt{n}}\sum_{j=1}^{n} f(\xi_j)$ is the normalized sum of independent, identically random variables, $P(|f(\xi_1)| \le 1) = 1$, $Ef(\xi_1) = 0$, $Ef(\xi_1)^2 \le \sigma^2$, then there exists some constant $\alpha > 0$ such that

$$P(|S_n(f)| > u) \le 2e^{-\alpha u^2/\sigma^2} \quad \text{if} \quad 0 < u < \sqrt{n}\sigma^2. \tag{6.1}$$

In Proposition 6.1 we shall give a good (Gaussian type) estimate on the probability $P\left(\sup_{f\in\mathcal{F}_{\bar\sigma}}|S_n(f)| > \frac{u}{\bar A}\right)$ with some parameter $\bar A > 1$, where $\mathcal{F}_{\bar\sigma}$ is an appropriate finite subset of a set of functions $\mathcal{F}$ satisfying the conditions of Theorem 4.1. (We introduced the number $\bar A$ because of some technical reasons. We can formulate with its help such a result which simplifies the reduction of the proof of Theorem 4.1 to the proof of another result formulated in Proposition 6.2.) We cannot give a good estimate for the above probability for all $u > 0$, we can do this only for such numbers u which are in an appropriate interval depending on the parameter σ appearing in condition (4.2) of Theorem 4.1 and the parameter $\bar{A}$ we chose in Proposition 6.1. This fact may explain why we could prove the estimate of Theorem 4.1 only for such numbers u which satisfy the condition imposed in formula (4.4). The choice of the set of functions $\mathscr{F}_{\bar{\sigma}} \subset \mathscr{F}$ depends of the number u appearing in the probability we want to estimate. It is such a subset of relatively small cardinality of $\mathscr{F}$ whose $L_2(\mu)$-norm distance from all elements of $\mathscr{F}$ is less than $\bar{\sigma} = \bar{\sigma}(u)$ with an appropriately defined number $\bar{\sigma}(u)$. With the help of Proposition 6.1 we want to reduce the proof of Theorem 4.1 to a result formulated in the subsequent Proposition 6.2. To do this we still need an upper bound on the cardinality of $\mathscr{F}_{\bar{\sigma}}$ and some upper and lower bounds on the value of $\bar{\sigma}(u)$. In Proposition 6.1 we shall formulate such results, too.

Proposition 6.1. *Let us have a countable, L_2-dense class of functions $\mathscr{F}$ with parameter $D \ge 1$ and exponent $L \ge 1$ with respect to some probability measure μ on a measurable space $(X, \mathscr{X})$ whose elements satisfy relations (4.1)–(4.3) with this probability measure μ on $(X, \mathscr{X})$ and some real number $0 < \sigma \le 1$. Take a sequence of independent, μ-distributed random variables $\xi_1, \dots, \xi_n$, $n \ge 2$, and define the normalized random sums $S_n(f) = \frac{1}{\sqrt{n}} \sum_{l=1}^{n} f(\xi_l)$, for all $f \in \mathscr{F}$. Let us fix some number $\bar{A} \ge 1$. There exists some number $M = M(\bar{A})$ such that with these parameters $\bar{A}$ and $M = M(\bar{A}) \ge 1$ the following relations hold.*

For all numbers $u > 0$ such that $n\sigma^2 \ge \left(\frac{u}{\sigma}\right)^2 \ge M(L \log \frac{2}{\sigma} + \log D)$ a number $\bar{\sigma} = \bar{\sigma}(u)$, $0 \le \bar{\sigma} \le \sigma \le 1$, and a collection of functions $\mathscr{F}_{\bar{\sigma}} = \{f_1, \dots, f_m\} \subset \mathscr{F}$ with $m \le D\bar{\sigma}^{-L}$ elements can be chosen in such a way that the union of the sets $\mathscr{D}_j = \{f \colon f \in \mathscr{F}, \int |f - f_j|^2 \, d\mu \le \bar{\sigma}^2\}$, $1 \le j \le m$, cover the set of functions $\mathscr{F}$, i.e. $\bigcup_{j=1}^{m} \mathscr{D}_j = \mathscr{F}$, and the normalized random sums $S_n(f)$, $f \in \mathscr{F}_{\bar{\sigma}}$, $n \ge 2$, satisfy the inequality

$$P\left(\sup_{f \in \mathscr{F}_{\bar{\sigma}}} |S_n(f)| \ge \frac{u}{\bar{A}} \right) \le 4 \exp\left\{ -\alpha \left(\frac{u}{10\bar{A}\sigma} \right)^2 \right\}$$

$$\text{under the condition } n\sigma^2 \ge \left(\tfrac{u}{\sigma}\right)^2 \ge M\left(L \log \tfrac{2}{\sigma} + \log D\right) \tag{6.2}$$

with the constants α in formula (6.1) and the exponent L and parameter D of the L_2-dense class $\mathscr{F}$. The inequality $\frac{1}{16}\left(\frac{u}{\bar{A}\bar{\sigma}}\right)^2 \ge n\bar{\sigma}^2 \ge \frac{1}{64}\left(\frac{u}{\bar{A}\sigma}\right)^2$ also holds with the number $\bar{\sigma} = \bar{\sigma}(u)$. If the number u satisfies also the inequality

$$n\sigma^2 \ge \left(\frac{u}{\sigma}\right)^2 \ge M\left(L^{3/2} \log \frac{2}{\sigma} + (\log D)^{3/2}\right) \tag{6.3}$$

with a sufficiently large number $M = M(\bar{A})$, then the relation $n\bar{\sigma}^2 \geq L \log n + \log D$ holds, too.

Remark. Under the conditions $L \geq 1$ and $D \geq 1$ of Proposition 6.1 the condition formulated in relation (6.3) (with a sufficiently large number $M = M(\bar{A})$) is stronger than the condition $(\frac{u}{\sigma})^2 \geq M(L \log \frac{2}{\sigma} + \log D)$ imposed in formula (6.2). To see this observe that although $(\log D)^{3/2} \leq \log D$ if $\log D \leq 1$, but this effect can be compensated by choosing a sufficiently large parameter M in formula (6.3) and exploiting that $L \log \frac{2}{\sigma} \geq \log 2$.

Proposition 6.1 helps to reduce the proof of Theorem 4.1 to the case when such classes of functions $\mathcal{F}$ are considered whose elements are such functions whose L_2-norm is bounded by a relatively small number $\bar{\sigma}$. In more detail, the proof of Theorem 4.1 can be reduced to a good estimate on the distribution of the supremum of random variables $\sup\limits_{f \in \mathcal{D}_j} |S_n(f - f_j)|$ for all classes $\mathcal{D}_j$, $1 \leq j \leq m$, by means of Proposition 6.1. To carry out such a reduction we also need the inequality $n\bar{\sigma}^2 \geq L \log n + \log D$ (or a slightly weaker version of it). This is the reason why we have finished Proposition 6.1 with the statement that this inequality holds under the condition (6.3). We also have to know that the number m of the classes $\mathcal{D}_j$ is not too large. Beside this, we need some estimates on the number $\bar{\sigma} = \bar{\sigma}(u)$ which is an upper bound for the L_2-norm of the functions $f - f_j$, $f \in \mathcal{D}_j$. To get such bounds for $\bar{\sigma}$ that we need in the applications of Proposition 6.1 we introduced a large parameter $\bar{A}$ in the formulation of Proposition 6.1 and imposed a condition with a sufficiently large number $M = M(\bar{A})$ in formula (6.3). This condition reappears in Theorem 4.1 in the conditions of the estimate (4.4).

Let me remark that one of the inequalities the number $\bar{\sigma}$ introduced in Proposition 6.1 satisfies has the consequence $u >$ const. $\sqrt{n}\bar{\sigma}^2$ with an appropriate constant. Hence to complete the proof of Theorem 4.1 we have to estimate the probability $P\left(\sup\limits_{f \in \mathcal{F}} S_n(f)| > u\right)$ also in such cases when the L_2 norm of the functions in $\mathcal{F}$ is bounded with such a number $\bar{\sigma}$ for which $u >$ const. $\sqrt{n}\bar{\sigma}^2$. On the other hand, we got an estimate in Proposition 6.1 if $u < \sqrt{n}\sigma^2$, (see formula (6.2)), and this is an inequality in the opposite direction. Hence to complete the proof of Theorem 4.1 with the help of Proposition 6.1 we need a result whose proof demands an essentially different method. Proposition 6.2 formulated below is such a result. I shall show that Theorem 4.1 is a consequence of Propositions 6.1 and 6.2. Proposition 6.1 is proved at the end of this chapter, while the proof of Proposition 6.2 is postponed to the next chapter.

Proposition 6.2. *Let us have a probability measure μ on a measurable space $(X, \mathcal{X})$ together with a sequence of independent and μ distributed random variables $\xi_1, \dots, \xi_n$, $n \geq 2$, and a countable, L_2-dense class of functions $f = f(x)$ on $(X, \mathcal{X})$ with some parameter $D \geq 1$ and exponent $L \geq 1$ which satisfies conditions (4.1), (4.2) and (4.3) with some $0 < \sigma \leq 1$ such that the inequality*

$n\sigma^2 > L\log n + \log D$ holds. Then there exists a threshold index $A_0 \ge 5$ such that the normalized random sums $S_n(f)$, $f \in \mathscr{F}$, introduced in Theorem 4.1 satisfy the inequality

$$P\left(\sup_{f\in\mathscr{F}} |S_n(f)| \ge An^{1/2}\sigma^2\right) \le e^{-A^{1/2}n\sigma^2/2} \quad \textit{if } A \ge A_0. \tag{6.4}$$

I did not try to find optimal parameters in formula (6.4). Even the coefficient $-A^{1/2}$ in the exponent at its right-hand side could be improved. The result of Proposition 6.2 is similar to that of Theorem 4.1. Both of them give an estimate on a probability of the form $P\left(\sup\limits_{f\in\mathscr{F}} |S_n(f)| \ge u\right)$ with some class of functions $\mathscr{F}$. The essential difference between them is that in Theorem 4.1 this probability is considered for $u \le n^{1/2}\sigma^2$ while in Proposition 6.2 the case $u = An^{1/2}\sigma^2$ with $A \ge A_0$ is taken, where A_0 is a sufficiently large positive number. Let us observe that in this case no good Gaussian type estimate can be given for the probabilities $P(S_n(f) \ge u)$, $f \in \mathscr{F}$. In this case Bernstein's inequality yields the bound $P(S_n(f) > An^{1/2}\sigma^2) = P\left(\sum\limits_{l=1}^n f(\xi_l) > uV_n\right) < e^{-\text{const.}\, An\sigma^2}$ with $u = A\sqrt{n}\sigma$ and $V_n = \sqrt{n}\sigma$ for each single function $f \in \mathscr{F}$ which takes part in the supremum of formula (6.4). The estimate (6.4) yields a slightly weaker estimate for the supremum of such random variables, since it contains the coefficient $A^{1/2}$ instead of A in the exponent of the estimate at the right-hand side. But also such a bound will be sufficient for us.

In Proposition 6.2 such a situation is considered when the irregularities of the summands provide a non-negligible contribution to the probabilities $P(|S_n(f)| \ge u)$, and the chaining argument applied in the proof of Theorem 4.2 does not give a good estimate on the probability at the left-hand side of (6.4). This is the reason why we separated the proof of Theorem 4.1 to two different statements given in Propositions 6.1 and 6.2.

In the proof of Theorem 4.1 Proposition 6.1 will be applied with a sufficiently large number $\bar{A} \ge 1$ and an appropriate number $M = M(\bar{A})$ appearing in the formulation of this result. Proposition 6.2 will be applied for the sets of functions $\mathscr{F} = \mathscr{F}_j = \left\{\frac{g-f_j}{2} \colon g \in \mathscr{D}_j\right\}$ and number $\sigma = \bar{\sigma}$, with the number $\bar{\sigma}$, functions f_j and sets of functions $\mathscr{D}_j$ introduced in Proposition 6.1 and with the parameter A_0 appearing in the formulation of Proposition 6.2. We can write

$$\begin{aligned} P\left(\sup_{f\in\mathscr{F}} |S_n(f)| \ge u\right) &\le P\left(\sup_{f\in\mathscr{F}_{\bar{\sigma}}} |S_n(f)| \ge \frac{u}{\bar{A}}\right) \\ &\quad + \sum_{j=1}^m P\left(\sup_{g\in\mathscr{D}_j} \left|S_n\left(\frac{f_j-g}{2}\right)\right| \ge \left(\frac{1}{2} - \frac{1}{2\bar{A}}\right)u\right), \end{aligned} \tag{6.5}$$

where m is the cardinality of the set of functions $\mathcal{F}_{\bar\sigma}$ appearing in Proposition 6.1, which is bounded by $m \le D\bar\sigma^{-L}$. We want to choose the number $\bar A$ in such a way that the inequality $(\frac12 - \frac{1}{2\bar A})u \ge A_0\sqrt{n}\bar\sigma^2$ holds, since in this case Proposition 6.2 with the choice $A = A_0$ yields a good estimate on the second term in (6.5). This inequality is equivalent to $n\bar\sigma^2 \le (\frac{1}{2A_0} - \frac{1}{2A_0\bar A})^2(\frac{u}{\bar\sigma})^2$. On the other hand, $(\frac{u}{4\bar A\bar\sigma})^2 \ge n\bar\sigma^2$ by Proposition 6.1, hence the desired inequality holds if $\frac{1}{2A_0} - \frac{1}{2A_0\bar A} \ge \frac{1}{4\bar A}$. Hence with the choice $\bar A = \max(1, \frac{A_0+2}{2})$ and a sufficiently large $M = M(\bar A)$ we can bound both terms at the right-hand side of (6.5) with the help of Propositions 6.1 and 6.2.

With such a choice of $\bar A$ we can write by Proposition 6.2

$$P\left(\sup_{g\in\mathcal{D}_j}\left|S_n\left(\frac{f_j-g}{2}\right)\right| \ge \left(\frac12 - \frac{1}{2\bar A}\right)u\right) \le P\left(\sup_{g\in\mathcal{D}_j}\left|S_n\left(\frac{f_j-g}{2}\right)\right| \ge A_0\sqrt{n}\bar\sigma^2\right)$$
$$\le e^{-A_0^{1/2}n\bar\sigma^2/2} \quad \text{for all } 1\le j\le m.$$

(Observe that the set of functions $\frac{f_j-g}{2}$, $g\in\mathcal{D}_j$, is an L_2-dense class with parameter D and exponent L.) Hence Proposition 6.1 together with the bound $m \le D\bar\sigma^{-L}$ and formula (6.5) imply that

$$P\left(\sup_{f\in\mathcal{F}}|S_n(f)| \ge u\right) \le 4\exp\left\{-\alpha\left(\frac{u}{10\bar A\sigma}\right)^2\right\} + D\bar\sigma^{-L}e^{-A_0^{1/2}n\bar\sigma^2/2}. \tag{6.6}$$

To get the estimate in Theorem 4.1 from inequality (6.6) we show that the inequality $n\bar\sigma^2 \ge L\log n + \log D$ (with $L\ge 1$, $D\ge 1$ and $n\ge 2$) which is valid under the conditions of Proposition 6.1 implies that $D\bar\sigma^{-L} \le e^{n\bar\sigma^2}$. Indeed, we have to show that $\log D + L\log\frac{1}{\bar\sigma} \le n\bar\sigma^2$. But we have $n\bar\sigma^2 \ge L\log n \ge \log n$, hence $\frac{1}{\bar\sigma} \le \sqrt{\frac{n}{\log n}} \le n$, thus $\log\frac{1}{\bar\sigma} \le \log n$, and $\log D + L\log\frac{1}{\bar\sigma} \le \log D + L\log n \le n\bar\sigma^2$, as we have claimed.

Proof. This inequality together with the inequality $n\bar\sigma^2 \ge \frac{1}{64}(\frac{u}{\bar A\sigma})^2$, proved in Proposition 6.1 imply that

$$D\bar\sigma^{-L}e^{-A_0^{1/2}n\bar\sigma^2/2} \le \exp\left\{-\left(\frac{A_0^{1/2}}{2}-1\right)n\bar\sigma^2\right\} \le \exp\left\{-\frac{(A_0^{1/2}-2)}{128\bar A^2}\left(\frac{u}{\sigma}\right)^2\right\}.$$

Hence relation (6.6) yields that

$$P\left(\sup_{f\in\mathcal{F}}|S_n(f)| \ge u\right) \le 4\exp\left\{-\frac{\alpha}{100\bar A^2}\left(\frac{u}{\sigma}\right)^2\right\} + \exp\left\{-\frac{(A_0^{1/2}-2)}{128\bar A^2}\left(\frac{u}{\sigma}\right)^2\right\},$$

and because of the relation $A_0 \ge 5$ this estimate implies Theorem 4.1. Let me remark that the condition $\sqrt{n}\sigma^2 \ge u \ge M\sigma(L^{3/4}\log^{1/2}\frac{2}{\sigma} + (\log D)^{3/4})$ appears in formula (4.4) because of condition (6.3) imposed in Proposition 6.1. (The parameter M in formula (4.4) can be chosen as twice the parameter M in (6.3).) In such a way we proved Theorem 4.1 with the help of Propositions 6.1 and 6.2. □

I finish this chapter with the proof of Proposition 6.1.

Proof of Proposition 6.1. Let us list the members of $\mathcal{F}$, as $f_1, f_2, \dots$, and choose for all $p = 0,1,2,\dots$ a set $\mathcal{F}_p = \{f_{a(1,p)}, \dots, f_{a(m_p,p)}\} \subset \mathcal{F}$ with $m_p \le D\,2^{2pL}\sigma^{-L}$ elements in such a way that $\inf\limits_{1\le j\le m_p} \int (f - f_{a(j,p)})^2\,d\mu \le 2^{-4p}\sigma^2$ for all $f \in \mathcal{F}$. For all indices $a(j,p)$, $p = 1,2,\dots$, $1 \le j \le m_p$, choose a predecessor $a(j',p-1)$, $j' = j'(j,p)$, $1 \le j' \le m_{p-1}$, in such a way that the functions $f_{a(j,p)}$ and $f_{a(j',p-1)}$ satisfy the relation $\int |f_{a(j,p)} - f_{a(j',p-1)}|^2\,d\mu \le \sigma^2 2^{-4(p-1)}$. Then we have $\int \left(\frac{f_{a(j,p)} - f_{a(j',p-1)}}{2}\right)^2 d\mu \le 4\sigma^2 2^{-4p}$ and

$$\sup_{x_j\in X,\,1\le j\le k} \left| \frac{f_{a(j,p)}(x_1,\dots,x_k) - f_{a(j',p-1)}(x_1,\dots,x_k)}{2} \right| \le 1.$$

Relation (6.1) yields that

$$\begin{aligned} P(A(j,p)) &= P\left(\frac{1}{2}|S_n(f_{a(j,p)} - f_{a(j',p-1)})| \ge \frac{2^{-(1+p)}u}{2\bar{A}}\right) \\ &\le 2\exp\left\{-\alpha\left(\frac{2^p u}{8\bar{A}\sigma}\right)^2\right\} \quad \text{if } n\sigma^2 \ge 2^{6p}\left(\frac{u}{16\bar{A}\sigma}\right)^2, \\ &\qquad 1 \le j \le m_p, \quad p = 1,2,\dots, \end{aligned} \tag{6.7}$$

and

$$\begin{aligned} P(B(s)) = P\left(|S_n(f_{s,0})| \ge \frac{u}{2\bar{A}}\right) &\le 2\exp\left\{-\alpha\left(\frac{u}{2\bar{A}\sigma}\right)^2\right\}, \quad 1 \le s \le m_0, \\ \text{if } n\sigma^2 &\ge \left(\frac{u}{2\bar{A}\sigma}\right)^2. \end{aligned} \tag{6.8}$$

Choose an integer $R = R(u)$, $R \ge 1$, by the inequality

$$2^{6(R+1)}\left(\frac{u}{16\bar{A}\sigma}\right)^2 > n\sigma^2 \ge 2^{6R}\left(\frac{u}{16\bar{A}\sigma}\right)^2,$$

define $\bar{\sigma}^2 = 2^{-4R}\sigma^2$ and $\mathcal{F}_{\bar{\sigma}} = \mathcal{F}_R$. (As $n\sigma^2 \ge \left(\frac{u}{\sigma}\right)^2$ and $\bar{A} \ge 1$ by our conditions, there exists such a number $R \ge 1$. The number R was chosen as the largest number p for which the second relation of formula (6.7) holds.) Then the cardinality m of the set $\mathcal{F}_{\bar\sigma}$ equals $m_R \le D2^{2RL}\sigma^{-L} = D\bar\sigma^{-L}$, and the sets $\mathcal{D}_j$ are $\mathcal{D}_j = \{f\colon f \in \mathcal{F}, \int (f_{a(j,R)} - f)^2\,d\mu \le 2^{-4R}\sigma^2\}$, $1 \le j \le m_R$, hence $\bigcup\limits_{j=1}^{m} \mathcal{D}_j = \mathcal{F}$. Beside this, with our choice of the number R inequalities (6.7) and (6.8) can be applied for $1 \le p \le R$. Hence the definition of the predecessor of an index (j,p) implies that $\left\{\omega\colon \sup\limits_{f\in\mathcal{F}_{\bar\sigma}} |S_n(f)(\omega)| \ge \frac{u}{\bar A}\right\} \subset \bigcup\limits_{p=1}^{R}\bigcup\limits_{j=1}^{m_p} A(j,p) \cup \bigcup\limits_{s=1}^{m_0} B(s)$, and

$$
\begin{aligned}
P&\left(\sup_{f\in\mathcal{F}_{\bar\sigma}} |S_n(f)| \ge \frac{u}{\bar A}\right)\\
&\le P\left(\bigcup_{p=1}^{R}\bigcup_{j=1}^{m_p} A(j,p) \cup \bigcup_{s=1}^{m_0} B(s)\right)\\
&\le \sum_{p=1}^{R}\sum_{j=1}^{m_p} P(A(j,p)) + \sum_{s=1}^{m_0} P(B(s))\\
&\le \sum_{p=1}^{\infty} 2D\,2^{2pL}\sigma^{-L} \exp\left\{-\alpha\left(\frac{2^p u}{8\bar A\sigma}\right)^2\right\} + 2D\sigma^{-L}\exp\left\{-\alpha\left(\frac{u}{2\bar A\sigma}\right)^2\right\}.
\end{aligned}
$$

If the relation $(\frac{u}{\sigma})^2 \ge M(L\log\frac{2}{\sigma} + \log D)$ holds with a sufficiently large constant M (depending on $\bar A$), and $\sigma \le 1$, then the inequalities

$$
D2^{2pL}\sigma^{-L}\exp\left\{-\alpha\left(\frac{2^p u}{8\bar A\sigma}\right)^2\right\} \le 2^{-p}\exp\left\{-\alpha\left(\frac{2^p u}{10\bar A\sigma}\right)^2\right\}
$$

hold for all $p = 1, 2, \dots$, and

$$
D\sigma^{-L}\exp\left\{-\alpha\left(\frac{u}{2\bar A\sigma}\right)^2\right\} \le \exp\left\{-\alpha\left(\frac{u}{10\bar A\sigma}\right)^2\right\}.
$$

Hence the previous estimate implies that

$$
\begin{aligned}
P\left(\sup_{f\in\mathcal{F}_{\bar\sigma}} |S_n(f)| \ge \frac{u}{\bar A}\right) &\le \sum_{p=1}^{\infty} 2\cdot 2^{-p}\exp\left\{-\alpha\left(\frac{2^p u}{10\bar A\sigma}\right)^2\right\}\\
&\quad + 2\exp\left\{-\alpha\left(\frac{u}{10\bar A\sigma}\right)^2\right\} \le 4\exp\left\{-\alpha\left(\frac{u}{10\bar A\sigma}\right)^2\right\},
\end{aligned}
$$

and relation (6.2) holds.

As $\sigma^2 = 2^{4R}\bar\sigma^2$ the inequality

$$2^{-4R}\cdot\frac{2^{6R}}{256}\left(\frac{u}{\bar A\sigma}\right)^2 \le n\bar\sigma^2 = 2^{-4R}n\sigma^2 \le 2^{-4R}\cdot\frac{2^{6(R+1)}}{256}\left(\frac{u}{\bar A\sigma}\right)^2 = \frac{1}{4}\cdot 2^{-2R}\left(\frac{u}{\bar A\bar\sigma}\right)^2$$

holds, and this implies (together with the relation $R\ge 1$) that

$$\frac{1}{64}\left(\frac{u}{\bar A\sigma}\right)^2 \le n\bar\sigma^2 \le \frac{1}{16}\left(\frac{u}{\bar A\bar\sigma}\right)^2,$$

as we have claimed. It remained to show that under the condition (6.3) $n\bar\sigma^2 \ge L\log n + \log D$.

This inequality clearly holds under the conditions of Proposition 6.1 if $\sigma \le n^{-1/3}$, since in this case $\log\frac{2}{\sigma} \ge \frac{\log n}{3}$, and $n\bar\sigma^2 \ge \frac{1}{64}(\frac{u}{\bar A\sigma})^2 \ge \frac{1}{64\bar A^2}M(L\log\frac{2}{\sigma} + \log D) \ge \frac{1}{192\bar A^2}M(L\log n + \log D)) \ge L\log n + \log D$ if $M \ge M_0(\bar A)$ with a sufficiently large number $M_0(\bar A)$.

If $\sigma \ge n^{-1/3}$, we can exploit that the inequality $2^{6R}\left(\frac{u}{\bar A\sigma}\right)^2 \le 256n\sigma^2$ holds because of the definition of the number R. It can be rewritten as

$$2^{-4R} \ge 2^{-16/3}\left[\frac{\left(\frac{u}{\bar A\sigma}\right)^2}{n\sigma^2}\right]^{2/3}.$$

Hence $n\bar\sigma^2 = 2^{-4R}n\sigma^2 \ge \frac{2^{-16/3}}{\bar A^{4/3}}(n\sigma^2)^{1/3}\left(\frac{u}{\sigma}\right)^{4/3}$. As $\log\frac{2}{\sigma} \ge \log 2 > \frac{1}{2}$ the inequalities $n\sigma^2 \ge n^{1/3}$ and $(\frac{u}{\sigma})^2 \ge M(L^{3/2}\log\frac{2}{\sigma} + (\log D)^{3/2}) \ge \frac{M}{2}(L^{3/2} + (\log D)^{3/2})$ hold. They yield that

$$\begin{aligned} n\bar\sigma^2 &\ge \frac{\bar A^{-4/3}}{50}(n\sigma^2)^{1/3}\left(\frac{u}{\sigma}\right)^{4/3} \ge \frac{\bar A^{-4/3}}{50}n^{1/9}\left(\frac{M}{2}\right)^{2/3}(L^{3/2} + (\log D)^{3/2})^{2/3} \\ &\ge \frac{M^{2/3}n^{1/9}(L + \log D)}{100\bar A^{4/3}} \ge L\log n + \log D \end{aligned}$$

if $M = M(\bar A)$ is chosen sufficiently large. □

Chapter 7
The Completion of the Proof of Theorem 4.1

This chapter contains the proof of Proposition 6.2 with the help of a symmetrization argument, and this completes the proof of Theorem 4.1. By symmetrization argument I mean the reduction of the investigation of sums of the form $\sum_j f(\xi_j)$ to sums of the form $\sum_j \varepsilon_j f(\xi_j)$, where ε_j are independent random variables, independent also of the random variables ξ_j, and $P(\varepsilon_j = 1) = P(\varepsilon_j = -1) = \frac{1}{2}$. First a symmetrization lemma is proved, and then such an inductive statement is formulated in Proposition 7.3 which implies Proposition 6.2. Proposition 7.3 will be proved with the help of the symmetrization lemma and a conditioning argument. To carry out such a program we shall need some estimates which follow from Hoeffding's inequality formulated in Theorem 3.4.

First I formulate the symmetrization lemma we shall apply.

Lemma 7.1 (Symmetrization Lemma). *Let Z_n and $\bar{Z}_n$, $n = 1, 2, \dots$, be two sequences of random variables independent of each other, and let the random variables $\bar{Z}_n$, $n = 1, 2, \dots$, satisfy the inequality*

$$P(|\bar{Z}_n| \le \alpha) \ge \beta \quad \textit{for all } n = 1, 2, \dots \tag{7.1}$$

with some numbers $\alpha > 0$ and $\beta > 0$. Then

$$P\left(\sup_{1 \le n < \infty} |Z_n| > u + \alpha\right) \le \frac{1}{\beta} P\left(\sup_{1 \le n < \infty} |Z_n - \bar{Z}_n| > u\right) \quad \textit{for all } u > 0.$$

Proof of Lemma 7.1. Put $\tau = \min\{n\colon |Z_n| > u + \alpha\}$ if there exists such an index n, and $\tau = 0$ otherwise. Then the event $\{\tau = n\}$ is independent of the sequence of random variables $\bar{Z}_1, \bar{Z}_2, \dots$ for all $n = 1, 2, \dots$, and because of this independence

$$P(\{\tau = n\}) \le \frac{1}{\beta} P(\{\tau = n\} \cap \{|\bar{Z}_n| \le \alpha\}) \le \frac{1}{\beta} P(\{\tau = n\} \cap \{|Z_n - \bar{Z}_n| > u\})$$

P. Major, *On the Estimation of Multiple Random Integrals and U-Statistics*,
Lecture Notes in Mathematics 2079, DOI 10.1007/978-3-642-37617-7_7,

for all $n = 1, 2, \dots$. Hence

$$
\begin{aligned}
P\left(\sup_{1\le n<\infty} |Z_n| > u+\alpha\right) &= \sum_{l=1}^{\infty} P(\tau = l) \\
&\le \frac{1}{\beta}\sum_{l=1}^{\infty} P(\{\tau = l\} \cap \{|Z_l - \bar Z_l| > u\}) \\
&\le \frac{1}{\beta}\sum_{l=1}^{\infty} P(\{\tau = l\} \cap \sup_{1\le n<\infty} |Z_n - \bar Z_n| > u\}) \\
&\le \frac{1}{\beta} P\left(\sup_{1\le n<\infty} |Z_n - \bar Z_n| > u\right).
\end{aligned}
$$

Lemma 7.1 is proved. □

We shall apply the following Lemma 7.2 which is a consequence of the Symmetrization Lemma 7.1.

Lemma 7.2. *Let us fix a countable class of functions $\mathcal{F}$ on a measurable space $(X, \mathcal{X})$ together with a real number $0 < \sigma < 1$. Consider a sequence of independent and identically distributed random variables $\xi_1, \dots, \xi_n$ with values in the space $(X, \mathcal{X})$ such that $Ef(\xi_1) = 0$, $Ef^2(\xi_1) \le \sigma^2$ for all $f \in \mathcal{F}$ together with another sequence $\varepsilon_1, \dots, \varepsilon_n$ of independent random variables with distribution $P(\varepsilon_j = 1) = P(\varepsilon_j = -1) = \frac{1}{2}$, $1 \le j \le n$, independent also of the random sequence $\xi_1, \dots, \xi_n$. Then*

$$
\begin{aligned}
&P\left(\frac{1}{\sqrt{n}} \sup_{f\in\mathcal{F}} \left|\sum_{j=1}^{n} f(\xi_j)\right| \ge A n^{1/2}\sigma^2\right) \\
&\quad \le 4P\left(\frac{1}{\sqrt{n}} \sup_{f\in\mathcal{F}} \left|\sum_{j=1}^{n} \varepsilon_j f(\xi_j)\right| \ge \frac{A}{3} n^{1/2}\sigma^2\right) \quad \text{if } A \ge \frac{3\sqrt{2}}{\sqrt{n}\sigma}. \qquad (7.2)
\end{aligned}
$$

Proof of Lemma 7.2. Let us construct an independent copy $\bar\xi_1, \dots, \bar\xi_n$ of the sequence $\xi_1, \dots, \xi_n$ in such a way that all three sequences $\xi_1, \dots, \xi_n$, $\bar\xi_1, \dots, \bar\xi_n$ and $\varepsilon_1, \dots, \varepsilon_n$ are independent. Define the random variables

$$
S_n(f) = \frac{1}{\sqrt{n}} \sum_{j=1}^{n} f(\xi_j) \quad \text{and} \quad \bar S_n(f) = \frac{1}{\sqrt{n}} \sum_{j=1}^{n} f(\bar\xi_j)
$$

for all $f \in \mathcal{F}$. The inequality

$$P\left(\sup_{f\in\mathscr{F}}|S_n(f)|>A\sqrt{n}\sigma^2\right)\le 2P\left(\sup_{f\in\mathscr{F}}|S_n(f)-\bar{S}_n(f)|>\frac{2}{3}A\sqrt{n}\sigma^2\right). \tag{7.3}$$

follows from Lemma 7.1 if it is applied for the countable set of random variables $Z_n(f)=S_n(f)$ and $\bar{Z}_n(f)=\bar{S}_n(f)$, $f\in\mathscr{F}$, and the numbers $u=\frac{2}{3}A\sqrt{n}\sigma^2$ and $\alpha=\frac{1}{3}A\sqrt{n}\sigma^2$, since the random fields $S_n(f)$ and $\bar{S}_n(f)$ are independent, and $P(|\bar{S}_n(f)|\le\alpha)>\frac{1}{2}$ for all $f\in\mathscr{F}$. Indeed, $\alpha=\frac{1}{3}A\sqrt{n}\sigma^2\ge\sqrt{2}\sigma$, $E\bar{S}_n(f)^2\le\sigma^2$, thus Chebishev's inequality implies that $P(|\bar{S}_n(f)|\le\alpha)\ge P(|\bar{S}_n(f)|\le\sqrt{2}\sigma)\ge\frac{1}{2}$ for all $f\in\mathscr{F}$.

Let us observe that the random field

$$S_n(f)-\bar{S}_n(f)=\frac{1}{\sqrt{n}}\sum_{j=1}^n\left(f(\xi_j)-f(\bar{\xi}_j)\right),\quad f\in\mathscr{F}, \tag{7.4}$$

and its randomized version

$$\frac{1}{\sqrt{n}}\sum_{j=1}^n\varepsilon_j\left(f(\xi_j)-f(\bar{\xi}_j)\right),\quad f\in\mathscr{F}, \tag{7.5}$$

have the same distribution. Indeed, even the conditional distribution of (7.5) under the condition that the values of the ε_j-s are prescribed agrees with the distribution of (7.4) for all possible values of the ε_j-s. This follows from the observation that the distribution of the random field (7.4) does not change if we exchange the random variables ξ_j and $\bar{\xi}_j$ for those indices j for which $\varepsilon_j=-1$ and do not change them for those indices j for which $\varepsilon_j=1$. On the other hand, the distribution of the random field obtained with such an exchange of its variables agrees with the conditional distribution of the random field defined in (7.5) under the condition that the random variables ε_j take these prescribed values.

The above relation together with formula (7.3) imply that

$$\begin{aligned}
&P\left(\frac{1}{\sqrt{n}}\sup_{f\in\mathscr{F}}\left|\sum_{j=1}^n f(\xi_j)\right|\ge An^{1/2}\sigma^2\right)\\
&\quad\le 2P\left(\frac{1}{\sqrt{n}}\sup_{f\in\mathscr{F}}\left|\sum_{j=1}^n\varepsilon_j\left[f(\xi_j)-\bar{f}(\xi_j)\right]\right|\ge\frac{2}{3}An^{1/2}\sigma^2\right)\\
&\quad\le 2P\left(\frac{1}{\sqrt{n}}\sup_{f\in\mathscr{F}}\left|\sum_{j=1}^n\varepsilon_j f(\xi_j)\right|\ge\frac{A}{3}n^{1/2}\sigma^2\right)
\end{aligned}$$

$$+2P\left(\frac{1}{\sqrt{n}}\sup_{f\in\mathscr{F}}\left|\sum_{j=1}^n \varepsilon_j f(\bar\xi_j)\right| \ge \frac{A}{3}n^{1/2}\sigma^2\right)$$

$$= 4P\left(\frac{1}{\sqrt{n}}\sup_{f\in\mathscr{F}}\left|\sum_{j=1}^n \varepsilon_j f(\xi_j)\right| \ge \frac{A}{3}n^{1/2}\sigma^2\right).$$

Lemma 7.2 is proved. □

First I try to explain briefly the method of proof of Proposition 6.2. A probability of the form $P\left(n^{-1/2}\sup\limits_{f\in\mathscr{F}}\left|\sum\limits_{j=1}^n f(\xi_j)\right| > u\right)$ has to be estimated. Lemma 7.2 enables us to replace this problem by the estimation of the probability

$$P\left(n^{-1/2}\sup_{f\in\mathscr{F}}\left|\sum_{j=1}^n \varepsilon_j f(\xi_j)\right| > \frac{u}{3}\right)$$

with some independent random variables ε_j, $P(\varepsilon_j = 1) = P(\varepsilon_j = -1) = \frac{1}{2}$, $j = 1,\dots,n$, which are also independent of the random variables ξ_j. We shall bound the conditional probability of the event appearing in this modified problem under the condition that each random variable ξ_j takes a prescribed value. This can be done with the help of Hoeffding's inequality formulated in Theorem 3.4 and the L_2-density property of the class of functions $\mathscr{F}$ we consider. We hope to get a sharp estimate in such a way which is similar to the result we got in the study of the Gaussian counterpart of this problem, because Hoeffding's inequality yields always a Gaussian type upper bound for the tail distribution of the random sum we are studying.

Nevertheless, there appears a problem when we try to apply such an approach. To get a good estimate on the conditional tail distribution of the supremum of the random sums we are studying with the help of Hoeffding's inequality we need a good estimate on the supremum of the conditional variances of the random sums we are studying, i.e. on the tail distribution of $\sup\limits_{f\in\mathscr{F}}\frac{1}{n}\sum\limits_{j=1}^n f^2(\xi_j)$. This problem is similar to the original one, and it is not simpler.

But a more detailed study shows that our approach to get a good estimate with the help of Hoeffding's inequality works. In comparing our original problem with the new, complementary problem we have to understand at which level we need a good estimate on the tail distribution of the supremum in the complementary problem to get a good tail distribution estimate at level u in the original problem. A detailed study shows that to bound the probability in the original problem with parameter u we have to estimate the probability $P\left(n^{-1/2}\sup\limits_{f\in\mathscr{F}'}\left|\sum\limits_{j=1}^n f(\xi_j)\right| > u^{1+\alpha}\right)$ with some new nice, appropriately defined L_2-dense class of bounded functions $\mathcal{F}'$ and some number $\alpha > 0$. We shall exploit that the number u is replaced by a larger number $u^{1+\alpha}$ in the new problem. Let us also observe that if the sum of bounded random variables is considered, then for very large numbers u the probability we investigate equals zero. On the basis of these observations an appropriate backward induction procedure can be worked out. In its n-th step we give a good upper bound on the probability $P\left(n^{-1/2}\sup\limits_{f\in\mathcal{F}}\left|\sum\limits_{j=1}^{n} f(\xi_j)\right| > u\right)$ if $u \ge T_n$ with an appropriately chosen number T_n, and try to diminish the number T_n in each step of this induction procedure. We can prove Proposition 6.2 as a consequence of the result we get by means of this backward induction procedure. To work out the details we introduce the following notion.

Definition of good tail behaviour for a class of normalized random sums. *Let us have some measurable space $(X, \mathcal{X})$ and a probability measure μ on it together with some integer $n \ge 2$ and real number $\sigma > 0$. Consider some class $\mathcal{F}$ of functions $f(x)$ on the space $(X, \mathcal{X})$, and take a sequence of independent and μ distributed random variables $\xi_1, \dots, \xi_n$ with values in the space $(X, \mathcal{X})$. Define the normalized random sums $S_n(f) = \frac{1}{\sqrt{n}}\sum\limits_{j=1}^{n} f(\xi_j)$, $f \in \mathcal{F}$. Given some real number $T > 0$ we say that the set of normalized random sums $S_n(f)$, $f \in \mathcal{F}$, has a good tail behaviour at level T (with parameters n and σ^2 which will be fixed in the sequel) if the inequality*

$$P\left(\sup_{f\in\mathcal{F}} |S_n(f)| \ge A\sqrt{n}\sigma^2\right) \le \exp\left\{-A^{1/2} n\sigma^2\right\} \tag{7.6}$$

holds for all numbers $A > T$.

Now I formulate Proposition 7.3 and show that Proposition 6.2 follows from it.

Proposition 7.3. *Let us fix a positive integer $n \ge 2$, a real number $0 < \sigma \le 1$ and a probability measure μ on a measurable space $(X, \mathcal{X})$ together with some numbers $L \ge 1$ and $D \ge 1$ such that $n\sigma^2 \ge L\log n + \log D$. Let us consider those countable L_2-dense classes $\mathcal{F}$ of functions $f = f(x)$ on the space $(X, \mathcal{X})$ with exponent L and parameter D for which all functions $f \in \mathcal{F}$ satisfy the conditions $\sup\limits_{x\in X}|f(x)| \le \frac{1}{4}$, $\int f(x)\mu(dx) = 0$ and $\int f^2(x)\mu(dx) \le \sigma^2$.*

Let a number $T > 1$ be such that for all classes of functions $\mathcal{F}$ which satisfy the above conditions the set of normalized random sums $S_n(f) = \frac{1}{\sqrt{n}}\sum\limits_{j=1}^{n} f(\xi_j)$, $f \in \mathcal{F}$, defined with the help of a sequence of independent μ distributed random variables $\xi_1, \dots, \xi_n$ have a good tail behaviour at level $T^{4/3}$. There is a universal constant $\bar{A}_0$ such that if $T \ge \bar{A}_0$, then the set of the above defined normalized sums, $S_n(f)$, $f \in \mathcal{F}$, have a good tail behaviour for all such classes of functions $\mathcal{F}$ not only at level $T^{4/3}$ but also at level T.

Proposition 6.2 simply follows from Proposition 7.3. To show this let us first observe that a class of normalized random sums $S_n(f)$, $f \in \mathscr{F}$, has a good tail behaviour at level $T_0 = \frac{1}{4\sigma^2}$ if this class of functions $\mathscr{F}$ satisfies the conditions of Proposition 7.3. Indeed, in this case

$$P\left(\sup_{f\in\mathscr{F}} |S_n(f)| \geq A\sqrt{n}\sigma^2\right) \leq P\left(\sup_{f\in\mathscr{F}} |S_n(f)| > \frac{\sqrt{n}}{4}\right) = 0$$

for all $A > T_0$. Then the repetitive application of Proposition 7.3 yields that a class of random sums $S_n(f)$, $f \in \mathscr{F}$, has a good tail behaviour at all levels $T \geq T_0^{(3/4)^j}$ with an index j such that $T_0^{(3/4)^j} \geq \bar{A}_0$ if the class of functions $\mathscr{F}$ satisfies the conditions of Proposition 7.3. Hence it has a good tail behaviour for $T = \bar{A}_0^{4/3}$ with the number $\bar{A}_0$ appearing in Proposition 7.3. If a class of functions $f \in \mathscr{F}$ satisfies the conditions of Proposition 6.2, then the class of functions $\bar{\mathscr{F}} = \left\{\bar{f} = \frac{f}{4} \colon f \in \mathscr{F}\right\}$ satisfies the conditions of Proposition 7.3, with the same parameters σ, L and D. (Actually some of the inequalities that must hold for the elements of a class of functions $\bar{\mathscr{F}}$ satisfying the conditions of Proposition 7.3 are valid with smaller parameters. But we did not change these parameters to satisfy also the condition $n\sigma^2 \geq L\log n + \log D$.) Hence the class of functions $S_n(\bar{f})$, $\bar{f} \in \bar{\mathscr{F}}$, has a good tail behaviour at level $T = \bar{A}_0^{4/3}$. This implies that the original class of functions $\mathscr{F}$ satisfies formula (6.4) in Proposition 6.2, and this is what we had to show.

Proof of Proposition 7.3. Fix a class of functions $\mathscr{F}$ which satisfies the conditions of Proposition 7.3 together with two independent sequences $\xi_1, \dots, \xi_n$ and $\varepsilon_1, \dots, \varepsilon_n$ of independent random variables, where ξ_j is μ-distributed, $P(\varepsilon_j = 1) = P(\varepsilon_j = -1) = \frac{1}{2}$, $1 \leq j \leq n$, and investigate the conditional probability

$$P(f, A|\xi_1, \dots, \xi_n) = P\left(\frac{1}{\sqrt{n}}\left|\sum_{j=1}^n \varepsilon_j f(\xi_j)\right| \geq \frac{A}{6}\sqrt{n}\sigma^2 \,\middle|\, \xi_1, \dots, \xi_n\right)$$

for all functions $f \in \mathscr{F}$, $A > T$ and values $(\xi_1, \dots, \xi_n)$ in the condition. By the Hoeffding inequality formulated in Theorem 3.4

$$P(f, A|\xi_1, \dots, \xi_n) \leq 2\exp\left\{-\frac{\frac{1}{36}A^2 n\sigma^4}{2\bar{S}^2(f, \xi_1, \dots, \xi_n)}\right\} \tag{7.7}$$

with

$$\bar{S}^2(f, x_1, \dots, x_n) = \frac{1}{n}\sum_{j=1}^n f^2(x_j), \quad f \in \mathscr{F}.$$

Let us introduce the set

$$H = H(A) = \left\{ (x_1, \dots, x_n) \colon \sup_{f \in \mathcal{F}} \bar{S}^2(f, x_1, \dots, x_n) \ge \left(1 + A^{4/3}\right) \sigma^2 \right\}. \quad (7.8)$$

I claim that

$$P((\xi_1, \dots, \xi_n) \in H) \le e^{-A^{2/3} n\sigma^2} \quad \text{if } A > T. \quad (7.9)$$

(The set H is the small exceptional set of those points $(x_1, \dots, x_n)$ for which we cannot give a good estimate for $P(f, A|\xi_1(\omega), \dots, \xi_n(\omega))$ with $\xi_1(\omega) = x_1, \dots, \xi_n(\omega) = x_n$ for some $f \in \mathcal{F}$.)

To prove relation (7.9) let us consider the functions $\bar{f} = \bar{f}(f)$, $\bar{f}(x) = f^2(x) - \int f^2(x)\mu(dx)$, and introduce the class of functions $\bar{\mathcal{F}} = \{\bar{f}(f) \colon f \in \mathcal{F}\}$. Let us show that the class of functions $\bar{\mathcal{F}}$ satisfies the conditions of Proposition 7.3, hence the estimate (7.6) holds for the class of functions $\bar{\mathcal{F}}$ if $A > T^{4/3}$.

The relation $\int \bar{f}(x)\mu(dx) = 0$ clearly holds. The condition $\sup|\bar{f}(x)| \le \frac{1}{8} < \frac{1}{4}$ also holds if $\sup|f(x)| \le \frac{1}{4}$, and $\int \bar{f}^2(x)\mu(dx) \le \int f^4(x)\mu(dx) \le \frac{1}{16}\int f^2(x)\,\mu(dx) \le \frac{\sigma^2}{16} < \sigma^2$ if $f \in \mathcal{F}$. It remained to show that $\bar{\mathcal{F}}$ is an L_2-dense class with exponent L and parameter D. For this goal we need a good estimate on $\int (\bar{f}(x) - \bar{g}(x))^2 \rho(dx)$, where $\bar{f}$, $\bar{g} \in \bar{\mathcal{F}}$, and ρ is an arbitrary probability measure.

Observe that

$$\begin{aligned}
&\int (\bar{f}(x) - \bar{g}(x))^2 \rho(dx) \\
&\quad \le 2\int (f^2(x) - g^2(x))^2 \rho(dx) + 2\int (f^2(x) - g^2(x))^2 \mu(dx) \\
&\quad \le 2(\sup(|f(x)| + |g(x)|)^2 \left(\int (f(x) - g(x))^2 (\rho(dx) + \mu(dx) \right) \\
&\quad \le \int (f(x) - g(x))^2 \bar{\rho}(dx)
\end{aligned}$$

for all $f, g \in \mathcal{F}$, $\bar{f} = \bar{f}(f)$, $\bar{g} = \bar{g}(g)$ and probability measure ρ, where $\bar{\rho} = \frac{\rho + \mu}{2}$. This means that if $\{f_1, \dots, f_m\}$ is an ε-dense subset of $\mathcal{F}$ in the space $L_2(X, \mathcal{X}, \bar{\rho})$, then $\{\bar{f}_1, \dots, \bar{f}_m\}$ is an ε-dense subset of $\bar{\mathcal{F}}$ in the space $L_2(X, \mathcal{X}, \rho)$, and not only $\mathcal{F}$, but also $\bar{\mathcal{F}}$ is an L_2-dense class with exponent L and parameter D.

Because of the conditions of Proposition 7.3 we can write for the number $A^{4/3} > T^{4/3}$ and the class of functions $\bar{\mathcal{F}}$ that

$$\begin{aligned}
&P((\xi_1, \dots, \xi_n) \in H) \\
&\quad = P\left(\sup_{f \in \mathcal{F}} \left(\frac{1}{n} \sum_{j=1}^{n} \bar{f}(f)(\xi_j) + \frac{1}{n} \sum_{j=1}^{n} Ef^2(\xi_j) \right) \ge \left(1 + A^{4/3}\right) \sigma^2 \right)
\end{aligned}$$

$$\le P\left(\sup_{\bar f\in\bar{\mathcal F}}\frac{1}{\sqrt n}\sum_{j=1}^n \bar f(\xi_j)\ge A^{4/3}n^{1/2}\sigma^2\right)\le e^{-A^{2/3}n\sigma^2},$$

i.e. relation (7.9) holds.

By formula (7.7) and the definition of the set H given in (7.8) the estimate

$$P(f,A|\xi_1,\dots,\xi_n)\le 2e^{-A^{2/3}n\sigma^2/144}\quad \text{if } (\xi_1,\dots,\xi_n)\notin H \tag{7.10}$$

holds for all $f\in\mathcal F$ and $A>T\ge 1$. (Here we used the estimate $1+A^{4/3}\le 2A^{4/3}$.) Let us introduce the conditional probability

$$P(\mathcal F,A|\xi_1,\dots,\xi_n)=P\left(\sup_{f\in\mathcal F}\frac{1}{\sqrt n}\left|\sum_{j=1}^n\varepsilon_j f(\xi_j)\right|\ge\frac{A}{3}\sqrt n\sigma^2\,\middle|\,\xi_1,\dots,\xi_n\right)$$

for all $(\xi_1,\dots,\xi_n)$ and $A>T$. We shall estimate this conditional probability with the help of relation (7.10) if $(\xi_1,\dots,\xi_n)\notin H$.

Given a vector $x^{(n)}=(x_1,\dots,x_n)\in X^n$, let us introduce the probability measure

$$\nu=\nu(x_1,\dots,x_n)=\nu(x^{(n)})\quad\text{on } (X,\mathcal X)$$

which is concentrated in the coordinates of the vector $x^{(n)}=(x_1,\dots,x_n)$, and $\nu(\{x_j\})=\frac1n$ for all points x_j, $j=1,\dots,n$. If $\int f^2(u)\nu(du)\le\delta^2$ for a function f, then $\left|\frac{1}{\sqrt n}\sum_{j=1}^n\varepsilon_j f(x_j)\right|\le n^{1/2}\int|f(u)|\nu(du)\le n^{1/2}\delta$. As a consequence, we can write that

$$\left|\frac{1}{\sqrt n}\sum_{j=1}^n\varepsilon_j f(x_j)-\frac{1}{\sqrt n}\sum_{j=1}^n\varepsilon_j g(x_j)\right|\le\frac{A}{6}\sqrt n\sigma^2$$

$$\text{if}\int (f(u)-g(u))^2\,d\nu(u)\le\left(\frac{A\sigma^2}{6}\right)^2. \tag{7.11}$$

Remark. We may assume in our proof that the distribution of the random variables ξ_j, $1\le j\le n$, are non-atomic, and as a consequence we can restrict our attention to such measures $\nu(x^{(n)})$ for which all coordinates of the vector $x^{(n)}$ are different. Otherwise we can define independent and uniformly distributed random variables on the interval $[0,1]$, $\eta_1,\dots,\eta_n$, which are also independent of the random variables ξ_j, $1\le j\le n$. With the help of these random variables η_j we can introduce the random variables $\tilde\xi_j=(\xi_j,\eta_j)$, $1\le j\le n$, and the class of functions $\tilde{\mathcal F}$ on the space $X \times [0,1]$ consisting of functions $\tilde f(x,y) = f(x)$, $f \in \mathcal{F}$, with $x \in X$ and $0 \le y \le 1$. It is not difficult to see that the random variables $\tilde\xi_j$ and the class of functions $\tilde{\mathcal{F}}$ satisfy the conditions of Proposition 7.3, and the distribution of the random variables $\tilde\xi_j$ is non-atomic. Hence we can apply Proposition 7.3 with such a choice, and this provides the statement of Proposition 7.3 in the original case, too.

Let us list the elements of the (countable) set $\mathcal{F}$ as $\mathcal{F} = \{f_1, f_2, \dots\}$, fix the number $\delta = \frac{A\sigma^2}{6}$, and choose for all vectors $x^{(n)} = (x_1,\dots,x_n) \in X^n$ a sequence of indices $p_1(x^{(n)}),\dots,p_m(x^{(n)})$ taking positive integer values with $m = \max(1, D\delta^{-L}) = \max(1, D(\frac{6}{A\sigma^2})^L)$ elements in such a way that $\inf\limits_{1\le l\le m} \int (f(u) - f_{p_l(x^{(n)})}(u))^2\,d\nu(x^{(n)})(u) \le \delta^2$ for all $f \in \mathcal{F}$ and $x^{(n)} \in X^n$ with the above defined measure $\nu(x^{(n)})$ on the space $(X, \mathcal{X})$. This is possible because of the L_2-dense property of the class of functions $\mathcal{F}$. (This is the point where the L_2-dense property of the class of functions $\mathcal{F}$ is exploited in its full strength.) In a complete proof of Proposition 7.3 we still have to show that we can choose the indices $p_j(x^{(n)})$, $1 \le j \le m$, as measurable functions of their argument $x^{(n)}$ on the space $(X^n, \mathcal{X}^n)$. We shall show this in Lemma 7.4 at the end of the proof.

Put $\xi^{(n)}(\omega) = (\xi_1(\omega),\dots,\xi_n(\omega))$. Because of relation (7.11), the choice of the number δ and the property of the functions $f_{p_l(x^{(n)})}(\cdot)$ we have

$$\left\{\omega\colon \sup_{f\in\mathcal{F}} \frac{1}{\sqrt n}\left|\sum_{j=1}^n \varepsilon_j(\omega) f(\xi_j(\omega))\right| \ge \frac{A}{3}\sqrt n\sigma^2\right\} \tag{7.12}$$

$$\subset \bigcup_{l=1}^m \left\{\omega\colon \frac{1}{\sqrt n}\left|\sum_{j=1}^n \varepsilon_j(\omega) f_{p_l(\xi^{(n)}(\omega))}(\xi_j(\omega))\right| \ge \frac{A}{6}\sqrt n\sigma^2\right\}.$$

We can estimate the conditional probability at the right-hand side of (7.12) under the condition that the vector $(\xi_1(\omega),\dots,\xi_n(\omega))$ takes such a prescribed value for which $(\xi_1(\omega),\dots,\xi_n(\omega)) \in H$. We get with the help of (7.12), inequality (7.10) and the definition of the quantity $P(f, A|\xi_1,\dots,\xi_n)$ before formula (7.7) that

$$\begin{aligned} P(\mathcal{F}, A|\xi_1,\dots,\xi_n) &\le \sum_{l=1}^m P(f_{p_l(\xi^{(n)})}, A|\xi_1,\dots,\xi_n) \\ &\le 2\max\left(1, D\left(\frac{6}{A\sigma^2}\right)^L\right) e^{-A^{2/3}n\sigma^2/144} \\ &\qquad \text{if } (\xi_1,\dots,\xi_n) \notin H \text{ and } A > T. \end{aligned} \tag{7.13}$$

If $A \ge \bar A_0$ with a sufficiently large constant $\bar A_0$, then this inequality together with Lemma 7.2 and the estimate (7.9) imply that

$$
\begin{aligned}
&P\left(\frac{1}{\sqrt{n}}\sup_{f\in\mathcal{F}}\left|\sum_{j=1}^n f(\xi_j)\right|\ge An^{1/2}\sigma^2\right)\\
&\quad\le 4P\left(\frac{1}{\sqrt{n}}\sup_{f\in\mathcal{F}}\left|\sum_{j=1}^n \varepsilon_j f(\xi_j)\right|\ge \frac{A}{3}n^{1/2}\sigma^2\right) \qquad (7.14)\\
&\quad\le \max\left(4, 8D\left(\frac{6}{A\sigma^2}\right)^L\right)e^{-A^{2/3}n\sigma^2/144}+4e^{-A^{2/3}n\sigma^2} \quad \text{if } A>T.
\end{aligned}
$$

(We may apply Lemma 7.2 if $A\ge A_0$ with a sufficiently large A_0, since $n\sigma^2\ge L\log n+\log D\ge\log 2$, hence $\sqrt{n}\sigma\ge\sqrt{\log 2}$, and the condition $A\ge\frac{3\sqrt{2}}{\sqrt{n}\sigma}$ demanded in relation (7.2) is satisfied.)

By the conditions of Proposition 7.3 the inequalities $n\sigma^2\ge L\log n+\log D$ hold with some $L\ge 1$, $D\ge 1$ and $n\ge 2$. This implies that $n\sigma^2\ge L\log 2\ge\frac12$, $(\frac{6}{A\sigma^2})^L\le(\frac{n}{2n\sigma^2})^L\le n^L=e^{L\log n}\le e^{n\sigma^2}$ if $A\ge\bar{A}_0$ with some sufficiently large constant $\bar{A}_0>0$, and $2D=e^{\log 2+\log D}\le e^{3n\sigma^2}$. Hence the first term at the right-hand side of (7.14) can be bounded by

$$
\max\left(4, 8D\left(\frac{6}{A\sigma^2}\right)^L\right)e^{-A^{2/3}n\sigma^2/144}\le e^{-A^{2/3}n\sigma^2/144}\cdot 4e^{4n\sigma^2}\le\frac12 e^{-A^{1/2}n\sigma^2}
$$

if $A\ge\bar{A}_0$ with a sufficiently large $\bar{A}_0$. The second term at the right-hand side of (7.14) can also be bounded as $4e^{-A^{2/3}n\sigma^2}\le\frac12 e^{-A^{1/2}n\sigma^2}$ with an appropriate choice of the number $\bar{A}_0$.

By the above calculation formula (7.14) yields the inequality

$$
P\left(\frac{1}{\sqrt{n}}\sup_{f\in\mathcal{F}}\left|\sum_{j=1}^n f(\xi_j)\right|\ge An^{1/2}\sigma^2\right)\le e^{-A^{1/2}n\sigma^2}
$$

if $A>T$, and the constant $\bar{A}_0$ is chosen sufficiently large. □

To complete the proof of Proposition 7.3 we still show in the following Lemma 7.4 that the functions $p_l(x^{(n)})$, $1\le l\le m$, we have introduced in the above argument can be chosen as measurable functions in the space $(X^n,\mathcal{X}^n)$. This implies that the expressions $f_{p_l(\xi^{(n)}(\omega))}(\xi_j(\omega))$ in formula (7.12) are $\mathcal{F}(\xi_1,\dots,\xi_n)$ measurable random variables. Hence the formulation of (7.13) is legitime, no measurability problem arises. We shall present Lemma 7.4 together with some generalizations in Lemmas 7.4A and 7.4B that we shall apply later in the proof of Propositions 15.3 and 15.4 which are multivariate versions of Proposition 7.3. We shall need these results in the proof of the multivariate version of Proposition 6.2.

We have formulated them not in their most general possible form, but in the form as we shall need them.

Lemma 7.4. *Let $\mathcal{F} = \{f_1, f_2, \dots\}$ be a countable and L_2-dense class of functions with some exponent $L > 0$ and parameter $D \ge 1$ on a measurable space $(X, \mathcal{X})$. Fix some positive integer n, and define for all $x^{(n)} = (x_1, \dots, x_n) \in X^n$ the probability measure $\nu(x^{(n)}) = \nu(x_1, \dots, x_n)$ on the space $(X, \mathcal{X})$ by the formula $\nu(x^{(n)})(x_j) = \frac{1}{n}$, $1 \le j \le n$. For a number $0 \le \varepsilon \le 1$ put $m = m(\varepsilon) = [D\varepsilon^{-L}]$, where $[\cdot]$ denotes integer part. For all $0 \le \varepsilon \le 1$ there exists $m = m(\varepsilon)$ measurable functions $p_l(x^{(n)})$, $1 \le l \le m$, on the measurable space $(X^n, \mathcal{X}^n)$ with positive integer values in such a way that $\inf\limits_{1 \le l \le m} \int (f(u) - f_{p_l(x^{(n)})}(u))^2 \nu(x^{(n)})(\,du) \le \varepsilon^2$ for all $x^{(n)} \in X^n$ and $f \in \mathcal{F}$.*

In the proof of Proposition 15.3 we need the following result.

Lemma 7.4A. *Let $\mathcal{F} = \{f_1, f_2, \dots\}$ be a countable and L_2-dense class of functions with some exponent $L > 0$ and parameter $D \ge 1$ on the k-fold product $(X^k, \mathcal{X}^k)$ of a measurable space $(X, \mathcal{X})$ with some $k \ge 1$. Fix some positive integer n, and define for all vectors $x^{(n)} = (x_l^{(j)}, 1 \le l \le n, 1 \le j \le k) \in X^{kn}$, where $x_l^{(j)} \in X$ for all j and l the probability measure $\rho(x^{(n)})$ in the space $(X^k, \mathcal{X}^k)$ by the formula $\rho(x^{(n)})(x_{l_j}^{(j)}, 1 \le j \le k, 1 \le l_j \le n) = \frac{1}{n^k}$ for all sequences $(x_{l_1}^{(1)}, \dots, x_{l_k}^{(k)})$, $1 \le j \le k$, $1 \le l_j \le n$, with coordinates of the vector $x^{(n)} = (x_l^{(j)}, 1 \le l \le n, 1 \le j \le k)$. For all $0 \le \varepsilon \le 1$ there exist $m = m(\varepsilon) = [D\varepsilon^{-L}]$ measurable functions $p_r(x^{(n)})$, $1 \le r \le m$, on the measurable space $(X^{kn}, \mathcal{X}^{kn})$ with positive integer values in such a way that $\inf\limits_{1 \le r \le m} \int (f(u) - f_{p_r(x^{(n)})}(u))^2 \rho(x^{(n)})(\,du) \le \varepsilon^2$ for all $x^{(n)} \in X^{kn}$ and $f \in \mathcal{F}$.*

In the proof of Proposition 15.4 the following result will be needed.

Lemma 7.4B. *Let $\mathcal{F} = \{f_1, f_2, \dots\}$ be a countable and L_2-dense class of functions with some exponent $L > 0$ and parameter $D \ge 1$ on the product space $(X^k \times Y, \mathcal{X}^k \times \mathcal{Y})$ with some measurable spaces $(X, \mathcal{X})$ and $(Y, \mathcal{Y})$ and integer $k \ge 1$. Fix some positive integer n, and define for all vectors $x^{(n)} = (x_l^{(j,1)}, x_l^{(j,-1)}, 1 \le l \le n, 1 \le j \le k) \in X^{2kn}$, where $x_l^{(j,\pm 1)} \in X$ for all j and l a probability measure $\alpha(x^{(n)})$ in the space $(X^k \times Y, \mathcal{X}^k \times \mathcal{Y})$ in the following way. Fix some probability measure ρ in the space $(Y, \mathcal{Y})$ and two ± 1 sequences $\varepsilon_1^{(k)} = (\varepsilon_{1,1}, \dots, \varepsilon_{k,1})$ and $\varepsilon_2^{(k)} = (\varepsilon_{1,2}, \dots, \varepsilon_{k,2})$ of length k. Define with their help first the following probability measures $\alpha_1(x^{(n)}) = \alpha_1(x^{(n)}, \varepsilon_1^{(k)}, \varepsilon_2^{(k)}, \rho)$ and $\alpha_2(x^{(n)}) = \alpha_2(x^{(n)}, \varepsilon_1^{(k)}, \varepsilon_2^{(k)}, \rho)$ in the space $(X^k \times Y, \mathcal{X}^k \times \mathcal{Y})$ for all $x^{(n)} \in \mathcal{X}^{2kn}$. Let $\alpha_1(x^{(n)})(\{x_{l_1}^{(1,\varepsilon_{1,1})}\} \times \cdots \times \{x_{l_k}^{(k,\varepsilon_{k,1})}\} \times B) = \frac{\rho(B)}{n^k}$ and $\alpha_2(x^{(n)})(\{x_{l_1}^{(1,\varepsilon_{1,2})}\} \times \cdots \times \{x_{l_k}^{(k,\varepsilon_{k,2})}\} \times B) = \frac{\rho(B)}{n^k}$ with $1 \le l_j \le n$ for all $1 \le j \le k$ and $B \in \mathcal{Y}$ if $x_{l_j}^{(j,\varepsilon_{j,1})}$ and $x_{l_j}^{(j,\varepsilon_{j,2})}$ are the appropriate coordinates of the vector $x^{(n)} \in X^{2kn}$. Put $\alpha(x^{(n)}) = \frac{\alpha_1(x^{(n)}) + \alpha_2(x^{(n)})}{2}$. For all $0 \le \varepsilon \le 1$ there exist $m = m(\varepsilon) = [D\varepsilon^{-L}]$ measurable* *functions $p_r(x^{(n)})$, $1 \le r \le m$, on the measurable space $(X^{2kn}, \mathscr{X}^{2kn})$ with positive integer values in such a way that $\inf\limits_{1\le r\le m} \int (f(u) - f_{p_r(x^{(n)})}(u))^2 \alpha(x^{(n)})(du) \le \varepsilon^2$ for all $x^{(n)} \in X^{2kn}$ and $f \in \mathscr{F}$.*

Proof of Lemma 7.4. Fix some $0 < \varepsilon \le 1$, put the number $m = m(\varepsilon)$ introduced in the lemma, and let us list the set of all vectors $(j_1, \dots, j_m)$ of length m with positive integer coordinates in some way. Define for all of these vectors $(j_1, \dots, j_m)$ the set $B(j_1, \dots, j_m) \subset X^n$ in the following way. The relation $x^{(n)} = (x_1, \dots, x_n) \in B(j_1, \dots, j_m)$ holds if and only if $\inf\limits_{1\le r\le m} \int (f(u) - f_{j_r}(u))^2 \, d\nu(x^{(n)})(u) \le \varepsilon^2$ for all $f \in \mathscr{F}$. Then all sets $B(j_1, \dots, j_m)$ are measurable, and $\bigcup\limits_{(j_1,\dots,j_m)} B(j_1, \dots, j_m) = X^n$ because $\mathscr{F}$ is an L_2-dense class of functions with exponent L and parameter D. Given a point $x^{(n)} = (x_1, \dots, x_n)$ let us choose the first vector $(j_1, \dots, j_m) = (j_1(x^{(n)}), \dots, j_m(x^{(n)}))$ in our list of vectors for which $x^{(n)} \in B(j_1, \dots, j_m)$, and define $p_l(x^{(n)}) = j_l(x^{(n)})$ for all $1 \le l \le m$ with this vector $(j_1, \dots, j_m)$. Then the functions $p_l(x^{(n)})$ are measurable, and the functions $f_{p_l(x^{(n)})}$, $1 \le l \le m$, defined with their help together with the probability measures $\nu(x^{(n)})$ satisfy the inequality demanded in Lemma 7.4. □

The proof of Lemmas 7.4A and 7.4B is almost the same. We only have to modify the definition of the sets $B(j_1, \dots, j_m)$ in a natural way. The space of arguments $x^{(n)}$ are the spaces X^{kn} and X^{2kn} in these lemmas, and we have to integrate with respect to the measures $\rho(x^{(n)})$ in the space X^k and with respect to the measures $\alpha(x^{(n)})$ in the space $X^k \times Y$ respectively. The sets $B(j_1, \dots, j_m)$ are measurable also in these cases, and the rest of the proof can be applied without any change.

Chapter 8
Formulation of the Main Results of This Work

Former chapters of this work contain estimates about the tail distribution of normalized sums of independent, identically distributed random variables and of the supremum of appropriate classes of such random sums. They were considered together with some estimates about the tail distribution of the integral of a (deterministic) function with respect to a normalized empirical distribution and of the supremum of such integrals. This two kinds of problems are closely related, and to understand them better it is useful to investigate them together with their natural Gaussian counterpart.

In this chapter I formulate the natural multivariate versions of these results. They will be proved in the subsequent chapters. To formulate them we have to introduce some new notions. I shall also discuss some new problems whose solutions help in their proof. I finish this chapter with a short overview about the content of the remaining part of this work.

I start this chapter with the formulation of two results, Theorems 8.1 and 8.2 together with some simple consequences. They yield a sharp estimate about the tail distribution of a multiple random integral with respect to a normalized empirical distribution and about the analogous problem when the tail distribution of the supremum of such integrals is considered. These results are the natural versions of the corresponding one-variate results about the tail behaviour of an integral or of the supremum of a class of integrals with respect to a normalized empirical distribution. They can be formulated with the help of the notions introduced before, in particular with the help of the notion of multiple random integrals with respect to a normalized empirical distribution introduced in formula (4.8).

To formulate the following two results, Theorems 8.3 and 8.4 and their consequences, which are the natural multivariate versions of the results about the tail distribution of partial sums of independent random variables, and of the supremum of such sums we have to make some preparations. First we introduce the so-called U-statistics which can be considered the natural multivariate generalizations of the sum of independent and identically distributed random variables. Beside this, observe that in the one-variate case we had a good estimation about the tail

P. Major, *On the Estimation of Multiple Random Integrals and U-Statistics*,
Lecture Notes in Mathematics 2079, DOI 10.1007/978-3-642-37617-7_8,

distribution of sums of independent random variables only if the summands had expectation zero. We have to find the natural multivariate version of this property. Hence we define the so-called degenerate U-statistics which can be considered as the natural multivariate counterparts of sums of independent and identically distributed random variables with zero expectation. Theorems 8.3 and 8.4 contain estimates about the tail-distribution of degenerate U-statistics and of the supremum of such expressions.

In Theorems 8.5 and 8.6, I formulate the Gaussian counterparts of the above results. They deal with multiple Wiener–Itô integrals with respect to a so-called white noise. The notion of white noise and multiple Wiener–Itô integrals with respect to it and their properties needed to have a good understanding of these results will be explained in Chap. 10. Still two results are discussed in this chapter. They are Examples 8.7 and 8.8, which state that the estimates of Theorems 8.5 and 8.3 are in a certain sense sharp.

To formulate the first two results of this chapter let us consider a sequence of independent and identically distributed random variables $\xi_1, \dots, \xi_n$ with values in a measurable space $(X, \mathcal{X})$. Let μ denote the distribution of the random variables ξ_j, and introduce the empirical distribution of the sequence $\xi_1, \dots, \xi_n$ defined in (4.5). Given a measurable function $f(x_1, \dots, x_k)$ on the k-fold product space $(X^k, \mathcal{X}^k)$ consider its integral $J_{n,k}(f)$ with respect to the k-fold product of the normalized empirical distribution $\sqrt{n}(\mu_n - \mu)$ defined in formula (4.8). In the definition of this integral the diagonals $x_j = x_l$, $1 \le j < l \le k$, were omitted from the domain of integration. The following Theorem 8.1 can be considered as the multiple integral version of Bernstein's inequality formulated in Theorem 3.1.

Theorem 8.1 (Estimate on the tail distribution of a multiple random integral with respect to a normalized empirical distribution). *Let us take a measurable function $f(x_1, \dots, x_k)$ on the k-fold product $(X^k, \mathcal{X}^k)$ of a measurable space $(X, \mathcal{X})$ with some $k \ge 1$ together with a non-atomic probability measure μ on $(X, \mathcal{X})$ and a sequence of independent and identically distributed random variables $\xi_1, \dots, \xi_n$ with distribution μ on $(X, \mathcal{X})$. Let the function f satisfy the conditions*

$$\|f\|_\infty = \sup_{x_j \in X,\, 1 \le j \le k} |f(x_1, \dots, x_k)| \le 1, \tag{8.1}$$

and

$$\|f\|_2^2 = \int f^2(x_1, \dots, x_k)\mu(dx_1)\dots\mu(dx_k) \le \sigma^2 \tag{8.2}$$

with some constant $0 < \sigma \le 1$. There exist some constants $C = C_k > 0$ and $\alpha = \alpha_k > 0$ such that the random integral $J_{n,k}(f)$ defined in formulas (4.5) and (4.8) satisfies the inequality

$$P(|k! J_{n,k}(f)| > u) \le C \max\left(e^{-\alpha(u/\sigma)^{2/k}}, e^{-\alpha(nu^2)^{1/(k+1)}}\right) \tag{8.3}$$

for all $u > 0$. The constants $C = C_k > 0$ and $\alpha = \alpha_k > 0$ in formula (8.3) depend only on the parameter k.

Theorem 8.1 can be reformulated in the following equivalent form.

Theorem 8.1′. *Under the conditions of Theorem 8.1*

$$P(|k!J_{n,k}(f)| > u) \le Ce^{-\alpha(u/\sigma)^{2/k}} \quad \textit{for all } 0 < u \le n^{k/2}\sigma^{k+1} \tag{8.4}$$

with a number σ, $0 \le \sigma \le 1$, satisfying relation in (8.2) and some universal constants $C = C_k > 0$, $\alpha = \alpha_k > 0$, depending only on the multiplicity k of the integral $J_{n,k}(f)$.

Theorem 8.1 clearly implies Theorem 8.1′, since in the case $u \le n^{k/2}\sigma^{k+1}$ the first term is larger than the second one in the maximum at the right-hand side of formula (8.3). On the other hand, Theorem 8.1′ implies Theorem 8.1 also if $u > n^{k/2}\sigma^{k+1}$. Indeed, in this case Theorem 8.1′ can be applied with $\bar\sigma = \left(un^{-k/2}\right)^{1/(k+1)} \ge \sigma$ if $u \le n^{k/2}$, since the condition $0 < \bar\sigma \le 1$ is satisfied. This yields that $P\left(|k!J_{n,k}(f)| > u\right) \le C\exp\left\{-\alpha\left(\frac{u}{\bar\sigma}\right)^{2/k}\right\} = C\exp\left\{-\alpha(nu^2)^{1/(k+1)}\right\}$ if $n^{k/2} \ge u > n^{k/2}\sigma^{k+1}$, and relation (8.3) holds in this case. If $u > 2^k n^{k/2}$, then $P(k!|J_{n,k}(f)| > u) = 0$, and if $n^{k/2} \le u < 2^k n^{k/2}$, then

$$\begin{aligned} P(|k!J_{n,k}(f)| > u) &\le P(|k!J_{n,k}(f)| > n^{k/2}) \\ &\le C\exp\left\{-\alpha((n\cdot n^{k/2})^2)^{1/(k+1)}\right\} \\ &\le C\exp\left\{-2^{-k}\alpha(nu^2)^{1/(k+1)}\right\}. \end{aligned}$$

Hence relation (8.3) holds (with a possibly different parameter α) in these cases, too.

Theorem 8.1 or Theorem 8.1′ state that the tail distribution $P(k!|J_{n,k}(f)| > u)$ of the k-fold random integral $k!J_{n,k}(f)$ can be bounded similarly to the probability $P(|\text{const.}\,\sigma\eta^k| > u)$, where η is a random variable with standard normal distribution, and the number $0 \le \sigma \le 1$ satisfies relation (8.2), provided that the level u we consider is less than $n^{k/2}\sigma^{k+1}$. As we shall see later (see Corollary 1 of Theorem 9.4), the value of the number σ^2 in formula (8.2) is closely related to the variance of $k!J_{n,k}(f)$. At the end of this chapter an example is given which shows that the condition $u \le n^{k/2}\sigma^{k+1}$ is really needed in Theorem 8.1′.

The next result, Theorem 8.2, is the generalization of Theorem 4.1′ for multiple random integrals with respect to a normalized empirical measure. In its formulation the notions of L_2-dense classes and countable approximability introduced in Chap. 4 are applied.

Theorem 8.2 (Estimate on the supremum of multiple random integrals with respect to an empirical distribution). *Let us have a non-atomic probability measure μ on a measurable space $(X, \mathscr{X})$ together with a countable and L_2-dense* *class $\mathcal{F}$ of functions $f = f(x_1, \dots, x_k)$ of k variables with some parameter $D \geq 2$ and exponent $L \geq 1$ on the product space $(X^k, \mathcal{X}^k)$ which satisfies the conditions*

$$\|f\|_\infty = \sup_{x_j \in X,\, 1 \leq j \leq k} |f(x_1, \dots, x_k)| \leq 1, \qquad \text{for all } f \in \mathcal{F} \tag{8.5}$$

and

$$\|f\|_2^2 = Ef^2(\xi_1, \dots, \xi_k) = \int f^2(x_1, \dots, x_k)\mu(dx_1)\dots\mu(dx_k) \leq \sigma^2 \quad \text{for all } f \in \mathcal{F} \tag{8.6}$$

with some constant $0 < \sigma \leq 1$. There exist some constants $C = C(k) > 0$, $\alpha = \alpha(k) > 0$ and $M = M(k) > 0$ depending only on the parameter k such that the supremum of the random integrals $k!J_{n,k}(f)$, $f \in \mathcal{F}$, defined by formula (4.8) satisfies the inequality

$$P\left(\sup_{f \in \mathcal{F}} |k!J_{n,k}(f)| \geq u\right) \leq C \exp\left\{-\alpha \left(\frac{u}{\sigma}\right)^{2/k}\right\}$$

for those numbers u for which

$$n\sigma^2 \geq \left(\frac{u}{\sigma}\right)^{2/k} \geq M(L^{3/2} \log \frac{2}{\sigma} + (\log D)^{3/2}), \tag{8.7}$$

where the numbers D and L agree with the parameter and exponent of the L_2-dense class $\mathcal{F}$.

The condition about the countable cardinality of the class $\mathcal{F}$ can be replaced by the weaker condition that the class of random variables $k!J_{n,k}(f)$, $f \in \mathcal{F}$, is countably approximable.

The condition given for the number u in formula (8.7) appears in Theorem 8.2 for a similar reason as the analogous condition formulated in (4.4) in its one-variate counterpart, Theorem 4.1. The lower bound is needed, since we have a good estimate in formula (8.7) only for $u \geq E \sup_{f \in \mathcal{F}} |k!J_{n,k}(f)|$. The upper bound appears, since we have a good estimate in Theorem 8.1′ only for $0 < u \leq n^{k/2}\sigma^{k+1}$. If a pair of numbers (u, σ) does not satisfy condition (8.7), then we may try to get an estimate by increasing the number σ or decreasing the number u.

To formulate such a version of Theorems 8.1 and 8.2 which corresponds to the results about sums of independent random variables in the case $k = 1$ the following notions will be introduced.

Definition of U-statistics. *Let us consider a function $f = f(x_1, \dots, x_k)$ on the k-th power $(X^k, \mathcal{X}^k)$ of a space $(X, \mathcal{X})$ together with a sequence of independent* *and identically distributed random variables* $\xi_1,\dots,\xi_n$, $n \ge k$, *which take their values in this space* $(X,\mathcal{X})$. *The expression*

$$I_{n,k}(f)=\frac{1}{k!}\sum_{\substack{(l_1,\dots,l_k)\colon 1\le l_j\le n,\ j=1,\dots,k,\\ l_j\ne l_{j'} \text{ if } j\ne j'}} f\left(\xi_{l_1},\dots,\xi_{l_k}\right) \tag{8.8}$$

is called a U-statistic of order k with the sequence $\xi_1,\dots,\xi_n$, *and kernel function* f.

Remark. In later calculations sometimes we shall work with U-statistics with kernel functions of the form $f(x_{u_1},\dots,x_{u_k})$ instead of $f(x_1,\dots,x_k)$, where $\{u_1,\dots,u_k\}$ is an arbitrary set with different elements. The U-statistic with such a kernel function will also be defined, and it equals the U-statistic with the original kernel function f defined in (8.8), i.e.

$$I_{n,k}(f(x_{u_1},\dots,x_{u_k}))=I_{n,k}(f(x_1,\dots,x_k)). \tag{8.9}$$

(Observe that if we define the function $f_\pi(x_1,\dots,x_k)=f(x_{\pi(1)},\dots,x_{\pi(k)})$ for all permutations π of the set $\{1,\dots,k\}$, then $I_{n,k}(f_\pi)=I_{n,k}(f)$, hence the above definition is legitimate.) Such a definition is natural, and it simplifies the notation in some calculations. A similar convention will be introduced about Wiener–Itô integrals in Chap. 10.

Some special U-statistics, called degenerate U-statistics, will also be introduced. They can be considered as the natural multivariate version of sums of identically distributed random variables with expectation zero. Degenerate U-statistics will be defined together with canonical kernel functions, because these two notions are closely related. For the sake of simpler notation in later discussions we shall allow general indexation of the variables in the definition of canonical functions, and we shall consider functions of the form $f(x_{l_1},\dots,x_{l_k})$ instead of $f(x_1,\dots,x_k)$.

Definition of degenerate U-statistics. *A U-statistic* $I_{n,k}(f)$ *of order* k *with a sequence of independent and identically distributed random variables* $\xi_1,\dots,\xi_n$ *is called degenerate if its kernel function* $f(x_1,\dots,x_k)$ *satisfies the relation*

$$E(f(\xi_1,\dots,\xi_k)|\xi_1=x_1,\dots,\xi_{j-1}=x_{j-1},\xi_{j+1}=x_{j+1},\dots,\xi_k=x_k)=0$$
$$\text{for all } 1\le j\le k \text{ and } x_s\in X,\ s\ne j.$$

Definition of a canonical function. *A function* $f(x_{l_1},\dots,x_{l_k})$ *taking values in the k-fold product of a measurable space* $(X,\mathcal{X})$ *is called a canonical function with respect to a probability measure* μ *on* $(X,\mathcal{X})$ *if*

$$\int f(x_{l_1},\dots,x_{l_{j-1}},u,x_{l_{j+1}},\dots,x_{l_k})\mu(du)=0$$
$$\text{for all } 1\le j\le k \text{ and } x_{l_s}\in X,\ s\ne j. \tag{8.10}$$

For the sake of more convenient notations in the subsequent part of this work we shall also speak of U-statistics of order zero. We shall write $I_{n,0}(c) = c$ for any constant c, and $I_{n,0}(c)$ will be called a degenerate U-statistic of order zero. A constant will be considered as a canonical function with zero arguments.

It is clear that a U-statistic $I_{n,k}(f)$ with kernel function f and independent μ-distributed random variables $\xi_1, \dots, \xi_n$ is degenerate if and only if its kernel function is canonical with respect to the probability measure μ. Let us also observe that

$$I_{n,k}(f) = I_{n,k}(\operatorname{Sym} f) \tag{8.11}$$

for all functions of k variables.

The next two results, Theorems 8.3 and 8.4, deal with degenerate U-statistics. Theorem 8.3 is the U-statistic version of Theorem 8.1, and Theorem 8.4 is the U-statistic version of Theorem 8.2. Actually Theorem 8.3 yields a sharper estimate than Theorems 8.1, because it contains more explicit and better universal constants. I shall return to this point later.

Theorem 8.3 (Estimate on the tail distribution of a degenerate U-statistic). *Let us have a measurable function $f(x_1, \dots, x_k)$ on the k-fold product $(X^k, \mathcal{X}^k)$, $k \ge 1$, of a measurable space $(X, \mathcal{X})$ together with a probability measure μ on $(X, \mathcal{X})$ and a sequence of independent and identically distributed random variables $\xi_1, \dots, \xi_n$, $n \ge k$, with distribution μ on $(X, \mathcal{X})$. Let us consider the U-statistic $I_{n,k}(f)$ of order k with this sequence of random variables $\xi_1, \dots, \xi_n$. Assume that this U-statistic is degenerate, i.e. its kernel function $f(x_1, \dots, x_k)$ is canonical with respect to the measure μ. Let us also assume that the function f satisfies conditions (8.1) and (8.2) with some number $0 < \sigma \le 1$. Then there exist some constants $A = A(k) > 0$ and $B = B(k) > 0$ depending only on the order k of the U-statistic $I_{n,k}(f)$ such that*

$$P(n^{-k/2}|k!I_{n,k}(f)| > u) \le A \exp\left\{ -\frac{u^{2/k}}{2\sigma^{2/k}\left(1 + B\left(un^{-k/2}\sigma^{-(k+1)}\right)^{1/k}\right)} \right\} \tag{8.12}$$

for all $0 \le u \le n^{k/2}\sigma^{k+1}$.

Let us also formulate the following simple corollary of Theorem 8.3.

Corollary of Theorem 8.3. *Under the conditions of Theorem 8.3 there exist some universal constants $C = C(k) > 0$ and $\alpha = \alpha(k) > 0$ that*

$$P(n^{-k/2}|k!I_{n,k}(f)| > u) \le C \exp\left\{ -\alpha \left(\frac{u}{\sigma}\right)^{2/k} \right\} \quad \text{for all } 0 \le u \le n^{k/2}\sigma^{k+1}. \tag{8.13}$$

The following estimate holds about the supremum of degenerate U-statistics.

Theorem 8.4 (Estimate on the supremum of degenerate U-statistics). *Let us have a probability measure μ on a measurable space $(X, \mathcal{X})$ together with a countable and L_2-dense class $\mathcal{F}$ of functions $f = f(x_1, \dots, x_k)$ of k variables with some parameter $D \ge 2$ and exponent $L \ge 1$ on the product space $(X^k, \mathcal{X}^k)$ which satisfies conditions (8.5) and (8.6) with some constant $0 < \sigma \le 1$. Let us take a sequence of independent μ distributed random variables $\xi_1, \dots, \xi_n$, $n \ge k$, and consider the U-statistics $I_{n,k}(f)$ with these random variables and kernel functions $f \in \mathcal{F}$. Let us assume that all these U-statistics $I_{n,k}(f)$, $f \in \mathcal{F}$, are degenerate, or in an equivalent form, all functions $f \in \mathcal{F}$ are canonical with respect to the measure μ. Then there exist some constants $C = C(k) > 0$, $\alpha = \alpha(k) > 0$ and $M = M(k) > 0$ depending only on the parameter k such that the inequality*

$$P\left(\sup_{f \in \mathcal{F}} n^{-k/2} |k! I_{n,k}(f)| \ge u\right) \le C \exp\left\{-\alpha \left(\frac{u}{\sigma}\right)^{2/k}\right\}$$

holds for those numbers u for which

$$n\sigma^2 \ge \left(\frac{u}{\sigma}\right)^{2/k} \ge M\left(L^{3/2} \log \frac{2}{\sigma} + (\log D)^{3/2}\right), \tag{8.14}$$

where the numbers D and L agree with the parameter and exponent of the L_2-dense class $\mathcal{F}$.

The condition about the countable cardinality of the class $\mathcal{F}$ can be replaced by the weaker condition that the class of random variables $n^{-k/2} I_{n,k}(f)$, $f \in \mathcal{F}$, is countably approximable.

Next I formulate a Gaussian counterpart of the above results. To do this I need some notions that will be introduced in Chap. 10. In that chapter the white noise with a reference measure μ will be defined. It is an appropriate set of jointly Gaussian random variables indexed by those measurable sets $A \in \mathcal{X}$ of a measure space $(X, \mathcal{X}, \mu)$ with a σ-finite measure μ for which $\mu(A) < \infty$. Its distribution depends on the measure μ which will be called the reference measure of the white noise.

In Chap. 10 it will also be shown that given a white noise μ_W with a non-atomic σ-additive reference measure μ on a measurable space $(X, \mathcal{X})$ and a measurable function $f(x_1, \dots, x_k)$ of k variables on the product space $(X^k, \mathcal{X}^k)$ such that

$$\int f^2(x_1, \dots, x_k) \mu(dx_1) \dots \mu(dx_k) \le \sigma^2 < \infty \tag{8.15}$$

a k-fold Wiener–Itô integral of the function f with respect to the white noise μ_W

$$Z_{\mu,k}(f) = \frac{1}{k!} \int f(x_1, \dots, x_k) \mu_W(dx_1) \dots \mu_W(dx_k) \tag{8.16}$$

can be defined, and the main properties of this integral will be proved there. It will be seen that Wiener–Itô integrals have a similar relation to degenerate U-statistics and multiple integrals with respect to normalized empirical measures as normally distributed random variables have to partial sums of independent random variables. Hence it is useful to find the analogues of the previous results to estimates about the tail distribution of Wiener–Itô integrals. This will be done in Theorems 8.5 and 8.6.

Theorem 8.5 (Estimate on the tail distribution of a multiple Wiener–Itô integral). *Let us fix a measurable space $(X,\mathcal{X})$ together with a σ-finite non-atomic measure μ on it, and let μ_W be a white noise with reference measure μ on $(X,\mathcal{X})$. If $f(x_1,\dots,x_k)$ is a measurable function on $(X^k,\mathcal{X}^k)$ which satisfies relation (8.15) with some $0<\sigma<\infty$, then*

$$P(|k!Z_{\mu,k}(f)|>u)\le C\exp\left\{-\frac{1}{2}\left(\frac{u}{\sigma}\right)^{2/k}\right\} \tag{8.17}$$

for all $u>0$ with some constants $C=C(k)$ depending only on k.

Theorem 8.6 (Estimate on the supremum of Wiener–Itô integrals). *Let $\mathcal{F}$ be a countable class of functions of k variables defined on the k-fold product $(X^k,\mathcal{X}^k)$ of a measurable space $(X,\mathcal{X})$ such that*

$$\int f^2(x_1,\dots,x_k)\mu(\,dx_1)\dots\mu(\,dx_k)\le\sigma^2 \quad \text{with some } 0<\sigma\le 1 \text{ for all } f\in\mathcal{F}$$

with some non-atomic σ-additive measure μ on $(X,\mathcal{X})$. Let us also assume that $\mathcal{F}$ is an L_2-dense class of functions in the space $(X^k,\mathcal{X}^k)$ with respect to the measure μ^k with some exponent $L\ge 1$ and parameter $D\ge 1$, where μ^k is the k-fold product of the measure μ. (The classes of L_2-dense classes with respect to a measure were defined in Chap. 4.)

Take a white noise μ_W on $(X,\mathcal{X})$ with reference measure μ, and define the Wiener–Itô integrals $Z_{\mu,k}(f)$ for all $f\in\mathcal{F}$. Fix some $0<\varepsilon\le 1$. The inequality

$$P\left(\sup_{f\in\mathcal{F}}|k!Z_{\mu,k}(f)|>u\right)\le CD\exp\left\{-\frac{1}{2}\left(\frac{(1-\varepsilon)u}{\sigma}\right)^{2/k}\right\} \tag{8.18}$$

holds for those numbers u which satisfy the inequality $u\ge ML^{k/2}\sigma\frac{1}{\varepsilon}(\log^{k/2}\frac{2}{\varepsilon}+\log^{k/2}\frac{2}{\sigma})$. Here $C=C(k)>0$, $M=M(k)>0$ are some universal constants depending only on the multiplicity k of the integrals.

Remark. Theorem 8.6 is the multivariate version of Theorem 4.2 about the tail distribution of the supremum of Gaussian random variables. In Theorem 4.2 we could get good estimates for such levels u which satisfy the inequality $u\ge \text{const.}\,\sigma\log^{1/2}\frac{2}{\sigma}$ with an appropriate constant, while in Theorem 8.6 we had a similar estimate under the condition $u\ge\text{const.}\sigma\log^{k/2}\frac{2}{\sigma}$ with an appropriate constant. In Chap. 4 we presented an example which shows that the above condition on the level u in Theorem 4.2 cannot be dropped. A similar example can be given about the necessity of the analogous condition in Theorem 8.6 with the help of the subsequent Example 8.7.

Put $f_{s,t}(u_1,\dots,u_k)=\prod_{j=1}^{k} f^0_{s,t}(u_j)$, where $f^0_{s,t}(u)$ denotes the indicator function of the interval $[s,t]$. Take the class of functions

$$\mathcal{F}=\mathcal{F}_\sigma=\{f_{s,t}\colon\ 0\le s<t\le 1,\ t-s\le\sigma^{2/k},\ s \text{ and } t \text{ are rational}\},$$

and define for all functions $f_{s,t}\in\mathcal{F}$ the k-fold Wiener–Itô integral

$$Z(f_{s,t})=\frac{1}{k!}\int f_{s,t}(u_1,\dots,u_k)W(du_1)\dots W(du_k).$$

Then $EZ(f_{s,t})^2\le\frac{\sigma^2}{k!}$ for all $f_{s,t}\in\mathcal{F}$, and it can be seen with the help of Example 8.7 similarly to the corresponding argument applied in Chap. 4 that there is some $c>0$ such that $P\left(\sup\limits_{f_{s,t}\in\mathcal{F}_\sigma} Z(f_{s,t})>c\sigma\log^{k/2}\frac{2}{\sigma}\right)\to 1$ as $\sigma\to 0$. Beside this, it can be seen that $\mathcal{F}$ is an L_2-dense class with respect to the Lebesgue measure. This implies that the lower bound imposed on u in Theorem 8.6 cannot be dropped. I omit the details of the proof.

Formula (8.18) yields an almost as good estimate for the supremum of Wiener–Itô integrals with the choice of a small $\varepsilon>0$ as formula (8.17) for a single Wiener–Itô integral. But the lower bound imposed on the number u in the estimate (8.18) depends on ε, and for a small number $\varepsilon>0$ it is large.

The subsequent result presented in Example 8.7 may help to understand why Theorems 8.3 and 8.5 are sharp. Its proof and the discussion of the question about the sharpness of Theorems 8.3 and 8.5 will be postponed to Chap. 13.

Example 8.7 (A converse estimate to Theorem 8.5). *Let us have a σ-finite measure μ on some measure space $(X,\mathcal{X})$ together with a white noise μ_W on $(X,\mathcal{X})$ with counting measure μ. Let $f_0(x)$ be a real valued function on $(X,\mathcal{X})$ such that $\int f_0(x)^2\mu(dx)=1$, and take the function $f(x_1,\dots,x_k)=\sigma f_0(x_1)\cdots f_0(x_k)$ with some number $\sigma>0$ together with the Wiener–Itô integral $Z_{\mu,k}(f)$ introduced in formula (8.16).*

Then the relation $\int f(x_1,\dots,x_k)^2\,\mu(dx_1)\dots\mu(dx_k)=\sigma^2$ holds, and the Wiener–Itô integral $Z_{\mu,k}(f)$ satisfies the inequality

$$P(|k!Z_{\mu,k}(f)|>u)\ge\frac{\bar C}{\left(\frac{u}{\sigma}\right)^{1/k}+1}\exp\left\{-\frac12\left(\frac{u}{\sigma}\right)^{2/k}\right\}\quad \text{for all } u>0 \tag{8.19}$$

with some constant $\bar C>0$.

The above results show that multiple integrals with respect to a normalized empirical distribution or degenerate U-statistics satisfy some estimates similar to those about multiple Wiener–Itô integrals, but they hold under more restrictive conditions. The difference between the estimates in these problems is similar to the difference between the corresponding results in Chap. 4 whose reason was explained there. Hence this will be only briefly discussed here.

The estimates of Theorems 8.1 and 8.3 are similar to that of Theorem 8.5. Moreover, for $0 \le u \le \varepsilon n^{k/2}\sigma^{k+1}$ with a small number $\varepsilon > 0$ Theorem 8.3 yields an almost as good estimate about degenerate U-statistics as Theorem 8.5 yields for a Wiener–Itô integral with the same kernel function f and underlying measure μ. Example 8.7 shows that the constant in the exponent of formula (8.17) cannot be improved, at least there is no possibility of an improvement if only the L_2-norm of the kernel function f is known. Some results discussed later indicate that neither the estimate of Theorem 8.3 can be improved. The main difference between Theorem 8.5 and the results of Theorem 8.1 or 8.3 is that in the latter case the kernel function f must satisfy not only an L_2 but also an L_∞ norm type condition, and the estimates of these results are formulated under the additional condition $u \le n^{k/2}\sigma^{k+1}$. It can be shown that the condition about the L_∞ norm of the kernel function cannot be dropped from the conditions of these theorems, and a version of Example 3.3 will be presented in Example 8.8 which shows that in the case $u \gg n^{k/2}\sigma^{k+1}$ the left-hand side of (8.12) may satisfy only a much weaker estimate. This estimate will be given only for $k = 2$, but with some work it can be generalized for general indices k.

Theorems 8.2, 8.4 and 8.6 show that for the tail distribution of the supremum of a not too large class of degenerate U-statistics or multiple integrals a similar upper bound can be given as for the tail distribution of a single degenerate U-statistic or multiple integral, only the universal constants may be worse in the new estimates. However, they hold only under the additional condition that the level at which the tail distribution of the supremum is estimated is not too low. A similar phenomenon appeared already in the results of Chap. 4. Moreover, such a restriction had to be imposed in the formulation of the results here and in Chap. 4 for the same reason.

In Theorems 8.2 and 8.4 an L_2-dense class of kernel functions was considered, and this meant that the class of random integrals or U-statistics we consider in this result is not too large. In Theorem 8.6 a similar, but weaker condition was imposed on the class of kernel functions. They had to satisfy a similar condition, but only for the reference measure μ of the white noise appearing in the Wiener–Itô integral. A similar difference appears in the comparison of Theorems 4.1 or 4.1′ with Theorem 4.2, and this difference has the same reason in the two cases.

Next I present the proof of the following Example 8.8 which is a multivariate version of Example 3.3. For the sake of simplicity I restrict my attention to the case $k = 2$.

Example 8.8 (A converse estimate to Theorem 8.3). *Let us take a sequence of independent and identically distributed random variables $\xi_1, \dots, \xi_n$ with values in the plane $X = R^2$ such that $\xi_j = (\eta_{j,1}, \eta_{j,2})$, $\eta_{j,1}$ and $\eta_{j,2}$ are independent random variables with the following distributions. The distribution of $\eta_{j,1}$ is defined with the help of a parameter σ^2, $0 < \sigma^2 \le \frac{1}{8}$, in the same way as the distribution of the random variables X_j in Example 3.3, i.e. $\eta_{j,1} = \bar\eta_{j,1} - E\bar\eta_{j,1}$ with $P(\bar\eta_{j,1} = 1) = \bar\sigma^2$,*

$P(\bar\eta_{j,1}=0)=1-\bar\sigma^2$, where $\bar\sigma^2$ is that solution of the equation $x^2-x+\sigma^2=0$, which is smaller than $\frac12$. The distribution of the random variables $\eta_{j,2}$ is given by the formula $P(\eta_{j,2}=1)=P(\eta_{j,2}=-1)=\frac12$ for all $1\le j\le n$. Introduce the function $f(x,y)=f((x_1,x_2),(y_1,y_2))=x_1y_2+x_2y_1$, $x=(x_1,x_2)\in R^2$, $y=(y_1,y_2)\in R^2$ if (x,y) is in the support of the distribution of the random vector (ξ_1,ξ_2), i.e. if x_1 and y_1 take the values $1-\bar\sigma^2$ or $-\bar\sigma^2$ and x_2 and y_2 take the values ±1. Put $f(x,y)=0$ otherwise. Define the U-statistic

$$I_{n,2}(f)=\frac12\sum_{1\le j,k\le n,\, j\ne k} f(\xi_j,\xi_k)=\frac12\sum_{1\le j,k\le n,\, j\ne k}(\eta_{j,1}\eta_{k,2}+\eta_{k,1}\eta_{j,2})$$

of order 2 with the above kernel function f and sequence of independent random variables $\xi_1,\dots,\xi_n$. Then $I_{n,2}(f)$ is a degenerate U-statistic such that $|\sup f(x,y)|\le 1$ and $Ef^2(\xi_j,\xi_j)=\sigma^2$.

If $u\ge B_1n\sigma^3$ with some appropriate constant $B_1>2$, $\bar B_2^{-1}n\ge u\ge \bar B_2 n^{-1/2}$ with a sufficiently large fixed number $\bar B_2>0$ and $\frac14\ge\sigma^2\ge\frac{1}{n^2}$, and n is a sufficiently large number, then the estimate

$$P(n^{-1}I_{n,2}(f)>u)\ge\exp\left\{-Bn^{1/3}u^{2/3}\log\left(\frac{u}{n\sigma^3}\right)\right\} \tag{8.20}$$

holds with some $B>0$.

Remark. In Theorem 8.3 we got the estimate $P(n^{-1}I_{n,2}(f)>u)\le e^{-\alpha u/\sigma}$ for the above defined degenerate U-statistic $I_{n,2}(f)$ if $0\precsim u\precsim n\sigma^3$. In the particular case $u=n\sigma^3$ we have the estimate $P(n^{-1}I_{n,2}(f)>n\sigma^3)\le e^{-\alpha n\sigma^2}$. On the other hand, the above example shows that in the case $u\gg n\sigma^3$ we can get only a weaker estimate. It is worth looking at the estimate (8.20) with fixed parameters n and u and to observe the dependence of the upper bound on the variance σ^2 of $I_{n,2}(f)$. In the case $\sigma^2=u^{2/3}n^{-2/3}$ we have the upper bound $e^{-\alpha n^{1/3}u^{2/3}}$. Example 8.8 shows that in the case $\sigma^2\ll u^{2/3}n^{-2/3}$ we can get only a relatively small improvement of this estimate. A similar picture appears as in Example 3.3 in the case $k=1$.

It is simple to check that the U-statistic introduced in the above example is degenerate because of the independence of the random variables $\eta_{j,1}$ and $\eta_{j,2}$ and the identity $E\eta_{j,1}=E\eta_{j,2}=0$. Beside this, $Ef(\xi_j,\xi_j)^2=\sigma^2$. In the proof of the estimate (8.20) the results of Chap. 3, in particular Example 3.3 can be applied for the sequence $\eta_{j,1}$, $j=1,2,\dots,n$. Beside this, the following result, known from the theory of large deviations will be applied. If $X_1,\dots,X_n$ are independent and identically distributed random variables, $P(X_1=1)=P(X_1=-1)=\frac12$, then for any number $0\le\alpha<1$ there exists some numbers $C_1=C_1(\alpha)>0$ and $C_2=C_2(\alpha)>0$ such that $P\left(\sum\limits_{j=1}^n X_j>u\right)\ge C_1e^{-C_2u^2/n}$ for all $0\le u\le\alpha n$.

Proof of Example 8.8. The inequality

$$P(n^{-1}I_{n,2}(f) > u) \tag{8.21}$$
$$\geq P\left(\left(\sum_{j=1}^{n}\eta_{j,1}\right)\left(\sum_{j=1}^{n}\eta_{j,2}\right) > 4nu\right) - P\left(\sum_{j=1}^{n}\eta_{j,1}\eta_{j,2} > 2nu\right)$$

holds. Because of the independence of the random variables $\eta_{j,1}$ and $\eta_{j,2}$ the first probability at the right-hand side of (8.21) can be bounded from below by bounding the multiplicative terms in it with $v_1 = 4n^{1/3}u^{2/3}$ and $v_2 = n^{2/3}u^{1/3}$. The first term will be estimated by means of Example 3.3. This estimate can be applied with the choice $y = v_1$, since the relation $v_1 \geq 4n\sigma^2$ holds if $u \geq B_1 n\sigma^3$ with $B_1 > 1$, and the remaining conditions $0 \leq \sigma^2 \leq \frac{1}{8}$ and $n \geq 4v_1 \geq 6$ also hold under the conditions of Example 8.8. The second term can be bounded with the help of the large-deviation result mentioned after the remark, since $v_2 \leq \frac{1}{2}n$ if $u \leq \bar{B}_2^{-1}n$ with a sufficiently large $\bar{B}_2 > 0$. In such a way we get the estimate

$$P\left(\left(\sum_{j=1}^{n}\eta_{j,1}\right)\left(\sum_{j=1}^{n}\eta_{j,2}\right) > 4nu\right) \geq P\left(\sum_{j=1}^{n}\eta_{j,1} > v_1\right)P\left(\sum_{j=1}^{n}\eta_{j,2} > v_2\right)$$
$$\geq C\exp\left\{-B_1 v_1 \log\left(\frac{v_1}{n\sigma^2}\right) - B_2\frac{v_2^2}{n}\right\}$$
$$\geq C\exp\left\{-B_3 n^{1/3}u^{2/3}\log\left(\frac{u}{n\sigma^3}\right)\right\}$$

with appropriate constants $B_1 > 1$, $B_2 > 0$ and $B_3 > 0$. On the other hand, by applying Bennett's inequality, more precisely its consequence given in formula (3.4) for the sum of the random variables $X_j = \eta_{j,1}\eta_{j,2}$ at level nu instead of level u we get the following upper bound for the second term at the right-hand side of (8.21).

$$P\left(\sum_{j=1}^{n}\eta_{j,1}\eta_{j,2} > 2nu\right) \leq \exp\left\{-Knu\log\frac{u}{\sigma^2}\right\}$$
$$\leq \exp\left\{-2B_4 n^{1/3}u^{2/3}\log\left(\frac{u}{n\sigma^3}\right)\right\},$$

since $E\eta_{j,1}\eta_{j,2} = 0$, $E\eta_{j,1}^2\eta_{j,2}^2 = \sigma^2$, $nu \geq B_1 n^2\sigma^3 \geq 2n\sigma^2$ because of the conditions $B_1 > 2$ and $n\sigma \geq 1$. Hence the estimate (3.4) (with parameter nu) can be applied in this case. Beside this, the constant B_4 can be chosen sufficiently large in the last inequality if the number n or the bound $\bar{B}_2$ in Example 8.8 us chosen sufficiently large. This means that this term is negligible small. The above estimates imply the statement of Example 8.8. □

Let me remark that under some mild additional restrictions the estimate (8.20) can be slightly sharpened, the term log can be replaced by $\log^{2/3}$ in the exponent of the right-hand side of (8.20). To get such an estimate some additional calculation is needed where the numbers v_1 and v_2 are replaced by $\bar{v}_1 = 4n^{1/3}u^{2/3}\log^{-1/3}\left(\frac{u}{n\sigma^3}\right)$ and $\bar{v}_2 = n^{2/3}u^{1/3}\log^{1/3}\left(\frac{u}{n\sigma^3}\right)$.

I finish this chapter with a short overview about the remaining part of this work. In our proofs we needed some results about U-statistics, and this is the main topic of Chap. 9. One of the results discussed there is the so-called Hoeffding decomposition of U-statistics to the linear combination of degenerate U-statistics of different order. We also needed some additional results which explain how some properties (e.g. a bound on the L_2 and L_∞ norm of a kernel function, the L_2-density property of a class $\mathcal{F}$ of kernel function) is inherited if we turn from the original U-statistics to the degenerate U-statistics appearing in their Hoeffding decomposition. Chapter 9 contains some results in this direction. Another important result in it is Theorem 9.4 which yields a decomposition of multiple integrals with respect to a normalized empirical distribution to the linear combination of degenerate U-statistics. This result is very similar to the Hoeffding decomposition of U-statistics. The main difference between them is that in the decomposition of multiple integrals much smaller coefficients appear. Theorem 9.4 makes possible to reduce the proof of Theorems 8.1 and 8.2 to the corresponding results in Theorems 8.3 and 8.4 about degenerate U-statistics.

The definition and the main properties of Wiener–Itô integrals needed in the proof of Theorems 8.5 and 8.6 are presented in Chap. 10. It also contains a result, called the diagram formula for Wiener–Itô integrals which plays an important role in our considerations. Beside this, we proved a limit theorem, where we expressed the limit of normalized degenerate U-statistics with the help of multiple Wiener–Itô integrals. This result may explain why it is natural to consider Theorem 8.5 as the natural Gaussian counterpart of Theorem 8.5, and Theorem 8.6 as the natural Gaussian counterpart of Theorem 8.6.

We could prove Bernstein's and Bennett's inequality by means of a good estimation of the exponential moments of the partial sums we were investigating. In the proof of their multivariate versions, in Theorems 8.3 and 8.5 this method does not work, because the exponential moments we have to bound in these cases may be infinite. On the other hand, we could prove these results by means of a good estimate on the high moments of the random variables whose tail distribution we wanted to bound. In the proof of Theorem 8.5 the moments of multiple Wiener–Itô integrals have to be bounded, and this can be done with the help of the diagram formula for Wiener–Itô integrals. In Chaps. 11 and 12 we proved that there is a version of the diagram formula for degenerate U-statistics, and this enables us to estimate the moments needed in the proof of Theorem 8.3. In Chap. 13 we proved Theorems 8.3, 8.5 and a multivariate version of the Hoeffding inequality. At the end of this chapter we still discussed some results which state that in certain cases when we have some useful additional information about the behaviour of the kernel function f beside the upper bound of their L_2 and L_∞ norm the estimates of Theorems 8.3 or 8.5 can be improved.

Chapter 14 contains the natural multivariate versions of the results in Chap. 6. In Chap. 6 Theorem 4.2 is proved about the supremum of Gaussian random variables and in Chap. 14 its multivariate version, Theorem 8.6. Both results are proved with the help of the chaining argument. On the other hand, the chaining argument is not strong enough to prove Theorem 4.1. But as it is shown in Chap. 6, it enables us to prove a result formulated in Proposition 6.1, and to reduce the proof of Theorem 4.1 with its help to a simpler result formulated in Proposition 6.2. One of the results in Chap. 14, Proposition 14.1, is a multivariate version of Proposition 6.1. We showed that the proof of Theorem 8.4 can be reduced with its help to the proof of a result formulated in Proposition 14.2, which can be considered a multivariate version of Proposition 6.2. Chapter 14 contains still another result. It turned out that it is simpler to work with so-called decoupled U-statistics introduced in this chapter than with usual U-statistics, because they have more independence properties. In Proposition 14.2′ a version of Proposition 14.2 is formulated about degenerate U-statistics, and it is shown with the help of a result of de la Peña and Montgomery–Smith that the proof of Proposition 14.2, and thus of Theorem 8.4 can be reduced to the proof of Proposition14.2′.

Proposition 14.2′ is proved similarly to its one-variate version, Proposition 6.2. The strategy of the proof is explained in Chap. 15. The main difference between the proof of the two propositions is that since the independence properties exploited in the proof of Proposition 6.2 hold only in a weaker form in this case, we have to apply a more refined and more difficult argument. In particular, we have to apply instead of the symmetrization lemma, Lemma 7.1, a more general version of it. We presented an appropriate version of this result in Lemma 15.2. It is hard to check the conditions of Lemma 15.2 when we try to apply it in the problems arising in the proof of Proposition 14.2′. This is the reason why we had to prove Proposition 14.2′ with the help of two inductive propositions, formulated in Propositions 15.3 and 15.4, while in the proof of Proposition 6.2 it was enough to prove a single result, presented in Proposition 7.3. We discuss the details of the problems and the strategy of the proof in Chap. 15. The proof of Propositions 15.3 and 15.4 is given in Chaps. 16 and 17. Chapter 16 contains the symmetrization arguments needed for us, and the proof is completed with its help in Chap. 17.

Finally in Chap. 18 we give an overview of this work, and explain its relation to some similar researches. The proof of some results is given in the appendix.

Chapter 9
Some Results About U-statistics

This chapter contains the proof of the Hoeffding decomposition theorem, an important result about U-statistics. It states that all U-statistics can be represented as a sum of degenerate U-statistics of different order. This representation can be considered as the natural multivariate version of the decomposition of a sum of independent random variable to the sum of independent random variables with expectation zero plus a constant (which can be interpreted as a random variable of zero variable). Some important properties of the Hoeffding decomposition will also be proved. In particular, it will be investigated how some properties of the kernel function of a U-statistic is inherited in the behaviour of the kernel functions of the U-statistics in its Hoeffding decomposition.

If the Hoeffding decomposition of a U-statistic is taken, then the L_2 and L_∞-norms of the kernel functions appearing in the U-statistics of the Hoeffding decomposition will be bounded by means of the corresponding norm of the kernel function of the original U-statistic. It will also be shown that if we take a class of U-statistics with an L_2-dense class of kernel functions (and the same sequence of independent and identically distributed random variables in the definition of each U-statistic) and consider the Hoeffding decomposition of all U-statistics in this class, then the kernel functions of the degenerate U-statistics appearing in these Hoeffding decompositions also constitute an L_2-dense class. Another important result of this chapter is Theorem 9.4. It yields a decomposition of a k-fold random integral with respect to a normalized empirical distribution to the linear combination of degenerate U-statistics. This result enables us to derive Theorem 8.1 from Theorem 8.3 and Theorem 8.2 from Theorem 8.4, and it is also useful in the proof of Theorems 8.3 and 8.4.

Let us first consider Hoeffding's decomposition. In the special case $k = 1$ it states that the sum $S_n = \sum\limits_{j=1}^{n} \xi_j$ of independent and identically distributed random variables can be rewritten as $S_n = \sum\limits_{j=1}^{n} (\xi_j - E\xi_j) + \left(\sum\limits_{j=1}^{n} E\xi_j \right)$, i.e. as the sum of

P. Major, *On the Estimation of Multiple Random Integrals and U-Statistics*, Lecture Notes in Mathematics 2079, DOI 10.1007/978-3-642-37617-7_9,

independent random variables with zero expectation plus a constant. We introduced the convention that a constant is the kernel function of a degenerate U-statistic of order zero, and $I_{n,0}(c) = c$ for a U-statistic of order zero. I wrote down the above trivial formula, because Hoeffding's decomposition is actually its adaptation to a more general situation. To understand this let us first see how to adapt the above construction to the case $k = 2$.

In this case a sum of the form $2I_{n,2}(f) = \sum\limits_{1\le j,k\le n, j\ne k} f(\xi_j, \xi_k)$ has to be considered. Write $f(\xi_j, \xi_k) = [f(\xi_j, \xi_k) - E(f(\xi_j, \xi_k)|\xi_k)] + E(f(\xi_j, \xi_k)|\xi_k) = f_1(\xi_j, \xi_k) + \bar{f_1}(\xi_k)$ with $f_1(\xi_j, \xi_k) = f(\xi_j, \xi_k) - E(f(\xi_j, \xi_k)|\xi_k)$, and $\bar{f_1}(\xi_k) = E(f(\xi_j, \xi_k)|\xi_k)$ to make the conditional expectation of $f_1(\xi_j, \xi_k)$ with respect to ξ_k equal zero. Repeating this procedure for the first coordinate we define $f_2(\xi_j, \xi_k) = f_1(\xi_j, \xi_k) - E(f_1(\xi_j, \xi_k)|\xi_j)$ and $\bar{f_2}(\xi_j) = E(f_1(\xi_j, \xi_k)|\xi_j)$. Let us also write $\bar{f_1}(\xi_k) = [\bar{f_1}(\xi_k) - E\bar{f_1}(\xi_k)] + E\bar{f_1}(\xi_k)$ and $\bar{f_2}(\xi_j) = [\bar{f_2}(\xi_j) - E\bar{f_2}(\xi_j)] + E\bar{f_2}(\xi_j)$. Simple calculation shows that $2I_{n,2}(f_2)$ is a degenerate U-statistics of order 2, and the identity $2I_{n,2}(f) = 2I_{n,2}(f_2) + I_{n,1}((n-1)(\bar{f_1} - E\bar{f_1})) + I_{n,1}((n-1)((\bar{f_2} - E\bar{f_2})) + n(n-1)E(\bar{f_1} + \bar{f_2})$ yields the decomposition of $I_{n,2}(f)$ into a sum of degenerate U-statistics of different orders.

Hoeffding's decomposition can be obtained by working out the details of the above argument in the general case. But it is simpler to calculate the appropriate conditional expectations by working with the kernel functions of the U-statistics. To carry out such a program we introduce the following notations.

Let us consider the k-fold product $(X^k, \mathcal{X}^k, \mu^k)$ of a measure space $(X, \mathcal{X}, \mu)$ with some probability measure μ, and define for all integrable functions $f(x_1, \dots, x_k)$ and indices $1 \le j \le k$ the projection $P_j f$ of the function f to its j-th coordinate, i.e. integration of the function f with respect to its j-th coordinate.

For the sake of simpler notations in our later considerations we shall define the operator P_j in a slightly more general setting. Let us consider a set $A = \{p_1, \dots, p_s\} \subset \{1, \dots, k\}$, put $X^A = X_{p_1} \times X_{p_2} \times \dots \times X_{p_s}$, $\mathcal{X}^A = \mathcal{X}_{p_1} \times \mathcal{X}_{p_2} \times \dots \times \mathcal{X}_{p_s}$, $\mu^A = \mu_{p_1} \times \mu_{p_2} \times \dots \times \mu_{p_s}$, take the product space $(X^A, \mathcal{X}^A, \mu^A)$ and if $j \in A$, then define the operator P_j as mapping a function on this product space to a function on the product space $(X^{A\setminus\{j\}}, \mathcal{X}^{A\setminus\{j\}})$ by the formula

$$(P_j f)(x_{p_1}, \dots, x_{p_{r-1}}, x_{p_{r+1}}, \dots, x_{p_s}) = \int f(x_{p_1}, \dots, x_{p_s})\mu(dx_j), \quad \text{if } j = p_r. \tag{9.1}$$

Let us also define the (orthogonal projection) operators $Q_j = I - P_j$ as $Q_j f = f - P_j f$ for all integrable functions f on the space $(X^A, \mathcal{X}^A, \mu^A)$, and $j \in A$, i.e. put

$$\begin{aligned}(Q_j f)(x_{p_1}, \dots, x_{p_s}) &= (I - P_j) f(x_{p_1}, \dots, x_{p_s}) \\ &= f(x_{p_1}, \dots, x_{p_s}) - \int f(x_{p_1}, \dots, x_{p_s})\mu(dx_j). \end{aligned} \tag{9.2}$$

In the definition (9.1) $P_j f$ is a function not depending on the coordinate x_j, but in the definition of Q_j we introduce the fictive coordinate x_j to make the expression $Q_j f = f - P_j f$ meaningful.

Remark. I shall use the following notation. $(P_j f)(x_{p_1},\dots,x_{p_{r-1}},x_{p_{r+1}},\dots,x_{p_s})$ will denote the value of the function $P_j f$ in the point $(x_{p_1},\dots,x_{p_{r-1}},x_{p_{r+1}},\dots,x_{p_s})$. On the other hand, I write sometimes $P_j f(x_{p_1},\dots,x_{p_s})$ (without parentheses) instead of $P_j f$ when it is more natural to write the function f together with its arguments. The same notation will be applied for the operator Q_j.

The following result holds.

Theorem 9.1 (The Hoeffding decomposition of U-statistics). *Let $f(x_1,\dots,x_k)$ be an integrable function on the k-fold product $(X^k,\mathcal{X}^k,\mu^k)$ of a space $(X,\mathcal{X},\mu)$ with a probability measure μ. It has a decomposition of the form*

$$f(x_1,\dots,x_k) = \sum_{V\subset\{1,\dots,k\}} f_V(x_{j_1},\dots,x_{j_{|V|}}) \tag{9.3}$$

with

$$f_V(x_{j_1},\dots,x_{j_{|V|}}) = \left(\prod_{j\in\{1,\dots,k\}\setminus V} P_j \prod_{j'\in V} Q_{j'}\right) f(x_1,\dots,x_k), \tag{9.4}$$

with $V=\{j_1,\dots,j_{|V|}\}$, $j_1<j_2<\dots<j_{|V|}$, for all $V\subset\{1,\dots,k\}$. Beside this, all functions f_V, $V\subset\{1,\dots,k\}$, defined in (9.3) are canonical with respect to the probability measure μ with $|V|$ arguments.

Let $\xi_1,\dots,\xi_n$ be a sequence of independent μ distributed random variables, and consider the U-statistics $I_{n,k}(f)$ and $I_{n,|V|}(f_V)$ corresponding to the kernel functions f, f_V defined in (9.3) and random variables $\xi_1,\dots,\xi_n$. Then

$$k!I_{n,k}(f) = \sum_{V\subset\{1,\dots,k\}} (n-|V|)(n-|V|-1)\cdots(n-k+1)|V|!I_{n,|V|}(f_V) \tag{9.5}$$

is a representation of $k!I_{n,k}(f)$ as a sum of degenerate U-statistics, where $|V|$ denotes the cardinality of the set V. (The product $(n-|V|)(n-|V|-1)\cdots(n-k+1)$ is defined as 1 for $V=\{1,\dots,k\}$, i.e. if $|V|=k$.) This representation is called the Hoeffding decomposition of $k!I_{n,k}(f)$.

Proof of Theorem 9.1. Write $f = \prod\limits_{j=1}^{k}(P_j+Q_j)f$. By carrying out the multiplications in this identity and applying the commutativity of the operators P_j and Q_j for different indices j we get formula (9.3). To show that the functions f_V in formula (9.3) are canonical let us observe that this property can be rewritten in the form $P_j f_V \equiv 0$ (in all points $(x_s,\ s\in V\setminus\{j\})$ if $j\in V$). Since $P_j=P_j^2$, and the identity $P_jQ_j = P_j - P_j^2 = 0$ holds for all $j\in\{1,\dots,k\}$ this relation follows from the above mentioned commutativity of the operators P_j and Q_j, as $P_j f_V = \left(\prod\limits_{s\in\{1,\dots,k\}\setminus V} P_s \prod\limits_{s\in V\setminus\{j\}} Q_s\right) P_j Q_j f = 0$. By applying identity (9.3) for all terms $f(\xi_{j_1},\dots,\xi_{j_k})$ in the sum defining the U-statistic $k!I_{n,k}(f)$ (see formula (8.8)) and then summing them up we get relation (9.5). □

In the Hoeffding decomposition we rewrote a general U-statistic in the form of a linear combination of degenerate U-statistics. In many applications of this result we still we have to know how the properties of the kernel function f of the original U-statistic are reflected in the properties of the kernel functions f_V of the degenerate U-statistics taking part in the Hoeffding composition. In particular, we need a good estimate on the L_2 and L_∞ norm of the functions f_V by means of the corresponding norm of the function f. Moreover, if we want to prove estimates on the tail distribution of the supremum of U-statistics $I_{n,k}(f)$ defined with the help of an L_2-dense class of kernel functions $\mathcal{F}$ with some exponent L and parameter D, then we may need a similar estimate on the classes of kernel functions $\mathcal{F}_V = \{f_V\colon f \in \mathcal{F}\}$ with functions f_V, $V \in \{1,\dots,k\}$ appearing in the Hoeffding decomposition of these functions. We have to show that this class of functions is also L_2-dense, and we also need a good bound on the exponent and parameter of this L_2-dense class. In the next result such statements will be proved.

Theorem 9.2 (Some properties of the Hoeffding decomposition). *Let us consider a square integrable function $f(x_1,\dots,x_k)$ on the k-fold product space $(X^k,\mathcal{X}^k,\mu^k)$ and take its decomposition defined in formula (9.3). The inequalities*

$$\int f_V^2(x_j,\, j\in V)\prod_{j\in V}\mu(dx_j) \le \int f^2(x_1,\dots,x_k)\mu(dx_1)\dots\mu(dx_k) \tag{9.6}$$

and

$$\sup_{x_j,\, j\in V} |f_V(x_j,\, j\in V)| \le 2^{|V|} \sup_{x_j,\, 1\le j\le k} |f(x_1,\dots,x_k)| \tag{9.7}$$

hold for all $V\subset\{1,\dots,k\}$. In particular,

$$f_\emptyset^2 \le \int f^2(x_1,\dots,x_k)\mu(dx_1)\dots\mu(dx_k) \quad \text{for } V=\emptyset.$$

Let us consider an L_2-dense class $\mathcal{F}$ of functions with some parameter $D\ge 1$ and exponent $L\ge 0$ on the space $(X^k,\mathcal{X}^k)$, take the decomposition (9.3) of all functions $f\in\mathcal{F}$, and define the classes of functions $\mathcal{F}_V = \{2^{-|V|}f_V\colon f\in\mathcal{F}\}$ for all $V\subset\{1,\dots,k\}$ with the functions f_V taking part in this decomposition. These classes of functions $\mathcal{F}_V$ are also L_2-dense with the same parameter D and exponent L for all $V\subset\{1,\dots,k\}$.

Theorem 9.2 will be proved as a consequence of Proposition 9.3 presented below. To formulate it first some notations will be introduced.

Let us consider the product $(Y \times Z, \mathcal{Y} \times \mathcal{Z})$ of two measurable spaces $(Y, \mathcal{Y})$ and $(Z, \mathcal{Z})$ together with a probability measure μ on $(Z, \mathcal{Z})$ and the operator

$$(Pf)(y) = (P_\mu f)(y) = \int f(y,z)\mu(dz), \quad y \in Y,\ z \in Z \tag{9.8}$$

defined for those $y \in Y$ for which the above integral is finite. Let I denote the identity operator on the space of functions on $Y \times Z$, i.e. let $(If)(y,z) = f(y,z)$, and introduce the operator $Q = Q_\mu = I - P = I - P_\mu$

$$\begin{aligned}(Q_\mu f)(y,z) &= ((I - P_\mu)f)(y,z) = f(y,z) - (P_\mu f)(y,z) \\ &= f(y,z) - \int f(y,z)\mu(dz),\end{aligned} \tag{9.9}$$

defined for those points $(y,z) \in Y \times Z$ whose first coordinate y is such that the expression $(P_\mu f)(y)$ is meaningful. (Here, and in the sequel a function $g(y)$ defined on the space $(Y, \mathcal{Y})$ will be sometimes identified with the function $\bar{g}(y,z) = g(y)$ on the space $(Y \times Z, \mathcal{Y} \times \mathcal{Z})$ which actually does not depend on the coordinate z.) The following result holds:

Proposition 9.3. *Let us consider the direct product $(Y \times Z, \mathcal{Y} \times \mathcal{Z})$ of two measurable spaces $(Y, \mathcal{Y})$ and $(Z, \mathcal{Z})$ together with a probability measure μ on the space $(Z, \mathcal{Z})$. Take the transformations P_μ and Q_μ defined in formulas (9.8) and (9.9). Given any probability measure ρ on the space $(Y, \mathcal{Y})$ consider the product measure $\rho \times \mu$ on $(Y \times Z, \mathcal{Y} \times \mathcal{Z})$. Then the transformations P_μ and Q_μ, as maps from the space $L_2(Y \times Z, \mathcal{Y} \times \mathcal{Z}, \mu \times \rho)$ to $L_2(Y, \mathcal{Y}, \rho)$ and $L_2(Y \times Z, \mathcal{Y} \times \mathcal{Z}, \rho \times \mu)$ respectively, have a norm less than or equal to 1, i.e.*

$$\int (P_\mu f)(y)^2 \rho(dy) \le \int f(y,z)^2 \rho(dy)\mu(dz), \tag{9.10}$$

and

$$\int (Q_\mu f)(y,z)^2 \rho(dy)\mu(dz) \le \int f(y,z)^2 \rho(dy)\mu(dz) \tag{9.11}$$

for all functions $f \in L_2(Y \times Z, \mathcal{Y} \times \mathcal{Z}, \rho \times \mu)$.

If $\mathcal{F}$ is an L_2-dense class of functions $f(y,z)$ in the product space $(Y \times Z, \mathcal{Y} \times \mathcal{Z})$, with some parameter $D \ge 1$ and exponent $L \ge 0$, then also the classes $\mathcal{F}_\mu = \{P_\mu f,\ f \in \mathcal{F}\}$ and $\mathcal{G}_\mu = \{\frac{1}{2} Q_\mu f = \frac{1}{2}(f - P_\mu f),\ f \in \mathcal{F}\}$ are L_2-dense classes with the same exponent L and parameter D in the spaces $(Y, \mathcal{Y})$ and $(Y \times Z, \mathcal{Y} \times \mathcal{Z})$ respectively.

The following corollary of Proposition 9.3 is formally more general, but it is a simple consequence of this result. Actually we shall need this corollary.

Corollary of Proposition 9.3. *Let us consider the product $(Y_1 \times Z \times Y_2, \mathcal{Y}_1 \times \mathcal{Z} \times \mathcal{Y}_2)$ of three measurable spaces $(Y_1, \mathcal{Y}_1)$, $(Z, \mathcal{Z})$ and $(Y_2, \mathcal{Y}_2)$ with a probability* *measure μ on the space $(Z, \mathcal{Z})$ and a probability measure ρ on $(Y_1 \times Y_2, \mathcal{Y}_1 \times \mathcal{Y}_2)$, and define the transformations*

$$(P_\mu f)(y_1, y_2) = \int f(y_1, z, y_2)\mu(dz), \quad y_1 \in Y_1,\ z \in Z,\ y_2 \in Y_2 \tag{9.12}$$

and

$$\begin{aligned}(Q_\mu f)(y_1, z, y_2) &= ((I - P_\mu)f)(y_1, z, y_2) = f(y_1, z, y_2) - (P_\mu f)(y_1, z, y_2)\\ &= f(y_1, z, y_2) - \int f(y_1, z, y_2)\mu(dz), \quad y_1 \in Y_1,\ z \in Z,\ y_2 \in Y_2 \end{aligned}\tag{9.13}$$

for the measurable functions f on the space $Y_1 \times Z \times Y_2$ integrable with respect the measure $\mu \times \rho$. Then

$$\int (P_\mu f)(y_1, y_2)^2 \rho(dy_1, dy_2) \le \int f(y, z)^2 (\rho \times \mu)(dy_1, dz, dy_2) \tag{9.14}$$

for all probability measures ρ on $(Y_1 \times Y_2, \mathcal{Y}_1 \times \mathcal{Y}_2)$, where $\rho \times \mu$ is the product of the probability measure ρ on $(Y_1 \times Y_2, \mathcal{Y}_1 \times \mathcal{Y}_2)$ and μ is a probability measure on $(Z, \mathcal{Z})$. Also the inequality

$$\int (Q_\mu f)(y_1, z, y_2)^2 \rho(dy_1, dy_2)\mu(dz) \le \int f(y_1, z, y_2)^2 \rho(dy_1, dy_2)\mu(dz) \tag{9.15}$$

holds for all functions $f \in L_2(Y \times Z, \mathcal{Y} \times \mathcal{Z}, \rho \times \mu)$.

If $\mathcal{F}$ is an L_2-dense class of functions $f(y_1, z, y_2)$ on the product space $(Y_1 \times Z \times Y_2, \mathcal{Y}_1 \times \mathcal{Z} \times Y_2)$, with some parameter $D \ge 1$ and exponent $L \ge 0$, then also the classes $\mathcal{F}_\mu = \{P_\mu f,\ f \in \mathcal{F}\}$ and $\mathcal{G}_\mu = \{\frac{1}{2} Q_\mu f = \frac{1}{2}(f - P_\mu f),\ f \in \mathcal{F}\}$ are L_2-dense classes with exponent L and parameter D in the spaces $(Y_1 \times Y_2, \mathcal{Y}_1 \times \mathcal{Y}_2)$ and $(Y_1 \times Z \times Y_2, \mathcal{Y}_1 \times \mathcal{Z} \times \mathcal{Y}_2)$ respectively.

This corollary is a simple consequence of Proposition 9.3 if we apply it with $(Y, \mathcal{Y}) = (Y_1 \times Y_2, \mathcal{Y}_1 \times \mathcal{Y}_2)$ and take the natural mapping $f((y_1, y_2), z) \to f(y_1, z, y_2)$ of a function from the space $(Y \times Z, \mathcal{Y} \times \mathcal{Z})$ to a function on $(Y_1 \times Z \times Y_2, \mathcal{Y}_1 \times \mathcal{Z} \times \mathcal{Y}_2)$. Beside this, we apply that measure on $(Y_1 \times Z \times Y_2, \mathcal{Y}_1 \times \mathcal{Z} \times \mathcal{Y}_2)$ which is the image of the product measure $\rho \times \mu$ with respect to the map induced by the above transformation on the space of measures.

Proposition 9.3, more precisely its corollary implies Theorem 9.2, since it implies that the operators P_s, Q_s, $1 \le s \le k$, applied in Theorem 9.2 do not increase the $L_2(\mu)$ norm of a function f, and it is also clear that the norm of P_s is bounded by 1, the norm of $Q_s = I - P_s$ is bounded by 2 as an operator from L_∞ spaces to L_∞ spaces. The corollary of Proposition 9.3 also implies that if $\mathcal{F}$ is an L_2-dense class of functions with parameter D and exponent L, then the same property holds for the classes of functions $\mathcal{F}_{P_s} = \{P_s f\colon f \in \mathcal{F}\}$ and

$\mathcal{F}_{Q_s} = \{\frac{1}{2}Q_s f\colon f \in \mathcal{F}\}$, $1 \le s \le k$. These relations together with the identity $f_V = \left(\prod_{s\in\{1,\dots,k\}\setminus V} P_s \prod_{s\in V} Q_s\right) f$ imply Theorem 9.2

Proof of Proposition 9.3. The Schwarz inequality yields that

$$(P_\mu f)(y)^2 \le \int f(y,z)^2 \mu(dz) \quad \text{for all } y \in Y,$$

and integrating this inequality with respect to the probability measure $\rho(dy)$ we get inequality (9.10). Also the inequality

$$\begin{aligned}\int (Q_\mu f)(y,z)^2 \rho(dy)\mu(dz) &= \int [f(y,z) - P_\mu f(y,z)]^2 \rho(dy)\mu(dz)\\ &\le \int f(y,z)^2 \rho(dy)\mu(dz)\end{aligned}$$

holds, and this is relation (9.11). This follows for instance from the observation that the functions $f(y,z) - (P_\mu f)(y,z)$ and $(P_\mu f)(y,z)$ are orthogonal in the space $L_2(Y \times Z, \mathcal{Y} \times \mathcal{Z}, \rho \times \mu)$.

Let us consider an arbitrary probability measure ρ on the space $(Y, \mathcal{Y})$. To prove that $\mathcal{F}_\mu$ is an L_2-dense class with parameter D and exponent L if the same relation holds for $\mathcal{F}$ we have to find for all $0 < \varepsilon \le 1$ a set $\{f_1,\dots,f_m\} \subset \mathcal{F}_\mu$, $1 \le j \le m$ with $m \le D\varepsilon^{-L}$ elements, such that $\inf\limits_{1\le j\le m} \int (f_j - f)^2\,d\rho \le \varepsilon^2$ for all $f \in \mathcal{F}_\mu$. But a similar property holds for $\mathcal{F}$ in the space $Y \times Z$ with the probability measure $\rho \times \mu$. This property together with the property of P_μ formulated in (9.10) imply that $\mathcal{F}_\mu$ is an L_2-dense class.

To prove that $\mathcal{G}_\mu$ is also L_2-dense with parameter D and exponent L under the same condition we have to find for all numbers $0 < \varepsilon \le 1$ and probability measures ρ on $Y \times Z$ a subset $\{g_1,\dots,g_m\} \subset \mathcal{G}_\mu$ with $m \le D\varepsilon^{-L}$ elements such that $\inf\limits_{1\le j\le m} \int (g_j - g)^2\,d\rho \le \varepsilon^2$ for all $g \in \mathcal{G}_\mu$.

To show this let us consider the probability measure $\tilde\rho = \frac{1}{2}(\rho + \bar\rho \times \mu)$ on $(Y \times Z, \mathcal{Y} \times \mathcal{Z})$, where $\bar\rho$ is the projection of the measure ρ to $(Y, \mathcal{Y})$, i.e. $\bar\rho(A) = \rho(A \times Z)$ for all $A \in \mathcal{Y}$, take a class of function $\mathcal{F}_0(\varepsilon,\tilde\rho) = \{f_1,\dots,f_m\} \subset \mathcal{F}$ with $m \le D\varepsilon^{-L}$ elements such that $\inf\limits_{1\le j\le m} \int (f_j - f)^2\,d\tilde\rho \le \varepsilon^2$ for all $f \in \mathcal{F}$, and put $\{g_1,\dots,g_m\} = \{\frac{1}{2}Q_\mu f_1,\dots,\frac{1}{2}Q_\mu f_m\}$. All functions $g \in \mathcal{G}_\mu$ can be written in the form $g = \frac{1}{2}Q_\mu f$ with some $f \in \mathcal{F}$, and there exists some function $f_j \in \mathcal{F}_0(\varepsilon,\tilde\rho)$ such that $\int (f - f_j)^2\,d\tilde\rho \le \varepsilon^2$. Hence to complete the proof of Proposition 9.3 it is enough to show that $\int \frac{1}{4}(Q_\mu f - Q_\mu \bar f)^2\,d\rho \le \int (f - \bar f)^2\,d\tilde\rho$ for all pairs $f, \bar f \in \mathcal{F}$. This inequality holds, since $\int \frac{1}{4}(Q_\mu f - Q_\mu \bar f)^2\,d\rho \le \int \frac{1}{2}(f - \bar f)^2\,d\rho + \int \frac{1}{2}(P_\mu f - P_\mu \bar f)^2\,d\rho$, and $\int (P_\mu f - P_\mu \bar f)^2\,d\rho = \int P_\mu (f - \bar f)^2\,d\bar\rho \le \int (f - \bar f)^2\,d(\bar\rho \times \mu)$

by formula (9.10). The above relations imply that $\int \frac{1}{4}(Q_\mu f - Q\mu \bar f)^2\, d\rho \le \int (f - \bar f)^2 \frac{1}{2} d\,(\rho + \bar\rho \times \mu) = \int (f - \bar f)^2 d\,\bar\rho$ as we have claimed. □

Now we shall discuss the relation between Theorems 8.1′ and 8.3 and between Theorems 8.2 and 8.4. First we show that Theorem 8.1 (or Theorem 8.1′) is equivalent to the estimate (8.13) in the corollary of Theorem 8.3 which is slightly weaker than the estimate (8.12) of Theorem 8.3. We also claim that Theorems 8.2 and 8.4 are equivalent. Both in Theorem 8.2 and in Theorem 8.4 we can restrict our attention to the case when the class of functions $\mathcal{F}$ is countable, since the case of countably approximable classes can be simply reduced to this situation. Let us remark that integration with respect to the measure $\mu_n - \mu$ in the definition (4.8) of the integral $J_{n,k}(f)$ yields some kind of normalization which is missing in the definition of the U-statistics $I_{n,k}(f)$. This is the cause why degenerate U-statistics had to be considered in Theorems 8.3 and 8.4. The deduction of the corollary of Theorem 8.3 from Theorems 8.1′ or of Theorem 8.4 from Theorem 8.2 is fairly simple if the underlying probability measure μ is non-atomic, since in this case the identity $I_{n,k}(f) = J_{n,k}(f)$ holds for a canonical function with respect to the measure μ. Let us remark that the non-atomic property of the measure μ is needed in this argument not only because of the conditions of Theorems 8.1′ and 8.2, but since in the proof of the above identity we need the identity $\int f(x_1,\dots,x_k)\mu(dx_j) \equiv 0$ in the case when the domain of integration is not the whole space X but the set $X \setminus \{x_1,\dots,x_{j-1},x_{j+1},\dots,x_k\}$.

The case of possibly atomic measures μ can be simply reduced to the case of non-atomic measures by means of the following enlargement of the space $(X,\mathcal{X},\mu)$. Let us introduce the product space $(\bar X,\bar{\mathcal{X}},\bar\mu) = (X,\mathcal{X},\mu) \times ([0,1],\mathcal{B},\lambda)$, where $\mathcal{B}$ is the σ-algebra and λ is the Lebesgue measure on $[0,1]$. Define the function $\bar f((x_1,u_1),\dots,(x_k,u_k)) = f(x_1,\dots,x_k)$ in this enlarged space. Then $I_{n,k}(f) = I_{n,k}(\bar f)$, the measure $\bar\mu = \mu \times \lambda$ is non-atomic, and $\bar f$ is canonical with respect to $\bar\mu$ if f is canonical with respect to μ. Hence the corollary of Theorem 8.3 and Theorem 8.4 can be derived from Theorems 8.1′ and 8.2 respectively by proving them first for their counterpart in the above constructed enlarged space with the above defined functions.

Also Theorems 8.1′ and 8.2 can be derived from Theorems 8.3 and 8.4 respectively, but this is a much harder problem. To do this let us observe that a random integral $J_{n,k}(f)$ can be written as a sum of U-statistics of different order, and it can also be expressed as a sum of degenerate U-statistics if Hoeffding's decomposition is applied for each U-statistic in this sum. Moreover, we shall show that the multiple integral of a function f of k variables with respect to a normalized empirical distribution can be decomposed to the linear combination of degenerate U-statistics with the same kernel functions f_V which appeared in Theorem 9.1 with relatively small coefficients. This is the content of the following Theorem 9.4. For the sake of a better understanding I shall reformulate it in a more explicit form in the special case $k = 2$ in Corollary 2 of Theorem 9.4 at the end of this chapter.

Theorem 9.4 (Decomposition of a multiple random integral with respect to a normalized empirical measure to a linear combination of degenerate U-statistics). *Let a non-atomic measure μ be given on a measurable space $(X, \mathcal{X})$ together with a sequence of independent, μ-distributed random variables $\xi_1, \dots, \xi_n$. Take a function $f(x_1, \dots, x_k)$ of k variables integrable with respect to the product measure μ^k on the product space $(X^k, \mathcal{X}^k)$, and consider the empirical distribution μ_n of the sequence $\xi_1, \dots, \xi_n$ introduced in (4.5) together with the k-fold random integral $J_{n,k}(f)$ of the function f defined in (4.8). The identity*

$$k!J_{n,k}(f) = \sum_{V \subset \{1,\dots,k\}} C(n,k,|V|)n^{-|V|/2}|V|!I_{n,|V|}(f_V) \tag{9.16}$$

holds with the set of (canonical) functions $f_V(x_j,\ j \in V)$ (with respect to the measure μ) defined in formula (9.3) together with some appropriate real numbers $C(n,k,p)$, $0 \le p \le k$, where $I_{n,|V|}(f_V)$ denotes the (degenerate) U-statistic of order $|V|$ with the random variables $\xi_1, \dots, \xi_n$ and kernel function f_V. The constants $C(n,k,p)$ in formula (9.16) satisfy the inequality $|C(n,k,p)| \le C(k)$ for all $n \ge k$ and $0 \le p \le k$ with some constant $C(k) < \infty$ depending only on the order k of the integral $J_{n,k}(f)$. The relations $\lim_{n\to\infty} C(n,k,p) = C(k,p)$ hold with some appropriate constant $C(k,p)$ for all $1 \le p \le k$, and $C(n,k,k) = 1$.

Remark. As the proof of Theorem 9.4 will show, the constant $C(n,k,p)$ in formula (9.16) is a polynomial of order $k-1$ of the argument $n^{-1/2}$ with some coefficients depending on the parameters k and p. As a consequence, $C(k,p)$ equals the constant term of this polynomial.

Theorems 8.1′ and 8.2 can be simply derived from Theorems 8.3 and 8.4 respectively with the help of Theorem 9.4. Indeed, to get Theorem 8.1′ observe that formula (9.16) implies the inequality

$$P(|k!J_{n,k}(f)| > u) \le \sum_{V \subset \{1,\dots,k\}} P\left(n^{-|V|/2}||V|!I_{n,|V|}(f_V)| > \frac{u}{2^k C(k)}\right) \tag{9.17}$$

with a constant $C(k)$ satisfying the inequality $p!C(n,k,p) \le k!C(k)$ for all coefficients $C(n,k,p)$, $1 \le p \le k$, in (9.16). Hence Theorem 8.1′ follows from Theorem 8.3 and relations (9.6) and (9.7) in Theorem 9.2 by which the L_2-norm of the functions f_V is bounded by the L_2-norm of the function f and the L_∞-norm of f_V is bounded by $2^{|V|}$-times the L_∞-norm or f. It is enough to estimate each term at the right-hand side of (9.17) by means of Theorem 8.3. It can be assumed that $2^k C(k) > 1$. Let us first assume that also the inequality $\frac{u}{2^k C(k)\sigma} \ge 1$ holds. In this case formula (8.4) in Theorem 8.1′ can be obtained by means of the estimation of each term at the right-hand side of (9.17). Observe that $\exp\left\{-\alpha\left(\frac{u}{2^k C(k)\sigma}\right)^{2/s}\right\} \le \exp\left\{-\alpha\left(\frac{u}{2^k C(k)\sigma}\right)^{2/k}\right\}$ for all $s \le k$ if $\frac{u}{2^k C(k)\sigma} \ge 1$.

In the other case, when $\frac{u}{2^k C(k)\sigma} \le 1$, formula (8.4) holds again with a sufficiently large $C > 0$, because in this case its right-hand side of (8.4) is greater than 1.

Theorem 8.2 can be similarly derived from Theorem 8.4 by observing that relation (9.17) remains valid if $|J_{n,k}(f)|$ is replaced by $\sup\limits_{f\in\mathscr{F}} |J_{n,k}(f)|$ and $|I_{n,|V|}(f_V)|$ by $\sup\limits_{f_V\in\mathscr{F}_V} |I_{n,|V|}(f_V)|$ in it, and we have the right to choose the constant M in formula (8.7) of Theorem 8.2 sufficiently large. The only difference in the argument is that beside formulas (9.6) and (9.7) the last statement of Theorem 9.2 also has to be applied in this case. It tells that if $\mathscr{F}$ is an L_2-dense class of functions on a space $(X^k, \mathscr{X}^k)$, then the classes of functions $\mathscr{F}_V = \{2^{-|V|} f_V \colon f \in \mathscr{F}\}$ are also L_2-dense classes of functions for all $V \subset \{1,\dots,k\}$ with the same exponent and parameter.

Before its proof I make some comments about the content of Theorem 9.4. The expression $J_{n,k}(f)$ was defined as a k-fold random integral with respect to the signed measure $\mu_n - \mu$, where the diagonals were omitted from the domain of integration. Formula (9.16) expresses the random integral $J_{n,k}(f)$ as a linear combination of degenerate U-statistics of different order. This is similar to the Hoeffding decomposition of the U-statistic $I_{n,k}(f)$ to the linear combination of degenerate U-statistics defined with the same kernel functions f_V. The main difference between these two formulas is that in the expansion (9.16) of $J_{n,k}(f)$ the terms $I_{n,|V|}(f_V)$ appear with small coefficients $C(n,k,|V|)|V|!\frac{1}{n^{|V|/2}}$. As we shall see, $E(C(n,k,|V|)|V|!\frac{1}{n^{|V|/2}} I_{n,V}(f_V))^2 < K$ with a constant $K < \infty$ not depending on n for each set $V \subset \{1,\dots,k\}$. This can be so interpreted that the sum at the right-hand side of (9.16) consists of such random variables $C(n,k,|V|)|V|!n^{-|V|/2} I_{n,V}(f_V)$ which are of constant magnitude. The smallness of these coefficients is related to fact that in the definition of $J_{n,k}(f)$ integration is taken with respect to the signed measure $\mu_n - \mu$ instead of the empirical measure μ_n, which means some kind of normalization. On the other hand, these coefficients $C(n,k,|V|)$ may have a non-zero limit as $n \to \infty$ also for $|V| < k$. In particular, the expansion (9.16) may contain a constant term $C(n,k,0) \ne 0$ such that even $\lim\limits_{n\to\infty} C(n,k,0) \ne 0$. In such a case also the expected value $EJ_{n,k}(f)$ does not equal zero. But even in such a case this expected value can be bounded by a finite number not depending on the sample size n. Next I show an example for a twofold random integral $J_{n,2}(f)$ such that $2EJ_{n,2}(f) = -1$.

Let us choose a sequence of independent random variables $\xi_1,\dots,\xi_n$ with uniform distribution on the unit interval, let μ_n denote its empirical distribution, let $f = f(x,y)$ denote the indicator function of the unit square, i.e. let $f(x,y) = 1$ if $0 \le x, y \le 1$, and $f(x,y) = 0$ otherwise. Let us consider the random integral $2J_{n,2}(f) = n\int_{x\ne y} f(x,y)(\mu_n(dx) - dx)(\mu_n(dy) - dy)$, and calculate its expected value $2EJ_{n,2}(f)$. By adjusting the diagonal $x = y$ to the domain of integration and taking out the contribution obtained in this way we get that $2EJ_{n,2}(f) = nE(\int_0^1 (\mu_n(dx) - \mu(dx))^2 - n^2 \cdot \frac{1}{n^2} = -1$. (The last term is the integral of the function $f(x,y)$ on the diagonal $x = y$ with respect to the product measure $\mu_n \times \mu_n$ which equals $(\mu_n - \mu) \times (\mu_n - \mu)$ on the diagonal.)

Now I turn to the proof of Theorem 9.4.

Proof of Theorem 9.4. Let us remark that for a canonical function g (with respect to the measure μ) of p variables the identity $n^{-p/2}p!I_{n,p}(g) = p!J_{n,p}(g)$ holds. (At this point we also exploit that μ is a non-atomic measure, which implies that the identity $\int g(x_1,\dots,x_p)\mu(dx_j) = 0$ for all $1 \le j \le p$ remains valid for arbitrary arguments x_u, $1 \le u \le p$, $u \ne j$, also if we omit finitely many points from the domain of integration.) This relation implies that if we calculate the (random) integral $p!J_{n,p}(g)$ for a canonical function g we do not change the value of this integral by replacing the measures $\mu_n(dx_j)-\mu(dx_j)$ by $\mu_n(dx_j)$ for all $1 \le j \le p$. The integral we get after such a replacement equals $p!n^{-1/2}I_{n,p}(g)$. Since all functions f_V appearing in formula (9.16) are canonical, the above relation between U-statistics and random integrals has the consequence that formula (9.16) can be rewritten in an equivalent form as

$$k!J_{n,k}(f) = \sum_{V\subset\{1,\dots,k\}} C(n,k,|V|)|V|!J_{n,|V|}(f_V). \tag{9.18}$$

Here we use the convention that a constant c is a canonical function of order zero, and $J_{n,0}(c) = c$. We shall prove identity (9.18) by means of induction with respect to the order k of the integral $k!J_{n,k}(f)$.

In the case $k = 1$ $f_{\{1\}}(x) = f(x) - \int f(x)\mu(dx)$, $f_\emptyset = \int f(x)\mu(dx)$, and

$$J_{n,1}(f_{\{1\}}) = \sqrt{n}\int (f(x) - f_\emptyset)(\mu_n(dx) - \mu(dx)) = J_{n,1}(f),$$

since $\int(\mu_n(dx) - \mu(dx)) = 0$. Hence formula (9.18) holds for $k = 1$ with $C(n,1,1) = 1$ and $C(n,1,0) = 0$. For $k = 0$ relation (9.18) holds with $C(n,0,0) = 1$ if the convention $f_V = f$ is applied for a function f of zero variables, i.e. if f is a constant function, and $V = \emptyset$. In the case $k \ge 2$ we can write by taking the identity (9.3) formulated in the Hoeffding decomposition Theorem 9.1, integrating it with respect to the product measure $\prod\limits_{j=1}^{k}(\mu_n(dx_j) - \mu(dx_j))$ and omitting the diagonals from the domain of integration that

$$k!J_{n,k}(f) = k!J_{n,k}(f_{\{1,\dots,k\}}) + \sum_{\bar V\subset\{1,\dots,k\},\, \bar V\ne\{1,\dots,k\}} k!J_{n,k}(f_{\bar V}). \tag{9.19}$$

Observe that in the case $\bar V \subset \{1,\dots,k\}$, $\bar V \ne \{1,\dots,k\}$ the function $f_{\bar V}$ has strictly less than k arguments, while the terms $J_{n,k}(f_{\bar V})$ at the right-hand side of (9.19) are random integrals of order k. We can rewrite these k-fold integrals as the linear combinations of random integrals of smaller multiplicity with the help of the following

Lemma 9.5. *Let us take a measure space $(X,\mathscr{X},\mu)$ with a non-atomic probability measure μ and an integrable function $f(x_1,\dots,x_{k-1})$ on its $k-1$-fold product, $(X^{k-1},\mathscr{X}^{k-1},\mu^{k-1})$, $k\ge 2$. Let us also take the operator $(P_l f)(x_j,\, j\in\{1,\dots,k-1\}\setminus\{l\})=\int f(x_1,\dots,x_{k-1})\mu(dx_l)$ for all $1\le l\le k-1$. Let us consider the function f also as a function $f(x_1,\dots,x_k)$ of k variables which does not depend on its last coordinate x_k. The identity*

$$k!J_{n,k}(f)=-n^{-1/2}(k-1)\cdot(k-1)!J_{n,k-1}(f)-\sum_{l=1}^{k-1}(k-2)!J_{n,k-2}(P_l f) \tag{9.20}$$

holds. The function $P_l f$ has arguments with indices $j\in\{1,\dots,k-1\}\setminus\{l\}$, and in the term $J_{n,k-2}(P_l f)$ in (9.20) we take integration with respect to

$$n^{(k-2)/2}\prod_{j\in\{1,\dots,k-1\}\setminus\{l\}}(d\mu_n(x_j)-\mu(dx_j)).$$

Proof of Lemma 9.5. Formula (9.20) is equivalent to the identity

$$\int' f(x_1,\dots,x_{k-1})(\mu_n(dx_1)-\mu(dx_1))\dots(\mu_n(dx_k)-\mu(dx_k))$$

$$=-\frac{k-1}{n}\int' f(x_1,\dots,x_{k-1})\prod_{s=1}^{k-1}(\mu_n(dx_s)-\mu(dx_s))$$

$$-\frac{1}{n}\sum_{l=1}^{k-1}\int'\left[\int f(x_1,\dots,x_{k-1})\mu(dx_l)\right]\prod_{1\le s\le k-1,\, s\neq l}(\mu_n(dx_s)-\mu(dx_s)).$$

The expressions at the two sides of this identity are linear combinations of terms of the form

$$\int' f(x_1,\dots,x_{k-1})\prod_{l\in V}\mu_n(dx_l)\prod_{l\in\{1,\dots,k-1\}\setminus V}\mu(dx_l)$$

with $V\subset\{1,\dots,k-1\}$. A term of this form with $|V|=p$ at the left-hand side of this identity has coefficient $(-1)^{k-p}(1-\frac{n-p}{n})=(-1)^{k-p}\frac{p}{n}$. To see this let us calculate the integral

$$\int' f(x_1,\dots,x_{k-1})\prod_{l\in V}\mu_n(dx_l)\prod_{l\in\{1,\dots,k-1\}\setminus V}\mu(dx_l)(\mu_n(dx_k)-\mu(dx_k))$$

by successive integration, and integrating with respect to the variable x_k in the last step. Then we integrate a constant function in the last step. Beside this, since the (random) measure μ_n is concentrated in n points with weights $\frac{1}{n}$, and in the integration $\int'$ we omit the diagonals from the domain of integration, we integrate with respect to a measure with total mass $\frac{n-p}{n}$ when we are integrating with respect to $\mu_n(dx_k)$. On the other hand, the first term at the right-hand side of the identity we want to prove has coefficient $(-1)^{(k-p)}\frac{k-1}{n}$ and the second term has coefficient $(-1)^{(k-p-1)}\frac{k-1-p}{n}$. Lemma 9.5 follows from these calculations. □

Lemma 9.5 was proved by means of elementary calculations. One may ask how its form can be found. It may be worth observing that there are some diagram formulas that play an important role in some subsequent proofs, and they also supply the identity formulated in Lemma 9.5 together with its proof.

In these diagram formulas the product of some random integrals or U-statistics are expressed by means of the sum of appropriately defined random integrals or U-statistics. In the subsequent part of this lecture note I discuss the diagram formula for Wiener–Itô integrals and U-statistics. I shall also mention that there is a diagram formula for the product of multiple integrals with respect to a normalized empirical distribution, and I shall indicate what its form looks like. An explicit formulation and proof of this result can be found in [35]. Lemma 9.5 can be obtained as a special case of this formula.

To get Lemma 9.5 with the help of the diagram formula take the function $e(x) \equiv 1$ on the space $(X, \mathscr{X})$. Then we have $J_{n,1}(e) = 0$ with probability one. Given a function $f(x_1, \dots, x_{k-1})$ write up the identity $J_{n,k-1}(f)J_{n,1}(e) = 0$ with probability one, and rewrite its left-hand side by means of the diagram formula. The identity we get in such a way agrees with Lemma 9.5. One of the terms in this identity is $k!J_{n,k}(f)$ which appears as the integral of the function $\bar{f}(x_1, \dots, x_k) = f(x_1, \dots, x_{k-1})e(x_k)$, and writing up all terms we get the desired formula.

Now I return to the proof of Theorem 9.4.

Completion of the proof of Theorem 9.4 with the help of Lemma 9.5. We shall prove the following slightly more general version of (9.18). If $f(x_j, j \in V)$ is an integrable function with arguments indexed by a set $V \subset \{1, \dots, k\}$, then

$$k!J_{n,k}(f) = \sum_{\bar{V} \subset V} C(n, k, |\bar{V}|, |V|)|\bar{V}|!J_{n,|\bar{V}|}(f_{\bar{V}}) \tag{9.21}$$

with some coefficients $C(n, k, p, q)$, $0 \le p \le q \le k$ such that $|C(n, k, p, q)| \le C_k < \infty$ for all arguments n and $0 \le p \le q \le k$, the limit $\lim_{n\to\infty} C(n, k, p, q) = C(k, p, q)$ exists, and $C(n, k, k, k) = 1$.

At the left-hand side of formulas (9.21) and (9.18) the same integral $J_{n,k}(f)$ of order k of a function f with less than or equal to k arguments is taken. (We define this integral by redefining its kernel function f as a function of k arguments by means of the introduction of some additional fictive coordinates.) At the right-hand side of these formulas the same canonical functions $f_{\bar{V}}$, $\bar{V} \subset \{1, \dots, k\}$, appear. They were introduced in the Hoeffding decomposition (9.3). But in (9.21) we take the integrals of the functions $f_{\bar{V}}$ only with respect to their 'real' coordinates with indices $l \in \bar{V} \subset V$. For the sake of simpler notations first we restrict our attention to the case $V=\{1,\dots,q\}$ with some $0\le q\le k$. (Actually, it can be seen with the help of the subsequent proof that we can choose $C(n,k,p,q)=C(n,k,p)$ with the constant $C(n,k,p)$ appearing in (9.16) or (9.18).)

We shall prove (9.21) by means of induction with respect to k. This relation holds for $k=0$, and to prove it for $k=1$ we still we have to check that it also holds in the special case when f is a function of zero variable, i.e. if it is a constant, and $V=\emptyset$. But relation (9.21) holds in this case with $C(n,1,0,0)=0$, since $J_{n,1}(f)=0$ if f is a variable of zero arguments, i.e. if it is a constant.

We shall prove relation (9.21) for general parameter k with the help of formula (9.19), Lemma 9.5 and formula (9.3) in the Hoeffding decomposition which gives the definition of the functions $f_{\tilde V}$ appearing in (9.19). I formulate a formally more general result than relation (9.20) which follows from Lemma 9.5 if we reindex the variables of the function f considered in it. I formulate this result, because this will be applied in our calculations.

Let us take a number $p\in\{1,\dots,k\}$, $k\ge 2$, and a function $f(x_j,\, j\in\{1,\dots,k\}\setminus\{p\})$, integrable with respect to the appropriate direct product of the measure μ together with the functions $P_l(f)=P_l(f)(x_j,\, j\in\{1,\dots,k\}\setminus\{l,p\})$ for all $l\in\{1,\dots,k\}\setminus\{p\}$ that we get by integrating the function f with respect to the measure $\mu(\,dx_l)$. The following modified version of (9.20) holds in this case.

$$k!J_{n,k}(f)=-n^{-1/2}(k-1)!(k-1)J_{n,k-1}(f)-\sum_{l\in\{1,\dots,k\}\setminus\{p\}}(k-2)!J_{n,k-2}(P_lf) \tag{9.22}$$

where $J_{n,k-1}(f)$ is the integral of the function f with respect to the measure

$$n^{(k-1)/2}\prod_{j\in\{1,\dots,k\}\setminus\{p\}}((\mu_n(\,dx_j)-\mu(\,dx_j))$$

and $J_{n,k-2}(P_lf)$ is the integral of the function P_lf with respect to the measure

$$n^{(k-2)/2}\prod_{j\in\{1,\dots,k\}\setminus\{p,l\}}((\mu_n(\,dx_j)-\mu(\,dx_j)).$$

(Naturally the diagonals are omitted from the domain of integration.)

First we prove (9.21) in the case $V=\{1,\dots,k\}$. We rewrite $k!J_{n,k}(f)$ by means of (9.19) as a sum of random integrals of order k with kernel functions $f_{\tilde V}$, $\tilde V\subset\{1,\dots,k\}$. We rewrite each term $k!J_{n,k}(f_{\tilde V})$ with $\tilde V\subset\{1,\dots,k\}$, $\tilde V\ne\{1,\dots,k\}$ in this sum (i.e. we do not consider the integral $k!J_{n,k}(f_{\{1,\dots,k\}})$) as a linear combination of multiple random integrals of the form $J_{n,k-1}(f_{\tilde V})$ and $J_{n,k-2}(P_lf_{\tilde V})$ of order $k-1$ and $k-2$ respectively with the help of identity (9.22). Then we can apply formula (9.21) for them because of our inductive hypothesis. Let us understand what kind of kernel functions appear in the integrals we get in such a way. If $\bar V\subset\tilde V$, then $(f_{\tilde V})_{\bar V}=f_{\bar V}$ by formula (9.3). On the other hand, $P_lf_{\tilde V}=$

$f_{\tilde{V}\setminus\{l\}}$, and in the expansion of $J_{n,k}(P_l f_{\tilde{V}})$ by means of (9.21) we get a linear combination of random integrals $J_{n,|\bar{V}|}(f_{\bar{V}})$ with $\bar{V} \subset \tilde{V} \setminus \{l\}$. By applying all these identities, summing them up, adding to them the term $J_{n,k}(f_{\{1,\dots,k\}})$ and applying formula (9.22) we get because of our inductive assumptions a representation $k! J_{n,k}(f) = \sum_{\bar{V} \subset V} C(n,k,\bar{V})|\bar{V}|! J_{n,|\bar{V}|}(f_{\bar{V}})$ (where $V = \{1,\dots,k\}$) of the random integral $k! J_{n,k}(f)$ with such coefficients $C(n,k,\bar{V})$ for which $|C(n,k,\bar{V})| \leq C(k)$ and the limit $C(\cdot,\bar{V}) = \lim_{n\to\infty} C(n,k,\bar{V})$ exists. We still have to show that these coefficients can be chosen in such a way that $C(n,k,\bar{V}) = C(n,k,|\bar{V}|)$, i.e. $C(n,k,\bar{V}_1) = C(n,k,\bar{V}_2)$ if $|\bar{V}_1| = |\bar{V}_2|$.

Given a set $\tilde{V} \subset \{1,\dots,k\}$, $\tilde{V} \neq \{1,\dots,k\}$, let us express the random integrals $J_{n,k-1}(f_{\tilde{V}})$ and $J_{n,k-2}(P_l f_{\tilde{V}})$ for all $p \in \{1,\dots,k\} \setminus \tilde{V}$ in the above way, and write $J_{n,k}(f_{\tilde{V}})$ and $J_{n,k}(P_l f_{\tilde{V}})$ as the average of these sums. Working with these expressions for $J_{n,k}(f_{\tilde{V}})$ and $J_{n,k}(P_l f_{\tilde{V}})$ it can be seen that our inductive assumption also holds with such coefficients $C(n,k,\bar{V})$ for which $C(n,k,\bar{V}_1) = C(n,k,\bar{V}_2)$ if $|\bar{V}_1| = |\bar{V}_2|$.

In the next step let us consider the case when $f = f(x_j, j \in V)$ with a set $V = \{1,\dots,q\}$ such that $0 \leq q < k$. I claim that in this case the identity $f_{\tilde{V}} \equiv 0$ holds for those sets $\tilde{V} \subset \{1,\dots,k\}$ for which $\tilde{V} \cap \{q+1,\dots,k\} \neq \emptyset$, and as a consequence $J_{n,k}(f_{\tilde{V}}) = 0$ with probability 1 for such sets $\tilde{V}$. First I show that relation (9.21) can be proved in the present case with the help of this relation similarly to the previous case.

In the present case formula (9.19) has the form $k! J_{n,k}(f) = \sum_{\tilde{V} \subset V} k! J_{n,k}(f_{\tilde{V}})$, and we can express each term $k! J_{n,k}(f_{\tilde{V}})$, $\tilde{V} \subset V$, in this sum by means of formula (9.22) by choosing $f_{\tilde{V}}$ as the function f and an integer p such that $q+1 \leq p \leq k$ (i.e. $p \in \{1,\dots,k\} \setminus V$) in it. In such a way we can write $k! J_{n,k}(f)$ as the linear combination of random integrals of the form $(k-1)! J_{n,k-1}(f_{\tilde{V}})$ and $(k-2)! J_{n,k-2}(P_l f_{\tilde{V}}) = (k-2)! J_{n,k-2}(f_{\tilde{V}\setminus\{l\}})$ with some sets $\tilde{V} \subset V$ and numbers $l \in \{1,\dots,k\} \setminus \{p\}$, where we took some number p such that $q+1 \leq p \leq k$. Then we can apply relation (9.21) for parameters $k-1$ and $k-2$ by our inductive hypothesis, and this enables us to write $J_{n,k}(f)$ as the linear combination of random integrals $|\bar{V}|! J_{n,|\bar{V}|}(f_{\bar{V}})$ with sets $\bar{V} \subset V$. Moreover, it can be seen similarly to the previous case (by writing the above identities for all $p \in \{1,\dots,k\} \setminus \tilde{V}$ and taking their average) that the coefficients in this linear combination can be chosen in such a way as it was demanded in formula (9.21).

To prove that $f_{\tilde{V}} \equiv 0$ if $\tilde{V} \cap \{q+1,\dots,k\} \neq \emptyset$ and $f = f(x_1,\dots,x_k)$ is the extension of a function $f = f(x_j, j \in \{1,\dots,q\})$ to X^k with the help of some "fictive" coordinates take a number $r \in \tilde{V} \cap \{q+1,\dots,k\}$, observe that $P_r f = f$ and $Q_r f \equiv 0$ for the operators P_r and Q_r defined in (9.1) and (9.2), since $r \notin V = \{1,\dots,q\}$. The definition of the function $f_{\tilde{V}}$ is given in formula (9.3). Observe that in the present case the operator Q_r and not the operator P_r appears in the formula defining $f_{\tilde{V}}$. Hence formula (9.3) and the exchangeability of the operators P_j and $Q_{j'}$ imply that $f_{\tilde{V}} \equiv 0$.

Formula (9.21) in the general case simply follows from the already proved results by a reindexation of the variables of the function f. Since (9.18) is a special case of (9.21) Theorem 9.4 is proved. □

Two corollaries of Theorem 9.4 will be formulated. The first one explains the content of conditions (8.2) and (8.6) in Theorems 8.1–8.4.

Corollary 1 of Theorem 9.4. *If $I_{n,k}(f)$ is a degenerate U-statistic of order k with some kernel function f, then*

$$\begin{aligned} E\left(n^{-k/2} I_{n,k}(f)\right)^2 &= \frac{n(n-1)\cdots(n-k+1)}{k!n^k} \int \operatorname{Sym} f^2(x_1,\dots,x_k)\mu(dx_1)\dots\mu(dx_k) \\ &\le \frac{1}{k!} \int f^2(x_1,\dots,x_k)\mu(dx_1)\dots\mu(dx_k), \end{aligned} \tag{9.23}$$

where μ is the distribution of the random variables taking part in the definition of the U-statistic $I_{n,k}(f)$, and $\operatorname{Sym} f$ *is the symmetrization of the function f. The k-fold multiple random integral $J_{n,k}(f)$ with an arbitrary square integrable kernel function f satisfies the inequality*

$$E J_{n,k}(f)^2 \le \bar{C}(k) \int f^2(x_1,\dots,x_k)\mu(dx_1)\dots\mu(dx_k)$$

with some constant $\bar{C}(k)$ depending only on the order k of the integral $J_{n,k}(f)$.

Proof of Corollary 1 of Theorem 9.4. The identity

$$E(n^{-k/2} I_{n,k}(f))^2 = \frac{1}{(k!)^2 n^k} \sum\nolimits' Ef(\xi_{l_1},\dots,\xi_{l_k}) f(\xi_{l'_1},\dots,\xi_{l'_k}) \tag{9.24}$$

holds, where the prime in $\sum'$ means that summation is taken for such pairs of k-tuples $(l_1,\dots,l_k)$, $(l'_1,\dots,l'_k)$, $1 \le l_j, l'_j \le n$, for which $l_j \ne l_{j'}$ and $l'_j \ne l'_{j'}$ if $j \ne j'$. On the other hand, the degeneracy of the U-statistic $I_{n,k}(f)$ implies that

$$Ef(\xi_{l_1},\dots,\xi_{l_k}) f(\xi_{l'_1},\dots,\xi_{l'_k}) = 0$$

if the two sets $\{l_1,\dots,l_k\}$ and $\{l'_1,\dots,l'_k\}$ differ. This can be seen by taking such an index l_j from the first k-tuple which does not appear in the second one, and by observing that the conditional expectation of the product we consider equals zero by the degeneracy condition of the U-statistic under the condition that the value of all random variables except that of ξ_{l_j} is fixed in this product. On the other hand,

$$
\begin{aligned}
&Ef(\xi_{l_1},\dots,\xi_{l_k})f(\xi_{l'_1},\dots,\xi_{l'_k})\\
&\qquad=\int f(x_1,\dots,x_k)f(x_{\pi(1)},\dots,x_{\pi(k)})\mu(\,dx_1)\dots\mu(\,dx_k)
\end{aligned}
$$

if $(l'_1,\dots,l'_k)=(\pi(l_1),\dots,\pi(l_k))$ with some $(\pi(1),\dots,\pi(k))\in\Pi_k$, where Π_k denotes the set of all permutations of the set $\{1,\dots,k\}$. By summing up the above identities for all pairs $(l_1,\dots,l_k)$ and $(l'_1,\dots,l'_k)$ and by applying formula (9.24) we get the identity at the left-hand side of formula (9.23). The second relation in (9.23) is obvious.

The bound for $J_{n,k}(f)$ follows from Theorem 9.4, formula (9.6) in Theorem 9.2 by which the L_2-norm of the functions f_V is not greater than the L_2-norm of the function f and the bound that formula (9.23) yields for the second moment of the degenerate U-statistics $n^{-|V|/2}I_{n,|V|}(f_V)$ appearing in the expansion (9.16). □

In Corollary 2 the decomposition (9.16) of a random integral $J_{n,2}(f)$ of order 2 is described in an explicit form. This result follows from the proof of Theorem 9.4.

Corollary 2 of Theorem 9.4. *Let the random integral $J_{n,2}(f)$ satisfy the conditions of Theorem 9.4. In this case formula (9.16) can be written in the following explicit form:*

$$
2J_{n,2}(f)=\frac{2}{n}I_{n,2}(f_{\{1,2\}})-\frac{1}{n}I_{n,1}(f_{\{1\}})-\frac{1}{n}I_{n,1}(f_{\{2\}})-f_\emptyset
$$

with the functions

$$
\begin{aligned}
f_{\{1,2\}}(x,y)&=f(x,y)-\int f(x,y)\mu(\,dx)-\int f(x,y)\mu(\,dy)\\
&\qquad+\int f(x,y)\mu(\,dx)\mu(\,dy),\\
f_{\{1\}}(x)&=\int f(x,y)\mu(\,dy)-\int f(x,y)\mu(\,dx)\mu(\,dy),\\
f_{\{2\}}(y)&=\int f(x,y)\mu(\,dx)-\int f(x,y)\mu(\,dx)\mu(\,dy),\quad \textit{and}\\
f_\emptyset&=\int f(x,y)\mu(\,dx)\mu(\,dy).
\end{aligned}
$$

Corollary 2 of Theorem 9.4 states that in the case $k=2$ formula (9.16) holds with $C(n,2,2)=1$, $C(n,2,1)=-\frac{1}{\sqrt{n}}$ and $C(n,2,0)=-1$.

Chapter 10
Multiple Wiener–Itô Integrals and Their Properties

In this chapter I present the definition of multiple Wiener–Itô integrals and some of their most important properties needed in the proof of the results formulated in Chap. 8. Wiener–Itô integrals provide a useful tool to handle non-linear functionals of Gaussian processes. To define them first I introduce the notion of the white noise with some reference measure. Then I define the multiple Wiener–Itô integrals with respect to a white noise with some non-atomic reference measure. A most important result in the theory of multiple Wiener–Itô integrals is the so-called diagram formula presented in Theorem 10.2A. This enables us to rewrite the product of two Wiener–Itô integrals in the form of a sum of Wiener–Itô integrals. The proof of the diagram formula is given in Appendix B. This result will be generalized in Theorem 10.2 to a formula about the representation of the product of finitely many Wiener–Itô integrals as a sum of Wiener–Itô integrals.

Another interesting result about Wiener–Itô integrals, formulated at the end of this chapter in Theorem 10.5 states that the class of random variables which can be written in the form of a sum of Wiener–Itô integrals of different order is sufficiently rich. All random variables with finite second moment which are measurable with respect to the σ-algebra generated by the (Gaussian) random variables appearing in the underlying white noise in the construction of multiple Wiener–Itô integrals can be written in such a form.

I shall also give a heuristic explanation of the diagram formula which may indicate why it has the form appearing in Theorem 10.2A. It also helps to find its analogue for (random) integrals with respect to the product of normalized empirical measures. Such a result will be useful later. A simple and useful consequence of Theorem 10.2A about the representation of the product of finitely many Wiener–Itô integrals in the form of a sum of Wiener–Itô integrals will be formulated in Theorem 10.2. This result will be also called the diagram formula. It has an important corollary about the calculation of the moments of Wiener–Itô integrals. Theorem 8.5 will be proved with the help of this corollary.

I shall give the proof of two other results about Wiener–Itô integrals in Appendix C. The first one, Theorem 10.3, is called Itô's formula for Wiener–

P. Major, *On the Estimation of Multiple Random Integrals and U-Statistics*,
Lecture Notes in Mathematics 2079, DOI 10.1007/978-3-642-37617-7_10,

Itô integrals, and it explains the relation between multiple Wiener–Itô integrals and Hermite polynomials of Gaussian random variables. This result is a relatively simple consequence of the diagram formula and some basic recursive relations about Hermite polynomials.

The other result proved in Appendix C, Theorem 10.4, is a limit theorem about a sequence of appropriately normalized degenerate U-statistics. Here the limit is presented in the form of a multiple Wiener–Itô integral. This result is interesting for us, because it helps to compare Theorems 8.3 and 8.1 with their one-variate counterpart, Bernstein's inequality. In the one-variate case Bernstein's inequality provides a comparison between the tail distribution of sums of independent random variables and the tail of the standard normal distribution. The normal distribution appears here in a natural way as the limit in the central limit theorem.

Theorem 8.3 yields a similar result about degenerate U-statistics. The upper bound for the tail-distribution of a degenerate U-statistic given in Theorem 8.3 or in its Corollary is similar to the bound of Theorem 8.5 about the tail-distribution of a Wiener–Itô integral with the same kernel function. On the other hand, by Theorem 10.4 this Wiener–Itô integral also appears as the limit of degenerate U-statistics with the same kernel function. This shows some similarity between Theorem 8.3, and its one-variate version, the Bernstein inequality. Theorem 8.1 which is an estimate of multiple integrals with respect to a normalized empirical distribution also has a similar interpretation.

My Lecture Note [32] contains a rather detailed description of Wiener–Itô integrals. But in that work the emphasis was put on the study of a slightly different version of it. The original version of this integral introduced in [26] was only briefly discussed there, and not all details were worked out. In particular, the diagram formula needed in this work was formulated and proved only for modified Wiener–Itô integrals. I shall discuss the difference between these random integrals together with the question why a modified version of Wiener–Itô integrals was studied in [32] at the end of this chapter.

To define multiple Wiener–Itô integrals first I introduce the notion of white noise.

Definition a white noise with some reference measure. *Let us have a σ-finite measure μ on a measurable space $(X, \mathcal{X})$. A white noise with reference measure μ is a Gaussian random field $\mu_W = \{\mu_W(A)\colon A \in \mathcal{X},\ \mu(A) < \infty\}$, i.e. a set of jointly Gaussian random variables indexed by the above sets A, which satisfies the relations $E\mu_W(A) = 0$ and $E\mu_W(A)\mu_W(B) = \mu(A \cap B)$ for all $A, B \in \mathcal{X}$ such that $\mu(A) < \infty$ and $\mu(B) < \infty$.*

I make some comments about this definition.

Remark. In the definition of a white noise sometimes also the property $\mu_W(A \cup B) = \mu_W(A) + \mu_W(B)$ with probability 1 if $A \cap B = \emptyset$, and $\mu(A) < \infty$, $\mu(B) < \infty$ is mentioned. But this condition can be omitted, because it follows from the remaining properties of the white noise. Indeed, simple calculation shows that $E(\mu_W(A \cup B) - \mu_W(A) - \mu_W(B))^2 = 0$ if $A \cap B = \emptyset$, hence $\mu_W(A \cup B) - \mu_W(A) - \mu_W(B) = 0$ with probability 1 in this case. It can also be observed that if some sets $A_1, \dots, A_k \in \mathscr{X}$, $\mu(A_j) < \infty$, $1 \le j \le k$, are disjoint, then the random variables $\mu_W(A_j)$, $1 \le j \le k$, are independent because of the uncorrelatedness of these jointly Gaussian random variables.

It is not difficult to see that for an arbitrary reference measure μ on a space $(X, \mathscr{X})$ a white noise μ_W with this reference measure really exists. This follows simply from Kolmogorov's fundamental theorem, by which if the finite dimensional distributions of a random field are defined in a consistent way, then there exists a random field with these finite dimensional distributions.

Now I turn to the definition of multiple Wiener–Itô integrals with respect to a white noise with some reference measure μ. First I introduce the class of functions whose Wiener–Itô integrals with respect to a white noise μ_W with a non-atomic reference measure μ will be defined.

Let us consider a measurable space $(X, \mathscr{X})$, a σ-finite, non-atomic measure μ on it and a white noise μ_W on $(X, \mathscr{X})$ with reference measure μ. Let us define the classes of functions $\mathscr{H}_{\mu,k}$, $k = 1, 2, \dots$, consisting of functions of k variables on $(X, \mathscr{X})$ by the formula

$$\mathscr{H}_{\mu,k} = \Bigg\{ f(x_1, \dots, x_k)\colon f(x_1, \dots, x_k) \text{ is an } \mathscr{X}^k \text{ measurable, real valued}$$
$$\text{function on } X^k, \text{ and} \int f^2(x_1, \dots, x_k)\mu(\,dx_1)\dots,\mu(\,dx_k) < \infty \Bigg\}. \tag{10.1}$$

We shall call a σ-finite measure μ on a measurable space $(X, \mathscr{X})$ non-atomic if for all sets $A \in \mathscr{X}$ such that $\mu(A) < \infty$ and all numbers $\varepsilon > 0$ there is a finite partition $A = \bigcup\limits_{s=1}^{N} B_s$ of the set A with the property $\mu(B_s) < \varepsilon$ for all $1 \le s \le N$. There is a formally weaker definition of a non-atomic measures by which a σ-finite measure μ is non-atomic if for all measurable sets A such that $0 < \mu(A) < \infty$ there is a measurable set $B \subset A$ with the property $0 < \mu(B) < \mu(A)$. But these two definitions of non-atomic measures are actually equivalent, although this equivalence is not trivial. I do not discuss this problem here, since it is a little bit outside from the direction of the present work. In our further considerations we shall work with the first definition of non-atomic measures.

I would also remark that non-atomic measures behave not completely so, as our first heuristic feeling would suggest. It is true that if μ is a non-atomic measure, then $\mu(\{a\}) = 0$ for all one-point sets $\{a\}$. But the reverse statement does not hold. There are (in some sense degenerate) measures μ for which each one-point set has zero μ measure, and which are nevertheless not non-atomic. I omit the discussion of this question.

The k-fold Wiener–Itô integrals of the functions $f \in \mathscr{H}_{\mu,k}$ with respect to the white noise μ_W will be defined in a rather standard way. First they will be defined for some simple functions, called elementary functions, then it will be shown that the integral for these elementary functions has an L_2 contraction property which makes possible to extend it to the class of all functions in $\mathcal{H}_{\mu,k}$.

Let us first introduce the following class of elementary functions $\bar{\mathcal{H}}_{\mu,k}$ of k variables. A function $f(x_1,\dots,x_k)$ on $(X^k,\mathcal{X}^k)$ belongs to $\bar{\mathcal{H}}_{\mu,k}$ if there exist finitely many disjoint measurable subsets $A_1,\dots,A_M$, $1\le M<\infty$, of the set X with finite μ-measure (i.e. $A_j\cap A_{j'}=\emptyset$ if $j\ne j'$, and $\mu(A_j)<\infty$ for all $1\le j\le M$) such that the function f has the form

$$f(x_1,\dots,x_k)=\begin{cases} c(j_1,\dots,j_k) \quad \text{if } (x_1,\dots,x_k)\in A_{j_1}\times\cdots\times A_{j_k} \text{ with} \\ \qquad \text{some indices } (j_1,\dots,j_k), \quad 1\le j_s\le M,\ 1\le s\le k, \\ \qquad \text{such that all numbers } j_1,\dots,j_k \text{ are different} \\ 0 \quad \text{if } (x_1,\dots,x_k)\notin \bigcup\limits_{\substack{(j_1,\dots,j_k)\colon 1\le j_s\le M,\ 1\le s\le k, \\ \text{and all } j_1,\dots,j_k \text{ are different.}}} A_{j_1}\times\cdots\times A_{j_k} \end{cases} \tag{10.2}$$

with some real numbers $c(j_1,\dots,j_k)$, $1\le j_s\le M$, $1\le s\le k$, defined for such arguments for which $j_1,\dots,j_k$ are different numbers. This means that the function f is constant on all k-dimensional rectangles $A_{j_1}\times\cdots\times A_{j_k}$ with different, non-intersecting edges, and it equals zero on the complementary set of the union of these rectangles. The property that the support of the function f is the union of rectangles with non-intersecting edges is sometimes interpreted so that the diagonals are omitted from the domain of integration of Wiener–Itô integrals.

The Wiener–Itô integral of an elementary function $f(x_1,\dots,x_k)$ of the form (10.2) with respect to a white noise μ_W with the (non-atomic) reference measure μ is defined by the formula

$$\begin{aligned} &\int f(x_1,\dots,x_k)\mu_W(dx_1)\dots\mu_W(dx_k) \\ &\quad=\sum_{\substack{1\le j_s\le M,\ 1\le s\le k \\ \text{all } j_1,\dots,j_k \text{ are different}}} c(j_1,\dots,j_k)\mu_W(A_{j_1})\cdots\mu_W(A_{j_k}). \end{aligned} \tag{10.3}$$

(The representation of the function f in (10.2) is not unique, the sets A_j can be divided into smaller disjoint sets, but the Wiener–Itô integral defined in (10.3) does not depend on the representation of the function f. This can be seen with the help of the additivity property $\mu_W(A\cup B)=\mu_W(A)+\mu_W(B)$ if $A\cap B=\emptyset$ of the white noise μ_W.) The notation

$$Z_{\mu,k}(f)=\frac{1}{k!}\int f(x_1,\dots,x_k)\mu_W(dx_1)\dots\mu_W(dx_k), \tag{10.4}$$

will be used in the sequel, and the expression $Z_{\mu,k}(f)$ will be called the normalized Wiener–Itô integral of the function f. Such a terminology will be applied also for the Wiener–Itô integrals of all functions $f\in\mathcal{H}_{\mu,k}$ to be defined later.

If f is an elementary function in $\bar{\mathcal{H}}_{\mu,k}$ defined in (10.2), then its normalized Wiener–Itô integral defined in (10.3) and (10.4) satisfies the relations

$$
\begin{aligned}
&Ek!Z_{\mu,k}(f) = 0,\\
&E(k!Z_{\mu,k}(f))^2 = \sum_{\substack{(j_1,\dots,j_k):\,1\le j_s\le M,\,1\le s\le k,\ \pi\in\Pi_k\\ \text{and all } j_1,\dots,j_k \text{ are different.}}} \sum c(j_1,\dots,j_k)c(j_{\pi(1)},\dots,j_{\pi(k)})\\
&\qquad\qquad E\mu_W(A_{j_1})\cdots\mu_W(A_{j_k})\mu_W(A_{j_{\pi(1)}})\cdots\mu_W(A_{j_{\pi(k)}})\\
&\qquad = k!\int \operatorname{Sym} f^2(x_1,\dots,x_k)\mu(\,dx_1)\dots\mu(\,dx_k)\\
&\qquad \le k!\int f^2(x_1,\dots,x_k)\mu(\,dx_1)\dots\mu(\,dx_k),
\end{aligned}
\tag{10.5}
$$

with $\operatorname{Sym} f(x_1,\dots,x_k) = \frac{1}{k!}\sum\limits_{\pi\in\Pi_k} f(x_{\pi(1)},\dots,x_{\pi(k)})$, where Π_k denotes the set of all permutations $\pi = \{\pi(1),\dots,\pi(k)\}$ of the set $\{1,\dots,k\}$.

The identities written down in (10.5) can be simply checked. The first relation follows from the identity $E\mu_W(A_{j_1})\cdots\mu_W(A_{j_k}) = 0$ for disjoint sets $A_{j_1},\dots,A_{j_k}$, which holds, since the expectation of the product of independent random variables with zero expectation is taken. The second identity follows similarly from the identity

$$
\begin{aligned}
&E\mu_W(A_{j_1})\cdots\mu_W(A_{j_k})\mu_W(A_{j'_1})\cdots\mu_W(A_{j'_k}) = 0\\
&\qquad \text{if the sets of indices } \{j_1,\dots,j_k\} \text{ and } \{j'_1,\dots,j'_k\} \text{ are different,}\\
&E\mu_W(A_{j_1})\cdots\mu_W(A_{j_k})\mu_W(A_{j'_1})\cdots\mu_W(A_{j'_k}) = \mu(A_{j_1})\cdots\mu(A_{j_k})\\
&\qquad \text{if } \{j_1,\dots,j_k\} = \{j'_1,\dots,j'_k\} \text{ i.e. if } j'_1 = j_{\pi(1)},\dots,j'_k = j_{\pi(k)}\\
&\qquad \text{with some permutation } \pi\in\Pi_k,
\end{aligned}
$$

which holds because of the facts that the μ_W measure of disjoint sets are independent with expectation zero, and $E\mu_W(A)^2 = \mu(A)$. The remaining relations in (10.5) can be simply checked.

It is not difficult to check that

$$
EZ_{\mu,k}(f)Z_{\mu,k'}(g) = 0 \tag{10.6}
$$

for all functions $f\in\bar{\mathcal{H}}_{\mu,k}$ and $g\in\bar{\mathcal{H}}_{\mu,k'}$ if $k\ne k'$, and

$$
Z_{\mu,k}(f) = Z_{\mu,k}(\operatorname{Sym} f) \tag{10.7}
$$

for all functions $f\in\bar{\mathcal{H}}_{\mu,k}$.

The definition of Wiener–Itô integrals can be extended to general functions $f \in \mathcal{H}_{\mu,k}$ with the help of formula (10.5). To carry out this extension we still have to know that the class of functions $\bar{\mathcal{H}}_{\mu,k}$ is a dense subset of the class $\mathcal{H}_{\mu,k}$ in the Hilbert space $L_2(X^k, \mathcal{X}^k, \mu^k)$, where μ^k is the k-th power of the reference measure μ of the white noise μ_W. I briefly explain how this property of $\bar{\mathcal{H}}_{\mu,k}$ can be proved. The non-atomic property of the measure μ is exploited at this point.

To prove this statement it is enough to show that the indicator function of any product set $A_1 \times \cdots \times A_k$ such that $\mu(A_j) < \infty$, $1 \le j \le k$, but the sets $A_1, \dots, A_k$ may be non-disjoint is in the $L_2(\mu^k)$ closure of $\bar{\mathcal{H}}_{\mu,k}$. In the proof of this statement it will be exploited that since μ is a non-atomic measure, the sets A_j can be represented for all $\varepsilon > 0$ and $1 \le j \le k$ as a finite union $A_j = \bigcup\limits_s B_{j,s}$ of disjoint sets $B_{j,s}$ with the property $\mu(B_{j,s}) < \varepsilon$. By means of these relations the product $A_1 \times \cdots \times A_k$ can be written in the form

$$A_1 \times \cdots \times A_k = \bigcup_{s_1,\dots,s_k} B_{1,s_1} \times \cdots \times B_{k,s_k} \tag{10.8}$$

with some sets B_{j,s_j} such that $\mu(B_{j,s_j}) < \varepsilon$ for all sets in this union. Moreover, we may assume, by refining the partitions of the sets A_j if this is necessary that any two sets B_{j,s_j} and $B_{j',s'_{j'}}$ in this representation are either disjoint, or they agree. Take such a representation of $A_1 \times \cdots \times A_k$, and consider the set we obtain by omitting those products $B_{1,s_1} \times \cdots \times B_{k,s_k}$ from the union at the right-hand side of (10.8) for which $B_{i,s_i} = B_{j,s_j}$ for some $1 \le i < j \le k$. The indicator function of the remaining set is in the class $\bar{\mathcal{H}}_{\mu,k}$. Hence it is enough to show that the distance between this indicator function and the indicator function of the set $A_1 \times \cdots \times A_k$ is less than const. ε in the $L_2(\mu^k)$ norm with some const. which may depend on the sets $A_1, \dots, A_k$, but not on ε. Indeed, by letting ε tend to zero we get from this relation that the indicator function of the set $A_1 \times A_2 \times \cdots \times A_k$ is in the closure of $\bar{\mathcal{H}}_{\mu,k}$ in the $L_2(\mu^k)$ norm.

Hence to prove the desired property of $\bar{\mathcal{H}}_{\mu,k}$ it is enough to prove the following statement. Take the representation (10.8) of $A_1 \times \cdots \times A_k$ (which depends on ε) and fix an arbitrary pair of integers i and j such that $1 \le i < j \le k$. Then the sum of the measures $\mu^k(B_{1,s_1} \times \cdots \times B_{k,s_k})$ of those sets $B_{1,s_1} \times \cdots \times B_{k,s_k}$ at the right-hand side of (10.8) for which $B_{i,s_i} = B_{j,s_j}$ is less than const. ε. To prove this estimate observe that the μ^k measure of such a set can be bounded by the μ^{k-1} measure of the set we obtain by omitting the i-th term from the product defining it in the following way:

$$\mu^k(B_{1,s_1} \times \cdots \times B_{k,s_k}) \le \varepsilon \mu^{k-1}(B_{1,s_1} \times \cdots \times B_{i-1,s_{i-1}} \times B_{i+1,s_{i+1}} \times \cdots \times B_{k,s_k}).$$

Let us sum up this inequality for all such sets $B_{1,s_1} \times \cdots \times B_{k,s_k}$ at the right-hand side of (10.8) for which $B_{i,s_i} = B_{j,s_j}$. The left-hand side of the inequality we get in such a way equals the quantity we want to estimate. The expression at its right-hand side is less than $\varepsilon \prod_{1\le s\le k, s\ne i} \mu(A_s)$, since ε-times the μ^{k-1} measure of such disjoint sets are summed up in it which are contained in the set $A_1\times\cdots\times A_{i-1}\times A_{i+1}\times\cdots\times A_k$. In such a way we get the estimate we wanted to prove.

Knowing that $\bar{\mathcal{H}}_{\mu,k}$ is a dense subset of $\mathcal{H}_{\mu,k}$ in $L_2(\mu^k)$ norm we can finish the definition of k-fold Wiener–itô integrals in the standard way. Given any function $f\in\mathcal{H}_{\mu,k}$ a sequence of functions $f_n\in\bar{\mathcal{H}}_{\mu,k}$, $n=1,2,\dots$, can be defined in such a way that

$$\int |f(x_1,\dots,x_k)-f_n(x_1,\dots,x_k)|^2\mu(\,dx_1)\dots\mu(\,dx_k)\to 0 \quad \text{as } n\to\infty.$$

By relation (10.5) the already defined Wiener–Itô integrals $Z_{\mu,k}(f_n)$ of the functions f_n, $n=1,2,\dots$, constitute a Cauchy sequence in the space of the square integrable random variables living on the probability space, where the white noise is given. (Observe that the difference of two functions from the class $\bar{\mathcal{H}}_{\mu,k}$ also belongs to this class.) Hence the limit $\lim\limits_{n\to\infty} Z_{\mu,k}(f_n)$ exists in L_2 norm, and this limit can be defined as the normalized Wiener–Itô integral $Z_{\mu,k}(f)$ of the function f. The definition of this limit does not depend on the choice of the approximating functions f_n, hence it is meaningful. It can be seen that relations (10.5) and (10.6) remain valid for all functions $f\in\mathcal{H}_{\mu,k}$. The following Theorem 10.1 describes the properties of multiple Wiener–Itô integrals. It contains already proved results. The only still non-discussed part of this Theorem is Property (f) of Wiener–Itô integrals. But it is easy to check this property by observing that onefold Wiener–Itô integrals are (jointly) Gaussian, they are measurable with respect to the σ-algebra generated by the white noise μ_W. Beside this, the random variable $\mu_W(A)$ for a set $A\in\mathcal{X}$, $\mu(A)<\infty$, equals the (onefold) Wiener–Itô integral of the indicator function of the set A.

Theorem 10.1 (Some properties of multiple Wiener–Itô integrals). *Let a white noise μ_W be given with some non-atomic, σ-additive reference measure on a measurable space $(X,\mathcal{X})$. Then the k-fold Wiener–Itô integrals of all functions in the class $\mathcal{H}_{\mu,k}$ introduced in formula (10.1) can be defined, and their normalized versions $Z_{\mu,k}(f)=\frac{1}{k!}\int f(x_1,\dots,x_k)\mu_W(\,dx_1)\dots\mu_W(dx_k)$ satisfy the following relations:*

(a) $Z_{\mu,k}(\alpha f+\beta g)=\alpha Z_{\mu,k}(f)+\beta Z_{\mu,k}(g)$ for all $f,g\in\mathcal{H}_{\mu,k}$ and real numbers α and β.

(b) If $A_1,\dots,A_k$ are disjoint sets, $\mu(A_j)<\infty$, then the function $f_{A_1,\dots,A_k}$ defined by the relation $f_{A_1,\dots,A_k}(x_1,\dots,x_k)=1$ if $x_1\in A_1,\dots,x_k\in A_k$, $f_{A_1,\dots,A_k}(x_1,\dots,x_k)=0$ otherwise, satisfies the identity

$$Z_{\mu,k}(f_{A_1,\dots,A_k}(x_1,\dots,x_k))=\frac{1}{k!}\mu_W(A_1)\cdots\mu_W(A_k).$$

(c)

$$EZ_{\mu,k}(f) = 0, \quad \textit{and} \quad EZ^2_{\mu,k}(f) = \frac{1}{k!}\|\mathrm{Sym}\, f\|_2^2 \le \frac{1}{k!}\|f\|_2^2$$

for all $f \in \mathcal{H}_{\mu,k}$, *where* $\|f\|_2^2 = \int f^2(x_1,\dots,x_k)\mu(dx_1)\dots\mu(dx_k)$ *is the square of the* L_2 *norm of a function* $f \in \mathcal{H}_{\mu,k}$.

(d) Relation (10.6) holds for all functions $f \in \mathcal{H}_{\mu,k}$ *and* $g \in \mathcal{H}_{\mu,k'}$ *if* $k \neq k'$.

(e) Relation (10.7) holds for all functions $f \in \mathcal{H}_{\mu,k}$.

(f) The Wiener–Itô integrals $Z_{\mu,1}(f)$ *of order* $k = 1$ *are jointly Gaussian. The smallest* σ*-algebra with respect to which they are all measurable agrees with the* σ*-algebra generated by the random variables* $\mu_W(A)$, $A \in \mathcal{X}$, $\mu(A) < \infty$, *of the white noise.*

We have defined Wiener–Itô integrals of order k for all $k = 1, 2, \dots$. For the sake of completeness let us introduce the class $\mathcal{H}_{\mu,0}$ for $k = 0$ which consists of the real constants (functions of zero variables), and put $Z_{\mu,0}(c) = c$. Because of relation (10.7) we could have restricted our attention to Wiener–Itô integrals with symmetric kernel functions. But at some points it was more convenient to work also with Wiener–Itô integrals of not necessarily symmetric functions.

Now I formulate the diagram formula for the product of two Wiener–Itô integrals. For this goal first I introduce some notations. Then I formulate the diagram formula with their help in Theorem 10.2A. To make this result more understandable I shall present after its formulation an example together with some pictures which may help to understand how to calculate the terms appearing in the diagram formula. A similar approach will be applied when the generalization of this result for the product of several Wiener–Itô integrals will be discussed, and also in the next chapter when a version of the diagram formula will be presented for the product of degenerate U-statistics.

To present the product of the multiple Wiener–Itô integrals of two functions $f(x_1,\dots,x_k) \in \mathcal{H}_{\mu,k}$ and $g(x_1,\dots,x_l) \in \mathcal{H}_{\mu,l}$ in the form of sums of Wiener–Itô integrals a class of diagrams $\Gamma = \Gamma(k,l)$ will be defined. The diagrams $\gamma \in \Gamma(k,l)$ have vertices $(1,1),\dots,(1,k)$ and $(2,1),\dots,(2,l)$, and edges $((1,j_1),(2,j_1')),\dots,((1,j_s),(2,j_s'))$ with some $1 \le s \le \min(k,l)$. The indices $j_1,\dots,j_s$ in the definition of the edges are all different, and the same relation holds for the indices $j_1',\dots,j_s'$. All diagrams γ with such properties belong to $\Gamma(k,l)$. The set of vertices of the form $(1,j)$, $1 \le j \le k$, will be called the first row, and the set of vertices of the form $(2,j')$, $1 \le j' \le l$, the second row of a diagram. We demanded that edges of a diagram can connect only vertices of different rows, and at most one edge may start from each vertex of a diagram.

Given a diagram $\gamma \in \Gamma(k,l)$ with the set of edges

$$E(\gamma) = \{(1,j_1),(2,j_1')),\dots,((1,j_s),(2,j_s')\}$$

let

$$V_1(\gamma) = \{(1,1),\dots,(1,k)\} \setminus \{(1,j_1),\dots,(1,j_s)\}$$

and

$$V_2(\gamma) = \{(2,1),\dots,(2,l)\} \setminus \{(2,j_1'),\dots,(2,j_s')\}$$

denote the set of those vertices in the first and in the second row of the diagram γ respectively from which no edge starts. Put $\alpha_\gamma((1,j)) = (2,j')$ if $((1,j),(2,j')) \in E(\gamma)$ and $\alpha_\gamma((1,j)) = (1,j)$ if the diagram γ contains no edge which is of the form $((1,j),(2,j')) \in E(\gamma)$. In words, the function $\alpha_\gamma(\cdot)$ is defined on the vertices of the first row of the diagram γ. It replaces a vertex to the vertex it is connected to by an edge of the diagram if there is such a vertex, and it does not change those vertices from which no edge starts. Put $|\gamma| = k + l - 2s$, i.e. let $|\gamma|$ equal the number of vertices in γ from which no edge starts. Given two functions $f(x_1,\dots,x_k) \in \mathcal{H}_{\mu,k}$ and $g(x_1,\dots,x_l) \in \mathcal{H}_{\mu,l}$ let us introduce their product

$$\begin{aligned}(f \circ g)&(x_{(1,1)},\dots,x_{(1,k)},x_{(2,1)},\dots,x_{(2,l)})\\ &= f(x_{(1,1)},\dots,x_{(1,k)})g(x_{(2,1)},\dots,x_{(2,l)})\end{aligned} \tag{10.9}$$

together with its transform

$$\begin{aligned}&\overline{(f \circ g)}_\gamma(x_{(1,j)}\colon (1,j) \in V_1(\gamma),\ x_{(2,j')}\colon 1 \le j' \le l)\\ &\quad = f(x_{\alpha_\gamma((1,1))},\dots,x_{\alpha_\gamma((1,k))})g(x_{(2,1)},\dots,x_{(2,l)}).\end{aligned} \tag{10.10}$$

(Here the function $f(x_1,\dots,x_k)$ is replaced by $f(x_{(1,1)},\dots,x_{(1,k)})$ and the function $g(x_1,\dots,x_l)$ by $g(x_{(2,1)},\dots,x_{(2,l)})$.) With the help of the above introduced sets $V_1(\gamma)$, $V_2(\gamma)$ and function $\alpha_\gamma(\cdot)$ let us introduce the functions $F_\gamma(f,g)$ as

$$\begin{aligned}&F_\gamma(f,g)(x_{(1,j)},x_{(2,j')}\colon (1,j) \in V_1(\gamma),\ (2,j') \in V_2(\gamma))\\ &\quad = \int \overline{(f \circ g)}_\gamma(x_{\alpha_\gamma((1,j))}\colon (1,j) \in V_1(\gamma),\ x_{(2,1)},\dots,x_{(2,l)})\\ &\qquad\qquad \prod_{(2,j') \in \{(2,1),\dots,(2,l)\} \setminus V_2(\gamma)} \mu(dx_{(2,j')})\end{aligned} \tag{10.11}$$

for all diagrams $\gamma \in \Gamma(k,l)$. In words: We take the product defined in (10.9), then if the index $(1,j)$ of a variable $x_{(1,j)}$ is connected with the index $(2,j')$ of some variable $x_{(2,j')}$ by an edge of the diagram γ, then we replace the variable $x_{(1,j)}$ by $x_{(2,j')}$ in this product. Finally we integrate the function obtained in such a way with respect to the arguments with indices $(2,j_1'),\dots,(2,j_s')$, i.e. with those vertices of the second row of the diagram γ from which an edge starts. It is clear that F_γ is a function of $|\gamma|$ variables. It depends on those coordinates whose indices are such vertices of γ from which no edge starts.

For the sake of simpler notations we shall also consider Wiener–Itô integrals with such kernel functions whose variables are more generally indexed. If the k-fold Wiener–Itô integral with a kernel function $f(x_1,\dots,x_k)$ is well-defined, then we shall say that the Wiener–Itô integral with kernel function $f(x_{u_1},\dots,x_{u_k})$, where

$\{u_1, \dots, u_k\}$ is an arbitrary set with k different elements, is also well defined, and it equals the Wiener–Itô integral with the original kernel function $f(x_1, \dots, x_k)$, i.e. we write

$$\int f(x_{u_1}, \dots, x_{u_k})\mu_W(dx_{u_1}) \dots \mu_W(dx_{u_k})$$
$$= \int f(x_1, \dots, x_k)\mu_W(dx_1) \dots \mu_W(dx_k). \tag{10.12}$$

(We have right to make such a convention, since the value of a Wiener–Itô integral does not change if we permute the indices of the variables of the kernel function in an arbitrary way. This follows e.g. from (10.7).) In particular, we shall speak about the Wiener–Itô integral of the function $F_\gamma(f_1, f_2)$ defined in (10.11) without reindexing its variables $x_{(1,j)}$ and $x_{(2,j')}$ "in the right way". Now we can formulate the diagram formula for the product of two Wiener–Itô integrals.

Theorem 10.2A (The diagram formula for the product of two Wiener–Itô integrals). *Let a non-atomic, σ-finite measure μ be given on a measurable space $(X, \mathcal{X})$ together with a white noise μ_W with reference measure μ, and take two functions $f(x_1, \dots, x_k) \in \mathcal{H}_{\mu,k}$ and $g(x_1, \dots, x_l) \in \mathcal{H}_{\mu,l}$. (The classes of functions $\mathcal{H}_{\mu,k}$ and $\mathcal{H}_{\mu,l}$ were introduced in (10.1).) Let us consider the class of diagrams $\Gamma(k, l)$ introduced above together with the functions $F_\gamma(f, g)$, $\gamma \in \Gamma(k, l)$, defined by formulas (10.9)–(10.11) with its help. They satisfy the inequality*

$$\|F_\gamma(f, g)\|_2 \le \|f\|_2 \|g\|_2 \quad \textit{for all } \gamma \in \Gamma(k, l), \tag{10.13}$$

where the L_2 norm of a (generally indexed) function $h(x_{u_1}, \dots, x_{u_s})$ is defined as

$$\|h\|_2^2 = \int h^2(x_{u_1}, \dots, x_{u_s})\mu(dx_{u_1}) \dots \mu\, dx_{u_s}).$$

Beside this, the product $k!Z_{\mu,k}(f)l!Z_{\mu,l}(g)$ of the Wiener–Itô integrals of the functions f and g (the notation $Z_{\mu,k}$ was introduced in (10.4)) satisfies the identity

$$(k!Z_{\mu,k}(f))(l!Z_{\mu,l}(g)) = \sum_{\gamma \in \Gamma(k,l)} |\gamma|!Z_{\mu,|\gamma|}(F_\gamma(f, g))$$
$$= \sum_{\gamma \in \Gamma(k,l)} |\gamma|!Z_{\mu,|\gamma|}(\operatorname{Sym} F_\gamma(f, g)). \tag{10.14}$$

The next example may help to understand how to apply the diagram formula. Take two Wiener–Itô integrals $2!Z_2(f) = \int f(x_1, x_2)\mu_W(dx_1)\mu_W(dx_2)$ and

$$3!Z_3(g) = \int g(x_1, x_2, x_3)\mu_W(dx_1)\mu_W(dx_2)\mu_W(dx_3)$$

with kernel functions $f(x_1,x_2)$ and $g(x_1,x_2,x_3)$. Let us understand how to calculate a term in the sum at the right-hand side of (10.14) which expresses the product $2!Z_2(f)3!Z_3(g)$ as a sum of Wiener–Itô integrals.

When we apply the diagram formula first we reindex the arguments of the functions f and g by the indices $(1,1)$, $(1,2)$ and $(2,1)$, $(2,2)$, $(2,3)$ respectively, and take the product of these reindexed functions. We get the function

$$(f\circ g)(x_{(1,1)},x_{(1,2)},x_{(2,1)},x_{(2,2)},x_{(2,3)}) = f(x_{(1,1)},x_{(1,2)})g(x_{(2,1)},x_{(2,2)},x_{(2,3)}).$$

We define the two rows of the diagrams we will be working with. The labels of their vertices agree with the indices of the arguments of the functions f and g. (See picture.)

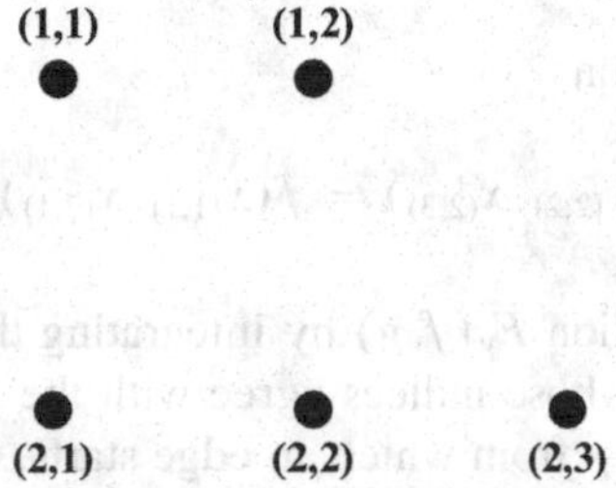

The vertices of the diagrams

We consider all diagrams γ in which vertices from the first and second row are connected by edges, and from each vertex there starts zero or one edge. We define with the help of these diagrams γ some function $F_\gamma(f,g)$ which will be the kernel functions of the Wiener–Itô integrals appearing in the diagram formula (10.14). Let us consider that diagram γ which contains one edge connecting the vertices $(1,2)$ and $(2,1)$.

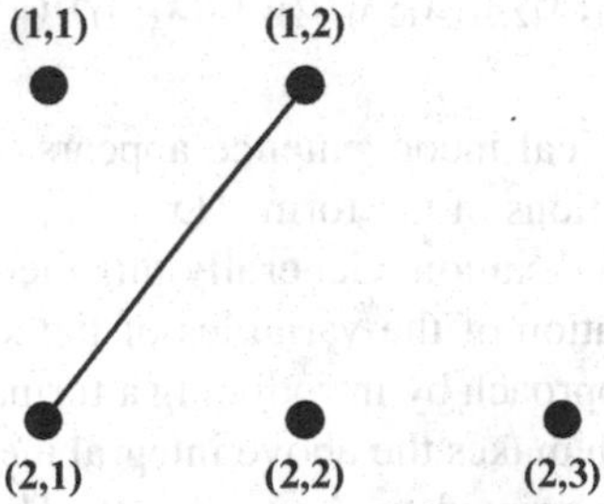

The diagram we consider

We make a relabelling of the vertices by replacing the label of the vertices from the first row from which an edge starts with the label of the vertex with which this vertex is connected. Then we make the same reindexation with the indices of the function $(f \circ g)$. In the present case the diagram we take is

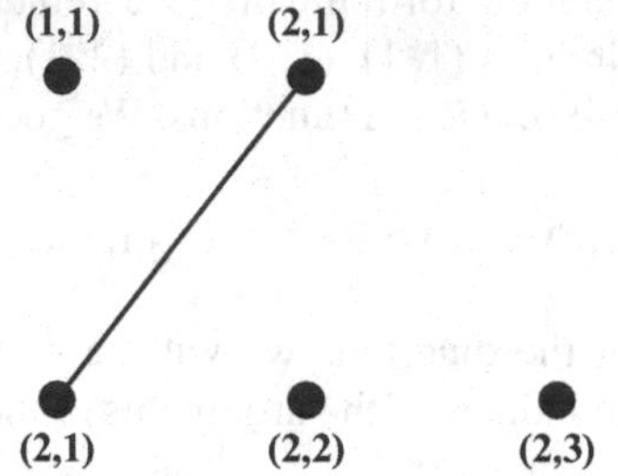

The reindexed version of our diagram

and we define the function

$$\overline{(f \circ g)}_\gamma (x_{(1,1)}, x_{(2,1)}, x_{(2,2)}, x_{(2,3)}) = f(x_{(1,1)}, x_{(2,1)}) g(x_{(2,1)}, x_{(2,2)}, x_{(2,3)}).$$

Finally we define the function $F_\gamma(f, g)$ by integrating the function $\overline{(f \circ g)}_\gamma$ with respect to those variables whose indices agree with the label of a vertex from the second row of the diagram γ from which an edge starts. (In the present case this is $x_{(2,1)}$.)

$$F_\gamma(f, g)(x_{(1,1)}, x_{(2,2)}, x_{(2,3)}) = \int \overline{(f \circ g)}_\gamma (x_{(1,1)}, x_{(2,1)}, x_{(2,2)}, x_{(2,3)}) \mu(dx_{(2,1)}).$$

We got a function of 3 variables, and the contribution of the above diagram γ to the diagram formula (10.14) is

$$3! Z_{\mu,3}(F_\gamma(f, g)) = \int F_\gamma(f, g)(x_{(1,1)}, x_{(2,2)}, x_{(2,3)}) \mu_W(dx_{(1,1)}) \mu_W(dx_{(2,2)}) \mu_W(dx_{(2,3)}).$$

In the last step some technical inconvenience appears. Originally we defined the Wiener–Itô integral of functions of the form $f(x_1, \dots, x_k)$, i.e. of functions whose variables have a different indexation. Generally this inconvenience is overcome in the literature by a reindexation of the variables of the kernel function $F_\gamma(f, g)$. I chose a slightly different approach by introducing a formally more general Wiener–Itô integral in (10.12) which makes the above integral meaningful.

Theorem 10.2A will be proved in Appendix B. The following consideration yields a heuristic explanation for it. Actually it can also be considered as a sketch of proof.

In the theory of general Itô integrals when stochastic processes are integrated with respect to a Wiener processes, one of the most basic results is Itô's formula about differentiation of functions of Itô integrals. It has a heuristic interpretation by means of the informal "identity" $(dW)^2 = dt$. In the case of general white noises this "identity" can be generalized as $(\mu_W(dx))^2 = \mu(dx)$. We present a rather informal "proof" of the diagram formula on the basis of this "identity" and the fact that the diagonals are omitted from the domain of integration in the definition of Wiener–Itô integrals.

In this "proof" we fix two numbers $k \ge 1$ and $l \ge 1$, and consider the product of two Wiener–Itô integrals of the functions f and g of order k and l. This product is a bilinear form of the functions f and g. Hence it is enough to check formula (10.14) for a sufficiently rich class of functions. It is enough to consider functions of the form $f(x_1,\dots,x_k) = I_{A_1}(x_1)\cdots I_{A_k}(x_k)$ and $g(x_1,\dots,x_l) = I_{B_1}(x_1)\cdots I_{B_l}(x_l)$ with disjoint sets $A_1,\dots,A_k$ and disjoint sets $B_1,\dots,B_l$, where $I_A(x)$ is the indicator function of a set A. (Here we have exploited that the functions f and g disappear in the diagonals.) Let us divide the sets A_j into the union of small disjoint sets $D_j^{(m)}$, $1 \le j \le k$ with some fixed number $1 \le m \le M$ in such a way that $\mu(D_j^{(m)}) \le \varepsilon$ with some fixed $\varepsilon > 0$, and the sets B_j into the union of small disjoint sets $F_j^{(m)}$, $1 \le j \le l$, with some fixed number $1 \le m \le M$, in such a way that $\mu(F_j^{(m)}) \le \varepsilon$ with some fixed $\varepsilon > 0$. Beside this, we also require that two sets $D_j^{(m)}$ and $F_{j'}^{(m')}$ should be either disjoint or they should agree. (The sets $D_j^{(m)}$ are disjoint for different indices, and the same relation holds for the sets $F_{j'}^{(m')}$.)

Then the identities

$$k!Z_{\mu,k}(f) = \prod_{j=1}^{k}\left(\sum_{m=1}^{M}\mu_W(D_j^{(m)})\right)$$

and

$$l!Z_{\mu,l}(g) = \prod_{j'=1}^{l}\left(\sum_{m'=1}^{M}\mu_W(F_{j'}^{(m')})\right),$$

hold, and the product of these two Wiener–Itô integrals can be written in the form of a sum by means of a term by term multiplication. Let us divide the terms of the sum we get in such a way into classes indexed by the diagrams $\gamma \in \Gamma(k,l)$ in the following way: Each term in this sum is a product of the form $\prod\limits_{j=1}^{k}\mu_W(D_j^{(m_j)})\prod\limits_{j'=1}^{l}\mu_W(F_{j'}^{(m'_j)})$. Let it belong to the class indexed by the diagram γ with edges $((1,j_1),(2,j_1'))$,..., and $((1,j_s),(2,j_s'))$ if the elements in the pairs $(D_{j_1}^{(m_{j_1})}, F_{j_1'}^{(m_{j_1'})})$,..., $(D_{j_s}^{(m_{j_s})}, F_{j_s'}^{(m_{j_s'})})$ agree, and otherwise all terms are different. Then letting $\varepsilon \to 0$ (and taking partitions of the sets D_j and $F_{j'}$ corresponding to the parameter ε) the sums of the terms in each class turn to integrals, and our calculation suggests the identity

$$(k!Z_{\mu,k}(f))(l!Z_{\mu,l}(g)) = \sum_{\gamma\in\Gamma(k,l)} \bar{Z}_\gamma(f,g) \tag{10.15}$$

with

$$\bar{Z}_\gamma(f,g) = \int f(x_{\alpha_\gamma((1,1))},\dots,x_{\alpha_\gamma((1,k))})g(x_{(2,1)},\dots,x_{(2,l)}) \tag{10.16}$$
$$\mu_W(dx_{\alpha_\gamma((1,1))})\dots\mu_W(dx_{\alpha_\gamma((1,k))})\mu_W(dx_{(2,1)})\dots\mu_W(dx_{(2,l)})$$

with the function $\alpha_\gamma(\cdot)$ introduced before formula (10.9). The indices $\alpha(1,j)$ of the arguments in (10.16) mean that in the case $\alpha_\gamma((1,j)) = (2,j')$ the argument $x_{(1,j)}$ has to be replaced by $x_{(2,j')}$. In particular,

$$\mu_W(dx_{\alpha_\gamma((1,j))})\mu_W(dx_{(2,j')}) = (\mu_W(dx_{(2,j')}))^2 = \mu(dx_{(2,j')})$$

in this case because of the "identity" $(\mu_W(dx))^2 = \mu(dx)$. Hence the above informal calculation yields the identity $\bar{Z}_\gamma(f,g) = |\gamma|!Z_{\mu,|\gamma|}(F_\gamma(f,g))$, and relations (10.15) and (10.16) imply formula (10.14).

A similar heuristic argument can be applied to get formulas for the product of integrals of normalized empirical distributions or (normalized) Poisson fields, only the starting "identity" $(\mu_W(dx))^2 = \mu(dx)$ changes in these cases, some additional terms appear in it, which modify the final result. I return to this question in the next chapter.

It is not difficult to generalize Theorem 10.2A with the help of some additional notations to a diagram formula about the product of finitely many Wiener–Itô integrals. We shall do this in Theorem 10.2. Then to understand this result better I present an example which shows how to calculate the terms in the sum expressing the product of three Wiener–Itô integrals as a sum of Wiener–Itô integrals.

We consider the product of the Wiener–Itô integrals $k_p!Z_{\mu,k_p}(f_p)$, $1 \le p \le m$, of $m \ge 2$ functions $f_p(x_1,\dots,x_{k_p}) \in \mathcal{H}_{\mu,k_p}$, of order $k_p \ge 1$, $1 \le p \le m$, and define a class of diagrams $\Gamma = \Gamma(k_1,\dots,k_m)$ in the following way.

The diagrams $\gamma \in \Gamma = \Gamma(k_1,\dots,k_m)$ have vertices of the form (p,r), $1 \le p \le m$, $1 \le r \le k_p$. The set of vertices $\{(p,r)\colon 1 \le r \le k_p\}$ with a fixed number p will be called the p-th row of the diagram γ. A diagram $\gamma \in \Gamma = \Gamma(k_1,\dots,k_m)$ may have some edges. All edges of a diagram connect vertices from different rows, and from each vertex there starts at most one edge. All diagrams satisfying these properties belong to $\Gamma(k_1,\dots,k_m)$. If a diagram γ contains an edge of the form $((p_1,r_1),(p_2,r_2))$ with $p_1 < p_2$, then (p_1,r_1) will be called the upper and (p_2,r_2) the lower end point of this edge. Let $E(\gamma) = \{((p_1^{(u)},r_1^{(u)}),(p_2^{(u)},r_2^{(u)})),\ p_1^{(u)} < p_2^{(u)},\ 1 \le u \le s\}$ denote the set of all edges of a diagram γ (the number of edges in γ was denoted by $s = |E(\gamma)|$), and let us also introduce the sets $V^u(\gamma) = \{((p_1^{(u)},r_1^{(u)}),\ 1 \le u \le s\}$, the set of all upper end points and $V^b(\gamma) = \{((p_2^{(u)},r_2^{(u)}),\ 1 \le u \le s\}$, the set of all lower end points of edges in a diagram γ. Let $V = V(\gamma) = \{(p,r)\colon 1 \le p \le m, 1 \le r \le k_p\}$ denote the set of all vertices of γ, and let $|\gamma| = k_1 + \dots + k_m - 2|E(\gamma)|$ denote the number of vertices in γ from which no edge starts. Vertices from which no edge starts will be called free vertices in the sequel. Let us also define the function $\alpha_\gamma((p,r))$ for a vertex (p,r)

of the diagram γ in the following way: $\alpha_\gamma((p,r)) = (\bar p, \bar r)$, if there is some pair of integers $(\bar p, \bar r)$ such that $((p,r),(\bar p,\bar r)) \in E(\gamma)$ and $p < \bar p$, i.e. $(p,r) \in V^u(\gamma)$ and $((p,r),(\bar p,\bar r)) \in E(\gamma)$, and put $\alpha_\gamma((p,r)) = (p,r)$ for $(p,r) \in V(\gamma) \setminus V^u(\gamma)$. In words, the function $\alpha_\gamma(\cdot)$ was defined on the set of vertices $V(\gamma)$ in such a way that it replaces the label of an upper end point of an edge with the label of the lower end point of this edge, and it does not change the labels of the remaining vertices of the diagram.

With the help of the above quantities the appropriate multivariate version of the functions given in (10.9)–(10.11) can be defined. Put

$$
\begin{aligned}
&(f_1 \circ f_2 \circ \cdots \circ f_m)(x_{(p,r)},\ 1 \le p \le m, 1 \le r \le k_p) \\
&\qquad = \prod_{p=1}^m f_p(x_{(p,1)}, \dots, x_{(p,k_p)}), \qquad (10.17)
\end{aligned}
$$

$$
\begin{aligned}
&\overline{(f_1 \circ f_2 \circ \cdots \circ f_m)_\gamma(x_{(p,r)},\ (p,r) \in V(\gamma) \setminus V^u(\gamma))} \\
&\qquad = \prod_{p=1}^m f_p(x_{\alpha_\gamma((p,1))}, \dots, x_{\alpha_\gamma((p,k_p))}), \qquad (10.18)
\end{aligned}
$$

and

$$
\begin{aligned}
&F_\gamma(f_1,\dots,f_m)(x_{(p,r)},\ (p,r) \in V(\gamma) \setminus (V^b(\gamma) \cup V^u(\gamma))) \qquad (10.19)\\
&\qquad = \int \overline{(f_1 \circ f_2 \circ \cdots \circ f_m)_\gamma(x_{(p,r)},\ (p,r) \in V(\gamma) \setminus V^u(\gamma))} \\
&\qquad\qquad \prod_{(p,r)\in V^b(\gamma)} \mu(dx_{(p,r)}).
\end{aligned}
$$

In words, first we replace the indices $1,\dots,k_p$ of the function $f_p(x_1,\dots,x_{k_p})$ by $(p,1),\dots,(p,k_p)$, and take the product of the functions f_p with these reindexed variables in (10.17). Then we replace those indices of the variables in this product which agree with the index of the upper end-point of an edge in γ with the index of the lower end-points of this edge in (10.18). Finally we integrate the function obtained in such a way with respect to those variables whose indices agree with the index of a lower end-point of an edge of γ in (10.19).

With the help of the above notations the diagram formula for the product of finitely many Wiener–Itô integrals can be formulated.

Theorem 10.2 (The diagram formula for the product of finitely many Wiener–Itô integrals). *Let a non-atomic, σ-finite measure μ be given on a measurable space $(X,\mathcal{X})$ together with a white noise μ_W with reference measure μ. Take $m \ge 2$ functions $f_p(x_1,\dots,x_{k_p}) \in \mathcal{H}_{\mu,k_p}$ with some order $k_p \ge 1$, $1 \le p \le m$. Let us consider the class of diagrams $\Gamma(k_1,\dots,k_m)$ introduced above together with the functions $F_\gamma(f_1,\dots,f_m)$, $\gamma \in \Gamma(k_1,\dots,k_m)$, defined by formulas (10.17)–(10.19) with its help. The L_2-norm of these functions satisfies the inequality*

$$\|F_\gamma(f_1,\dots,f_m)\|_2 \le \prod_{p=1}^m \|f_p\|_2 \quad \textit{for all } \gamma\in\Gamma(k_1,\dots,k_m). \tag{10.20}$$

Beside this, the product $\prod\limits_{p=1}^m k_p!Z_{\mu,k_p}(f_p)$ *of the Wiener–Itô integrals of the functions* f_p, $1\le p\le m$, *satisfies the identity*

$$\prod_{p=1}^m k_p!Z_{\mu,k_p}(f_p) = \sum_{\gamma\in\Gamma(k_1,\dots,k_m)} |\gamma|!Z_{\mu,|\gamma|}(F_\gamma(f_1,\dots,f_m)) \tag{10.21}$$
$$= \sum_{\gamma\in\Gamma(k_1,\dots,k_m)} |\gamma|!Z_{\mu,|\gamma|}(\operatorname{Sym} F_\gamma(f_1,\dots,f_m)).$$

To understand the notations of the above result better let us take the product of three Wiener–Itô integrals $2!Z_{\mu,2}(f_2)4!Z_{\mu,4}(f_2)3!Z_{\mu,3}(f_3)$ with kernel functions $f_1(x_1,x_2)$, $f_2(x_1,x_2,x_3,x_4)$ and $f_3(x_1,x_2,x_3)$ and see how to calculate a term in the sum of diagram formula (10.21) which expresses this product as a sum of Wiener–Itô integrals.

Let us first define the rows of the diagrams we shall working with together with their labelling. There will be three rows with labels $(1,1)$, $(1,2)$, then with $(2,1)$, $(2,2)$, $(2,3)$, $(2,4)$ and finally with $(3,1)$, $(3,2)$, $(3,3)$. We consider all possible diagrams which are graphs containing these vertices and edges connecting vertices from different rows with the restriction that from each vertex there can start at most one edge. We define with the help of all diagrams a function which will be the kernel-function of a Wiener–Itô integral appearing in the diagram formula (10.21). Let us consider for instance the diagram γ containing the edges $((1,1),(3,2))$, $((1,2),(2,2))$ and $((2,4),(3,3))$, (see picture).

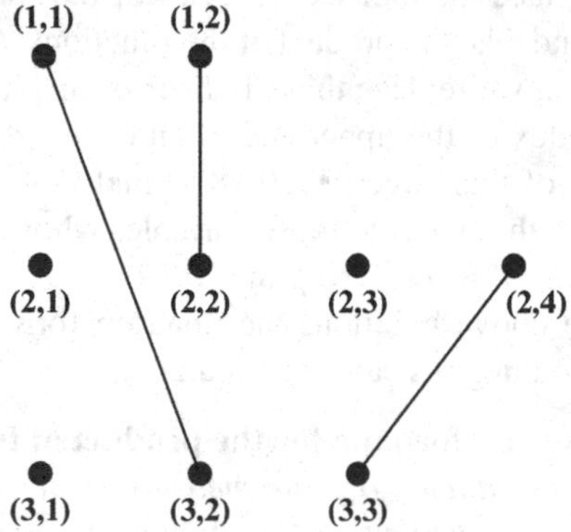

The diagram we consider

Let us relabel the vertices of the diagram γ by relabelling the upper vertices of each edge by the lower vertex of this edge.

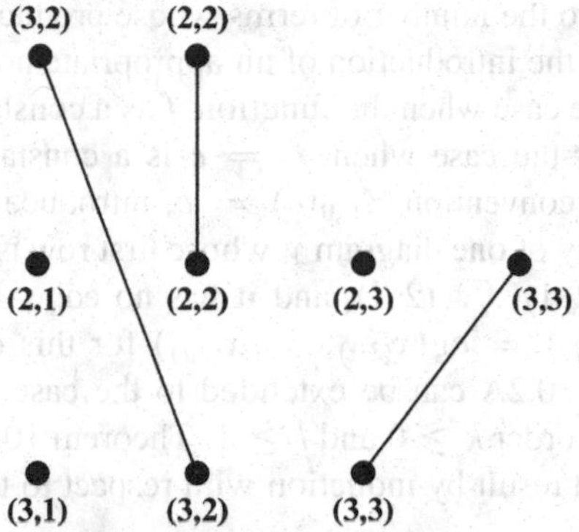

The relabelled version of our diagram

We take the product of our functions with the indexation of the variables corresponding the labels of the diagrams. Then we reindex these variables corresponding to the relabelling of our diagram γ, i.e. define first the function

$$\begin{aligned}(f_1 \circ f_2 \circ f_3)&(x_{(1,1)}, x_{(1,2)}, x_{(2,1)}, x_{(2,2)}, x_{(2,3)}, x_{(2,4)}, x_{(3,1)}, x_{(3,2)}, x_{(3,3)}) \\ &= f_1(x_{(1,1)}, x_{(1,2)}) f_2(x_{(2,1)}, x_{(2,2)}, x_{(2,3)}, x_{(2,4)}) f_3(x_{(3,1)}, x_{(3,2)}, x_{(3,3)})\end{aligned}$$

and then

$$\begin{aligned}\overline{(f_1 \circ f_2 \circ f_3)_\gamma}&(x_{(2,1)}, x_{(2,2)}, x_{(2,3)}, x_{(3,1)}, x_{(3,2)}, x_{(3,3)}) \\ &= f_1(x_{(3,2)}, x_{(2,2)}) f_2(x_{(2,1)}, x_{(2,2)}, x_{(2,3)}, x_{(3,3)}) f_3(x_{(3,1)}, x_{(3,2)}, x_{(3,3)}).\end{aligned}$$

Then we integrate the function $\overline{(f_1 \circ f_2 \circ f_3)_\gamma}$ with respect to the variables whose indices correspond to the labels of those vertices which are the lower labels of some edge. In our cases these are the indices $(2,2)$, $(3,2)$ and (3.3). This means that we define the function

$$\begin{aligned}F_\gamma&(f_1 \circ f_2 \circ f_3)(x_{(2,1)}, x_{(2,3)}, x_{(3,1)}) \\ &= \int \overline{(f_1 \circ f_2 \circ f_3)_\gamma}(x_{(2,1)}, x_{(2,2)}, x_{(2,3)}, x_{(3,1)}, x_{(3,2)}, x_{(3,3)}) \\ &\qquad \mu(dx_{(2,2)})\mu(dx_{(3,2)})\mu(dx_{3,3}).\end{aligned}$$

The function $F_\gamma(f_1, f_2, f_3)$ is a function of three variables, and the contribution of the diagram γ to the sum at the right-hand side of (10.17) equals

$3!Z_{\mu,3}(F_\gamma(f_1, f_2, f_3))$ with the above defined kernel function $F_\gamma(f_1, f_2, f_3)$. In the definition of this integral we apply again the convention described in (10.12).

Theorem 10.2 can be relatively simply derived from Theorem 10.2A by means of induction with respect to the number of terms whose product we consider. We still have to check that with the introduction of an appropriate notation Theorem 10.2A remains valid also in the case when the function f is a constant.

Let us also consider the case when $f = c$ is a constant, and $g \in \mathcal{H}_{\mu,l}$. In this case we apply the convention $Z_{\mu,0}(c) = c$, introduce the class of diagrams $\Gamma(0, l)$ that consists only of one diagram γ whose first row is empty, its second row contains the vertices $(2, 1), \dots, (2, l)$, and it has no edge. Beside this, we define $F_\gamma(c, g)(x_{(2,1)}, \dots, x_{(2,l)}) = cg(x_{(2,1)}, \dots, x_{(2,l)})$ for this diagram γ. With such a convention Theorem 10.2A can be extended to the case of the product of two Wiener–Itô integrals of order $k \geq 0$ and $l \geq 1$. Theorem 10.2 can be derived from this slightly generalized result by induction with respect to the number of terms m in the product.

I explain only briefly the proof of Theorem 10.2 which is similar to the proof of Theorem 11.2 about the product of degenerate U-statistics given in Chaps. 11 and 12, only some technical difficulties disappear in this case.

We can define, similarly to the corresponding definition in Chap. 11 where the diagram formula for the products of U-statistics will be formulated such a diagram $\gamma_{pr} \in \Gamma(k_1, \dots, k_{m-1})$ for all $\gamma \in \Gamma(k_1, \dots, k_m)$ which is actually the restriction of the diagram γ to its first $m - 1$ rows. Beside this, we can define a diagram $\gamma_{cl} \in \Gamma(|\gamma_{pr}|, k_m)$, where $|\gamma_{pr}|$ denotes the number of free vertices of γ_{pr} in the following way. This diagram consists of two rows with $|\gamma_{pr}|$ and k_m vertices respectively. It contains those edges of γ (after a reenumeration of the free vertices of γ_{pr} with the numbers $1, 2, \dots, |\gamma_{pr}|$) whose lower end points are in the m-th row of γ. It can be seen that $F_\gamma(f_1, \dots, f_m) = F_{\gamma_{cl}}(F_{\gamma_{pr}}(f_1, \dots, f_{m-1}), f_m)$, and there is such a one to one correspondence $(\bar\gamma, \hat\gamma) \leftrightarrow \gamma$ between the pairs of diagrams $(\bar\gamma, \hat\gamma)$, $\bar\gamma \in \Gamma(k_1, \dots, k_{m-1})$, $\hat\gamma \in \Gamma(|\bar\gamma|, k_m)$ and diagrams $\gamma \in \Gamma(k_1, \dots, k_m)$ for which $\bar\gamma = \gamma_{pr}$ and $\hat\gamma = \gamma_{cl}$.

To prove the diagram formula for a product of the form $\prod\limits_{p=1}^{m} k_p! Z_{\mu,k_p}(f_p)$ first we express the product $\prod\limits_{p=1}^{m-1} k_p! Z_{\mu,k_p}(f_p)$ with the help of the diagram formula by exploiting that by our inductive hypothesis it can be applied for the parameter $m-1$. In such a way we can rewrite the above product as a sum of Wiener–Itô integrals with such kernel functions which can be calculated with the help of the restrictions γ_{pr} to the first $m - 1$ rows of the diagrams $\gamma \in \Gamma(k_1, \dots, k_m)$. Then by multiplying each term of this sum by $k_m! Z_{\mu,k_m}(f_m)$, calculating these products with the help of Theorem 10.2A and summing up the expressions we get in such a way we can rewrite the product at the left-hand side of (10.21) as a sum of Wiener–Itô integrals. It can be seen with the help of the properties of the diagrams $\gamma \in \Gamma(k_1, \dots, k_m)$

mentioned in the previous paragraph that the identity we get in such a way is equivalent to formula (10.21) in Theorem 10.2. □

By statement (c) of Theorem 10.1 all Wiener–Itô integrals of order $k \geq 1$ have expectation zero. This fact together with Theorem 10.2 enable us to compute the expectation of a product of Wiener–Itô integrals. Theorem 10.2 makes possible to rewrite the product of Wiener–Itô integrals as a sum of Wiener–Itô integrals. Then its expectation can be calculated by taking the expected value of each term and summing them up. Only Wiener–Itô integrals of order zero yield a non-zero contribution to this expectation. These terms agree with the integrals of kernel functions $F_\gamma(f_1, \dots, f_m)$ corresponding to diagrams with no free vertices. In the next corollary I write down the result we got in this way.

Corollary of Theorem 10.2 about the expectation of a product of Wiener–Itô integrals. *Let a non-atomic σ-finite measure μ be given on a measurable space $(X, \mathcal{X})$ together with a white noise μ_W with reference measure μ. Take $m \geq 2$ functions $f_p(x_1, \dots, x_{k_p}) \in \mathcal{H}_{\mu,k_p}$, and consider their Wiener–Itô integrals $Z_{\mu,k_p}(f_p)$, $1 \leq p \leq m$. The expectation of the product of these random variables satisfies the identity*

$$E\left(\prod_{p=1}^m k_p! Z_{\mu,k_p}(f_p)\right) = \sum_{\gamma \in \bar{\Gamma}(k_1,\dots,k_m)} F_\gamma(f_1, \dots, f_m), \tag{10.22}$$

where $\bar{\Gamma}(k_1, \dots, k_m)$ denotes the set of those *diagrams* $\gamma \in \Gamma(k_1, \dots, k_m)$ *which have no free vertices, i.e.* $|\gamma| = 0$. *Such diagrams will be called closed diagrams in the sequel. (If $\bar{\Gamma}(k_1, \dots, k_m)$ is empty, then the sum at the right-hand side of (10.22) equals zero.) The functions $F_\gamma(f_1, \dots, f_m)$ for $\gamma \in \bar{\Gamma}(k_1, \dots, k_m)$ are constants, and they satisfy the inequality*

$$|F_\gamma(f_1, \dots, f_m)| \leq \prod_{p=1}^m \|f_p\|_2 \quad \textit{for all } \gamma \in \bar{\Gamma}(k_1, \dots, k_m). \tag{10.23}$$

Proof of the Corollary. Relation (10.22) is a straight consequence of formula (10.21), part (c) of Theorem 10.1 and the identity $Z_{\mu,0}(F_\gamma(f_1, \dots, f_m)) = F_\gamma(f_1, \dots, f_m)$, if $|\gamma| = 0$. Relation (10.23) follows from (10.20). □

The next result I formulate is Itô's formula for multiple Wiener–Itô integrals. It can also be considered as a consequence of the diagram formula. It will be proved in Appendix C.

Theorem 10.3 (Itô's formula for multiple Wiener–Itô integrals)). *Let a non-atomic, σ-finite measure μ be given on a measurable space $(X, \mathcal{X})$ together with a white noise μ_W with reference measure μ. Let us take some real valued, orthonormal functions $\varphi_1(x), \dots, \varphi_m(x)$ on the measure space $(X, \mathcal{X}, \mu)$. Let $H_k(u)$ denote the k-th Hermite polynomial with leading coefficient 1. Take the* *onefold Wiener–Itô integrals* $\eta_p = Z_{\mu,1}(\varphi_p)$, $1 \le p \le m$, *and introduce the random variables* $H_{k_p}(\eta_p)$, $1 \le p \le m$, *with some integers* $k_p \ge 1$, $1 \le p \le m$. *Put* $K_p = \sum\limits_{j=1}^{p} k_r$, $1 \le p \le m$, $K_0 = 0$. *Then* $\eta_1, \dots, \eta_m$ *are independent, standard normal random variables, and the identity*

$$\prod_{p=1}^{m} H_{k_p}(\eta_p) = K_m! Z_{\mu,K_m}\left(\prod_{p=1}^{m}\left(\prod_{j=K_{p-1}+1}^{K_p} \varphi_p(x_j)\right)\right) \tag{10.24}$$

$$= K_m! Z_{\mu,K_m}\left(\operatorname{Sym}\left(\prod_{p=1}^{m}\left(\prod_{j=K_{p-1}+1}^{K_p} \varphi_p(x_j)\right)\right)\right)$$

holds. In particular, if $\varphi(x)$ *is a real valued function such that* $\int \varphi^2(x)\mu(dx) = 1$, *then*

$$H_k\left(\int \varphi(x)\mu_W(dx)\right) = \int \varphi(x_1)\cdots\varphi(x_k)\mu_W(dx_1)\dots\mu_W(dx_k). \tag{10.25}$$

I also formulate a limit theorem about the distribution of normalized degenerate U-statistics that will be proved in Appendix C. The limit distribution in this result is given by means of multiple Wiener–Itô integrals.

Theorem 10.4 (Limit theorem about normalized degenerate U-statistics). *Let us consider a sequence of degenerate* U*-statistics* $I_{n,k}(f)$ *of order* k, $n = k, k+1, \dots$, *defined in (8.8) with the help of a sequence of independent and identically distributed random variables* $\xi_1, \xi_2, \dots$ *taking values in a measurable space* $(X, \mathcal{X})$ *with a non-atomic distribution* μ *and a kernel function* $f(x_1, \dots, x_k)$, *canonical with respect to the measure* μ, *defined on the* k*-fold product* $(X^k, \mathcal{X}^k)$ *of the space* $(X, \mathcal{X})$ *for which* $\int f^2(x_1, \dots, x_k)\mu(dx_1)\dots\mu(dx_k) < \infty$. *Then the sequence of normalized* U*-statistics* $n^{-k/2} I_{n,k}(f)$ *converges in distribution, as* $n \to \infty$, *to the* k*-fold Wiener–Itô integral*

$$Z_{\mu,k}(f) = \frac{1}{k!}\int f(x_1, \dots, x_k)\mu_W(dx_1)\dots\mu_W(dx_k)$$

with kernel function $f(x_1, \dots, x_k)$ *and a white noise* μ_W *with reference measure* μ.

Remark. The limit behaviour of degenerate U-statistics $I_{n,k}(f)$ with an atomic measure μ which satisfy the remaining conditions of Theorem 10.4 can be described in the following way. Take the probability space $(U, \mathcal{U}, \lambda)$, where $U = [0,1]$, $\mathcal{U}$ is the Borel σ-algebra and λ is the Lebesgue measure on it. Introduce a sequence of independent random variables $\eta_1, \eta_2, \dots$ with uniform distribution on the interval $[0,1]$, which is independent also of the sequence $\xi_1, \xi_2, \dots$. Define the product space $(\tilde{X}, \tilde{\mathcal{X}}, \tilde{\mu}) = (X \times U, \mathcal{X} \times \mathcal{U}, \mu \times \lambda)$ together with the function $\tilde{f}(\tilde{x}_1, \dots, \tilde{x}_k) =$

$\tilde{f}((x_1,u_1),\dots,(x_k,u_k)) = f(x_1,\dots,x_k)$ with the notation $\tilde{x} = (x,u) \in X \times U$, and $\tilde{\xi}_j = (\xi_j,\eta_j)$, $j = 1,2,\dots$. Then $I_{n,k}(f) = I_{n,k}(\tilde{f})$ (with the above defined function $\tilde{f}$ and $\tilde{\mu}$ distributed random variables $\tilde{\xi}_j$). Beside this, Theorem 10.4 can be applied for the degenerate U-statistics $I_{n,k}(\tilde{f})$, $n = 1,2,\dots$.

In the next result I give an interesting representation of the Hilbert space consisting of the square integrable functions measurable with respect to a white noise μ_W. An isomorphism will be given with the help of Wiener–Itô integrals between this Hilbert space and the so-called Fock space to be defined below. To formulate this result first some notations will be introduced.

Let $\mathscr{H}^0_{\mu,k} \subset \mathscr{H}_{\mu,k}$ denote the class of symmetric functions in the space $\mathscr{H}_{\mu,k}$, $k = 0,1,2,\dots$, i.e. $f \in \mathscr{H}_{\mu,k}$ is in its subspace $\mathscr{H}^0_{\mu,k}$ if and only if $f(x_1,\dots,x_k) = \operatorname{Sym} f(x_1,\dots,x_k)$. Let us introduce for all $k = 0,1,2,\dots$ the Hilbert space $\mathscr{G}_k$ consisting of those random variables η (on the probability space where the white noise μ_W is defined) which can be written in the form

$$\eta = Z_{\mu,k}(f) = \frac{1}{k!}\int f(x_1,\dots,x_k)\mu_W(dx_1)\dots\mu_W(dx_k)$$

with some $f \in \mathscr{H}^0_{k,\mu}$.

It follows from parts (a) and (c) of Theorem 10.1 that the map $f \to Z_{\mu,k}(f)$ is a linear transformation of $\mathscr{H}^0_{\mu,k}$ to $\mathscr{G}_k$, and $\frac{1}{k!}\|f\|_2^2 = EZ^2_{\mu,k}(f)$ for all $f \in \mathscr{H}^0_{\mu,k}$, where $\|f\|_2$ denotes the usual L_2-norm of the function f with respect to the k-fold power of the measure μ. By the definition of Wiener–Itô integrals the set $\mathscr{G}_1$ consists of jointly Gaussian random variables with expectation zero. The spaces $\mathscr{H}_{\mu,0}$ and $\mathscr{G}_0$ consist of the real constants. Let us define the space $\operatorname{Exp}(\mathscr{H}_\mu)$ of infinite sequences $f = (f_0, f_1,\dots)$, $f_k \in \mathscr{H}^0_{\mu,k}$, $k = 0,1,2,\dots$, such that $\|f\|_2^2 = \sum\limits_{k=0}^{\infty} \frac{1}{k!}\|f_k\|_2^2 < \infty$. The space $\operatorname{Exp}(\mathscr{H}_\mu)$ with the natural addition and multiplication by a constant and the above introduced norm $\|f\|_2$ for $f \in \operatorname{Exp}(\mathscr{H}_\mu)$ is a Hilbert space which is called the Fock space in the literature.

Let $\mathscr{G}$ denote the class of random variables of the form

$$Z(f) = \sum_{k=0}^{\infty} Z_{\mu,k}(f_k), \quad f = (f_0, f_1, f_2,\dots) \in \operatorname{Exp}(\mathscr{H}_\mu).$$

The next result describes the structure of the space of random variables $\mathscr{G}$. It is useful for a better understanding of Wiener–Itô integrals, but it will be not used in the sequel. In its proof I shall refer to some basic measure theoretical results.

Theorem 10.5 (Isomorphism of the space of square integrable random variables measurable with respect to a white noise with a Fock space). *Let a non-atomic, σ-finite measure μ be given on a measurable space $(X, \mathscr{X})$ together with a white noise μ_W with reference measure μ. Let us consider the class of* *functions* $\mathcal{H}^0_{\mu,k}$, $k = 0,1,2,\ldots$, *and* $\mathrm{Exp}\,(\mathcal{H}_\mu)$ *together with the spaces of random variables* $\mathcal{G}_k$, $k = 0,1,2,\ldots$, *and* $\mathcal{G}$ *defined above. The transformation* $Z\colon Z(f) = \sum_{k=0}^{\infty} Z_{\mu,k}(f_k)$, $f = (f_0, f_1, f_2, \ldots) \in \mathrm{Exp}\,(\mathcal{H}_\mu)$, *is a unitary transformation from the Hilbert space* $\mathrm{Exp}\,(\mathcal{H}_\mu)$ *to* $\mathcal{G}$. *The Hilbert space* $\mathcal{G}$ *consists of all random variables with finite second moment, measurable with respect to the* σ*-algebra generated by the random variables* $\mu_W(A)$, $A \in \mathcal{X}$, $\mu(A) < \infty$. *This* σ*-algebra agrees with the* σ*-algebra generated by the random variables* $Z_{\mu,1}(f_1)$, $f_1 \in \mathcal{H}^0_{\mu,1}$.

Remark. For the sake of simpler notations we restrict our attention to the case when the measure space $(X, \mathcal{X}, \mu)$ is such that the Hilbert space of square integrable functions on this space is separable. This condition is satisfied in all interesting cases.

Proof of Theorem 10.5. Properties (a) and (c) in Theorem 10.1 imply that the transformation $f_k \to Z_{\mu,k}(f_k)$ is a linear transformation of $\mathcal{H}^0_{\mu,k}$ to $\mathcal{G}_k$, and $\frac{1}{k!}\|f_k\|_2^2 = EZ_{\mu,k}(f)^2$. Beside this, $EZ_{\mu,k}(f)Z_{\mu,k'}(f'_{k'}) = 0$ if $f_k \in \mathcal{H}^0_{\mu,k}$, and $f'_{k'} \in \mathcal{H}^0_{\mu,k'}$ with $k \neq k'$ by properties (d) and (c). (The latter property is needed to guarantee this relation also holds if $k = 0$ or $k' = 0$.) It follows from these relations that the map $Z\colon Z(f) = \sum_{k=0}^{\infty} Z_{\mu,k}(f_k)$, $f = (f_0, f_1, f_2, \ldots) \in \mathrm{Exp}\,(\mathcal{H}_\mu)$ is an isomorphism between the Hilbert spaces $\mathrm{Exp}\,(\mathcal{H}_\mu)$ and $\mathcal{G}$.

It remained to show that $\mathcal{G}$ contains all random variables with finite second moment, measurable with respect to the corresponding σ-algebra. Let $g_j(u)$, $j = 1,2,\ldots$, be an orthonormal basis in $\mathcal{H}^0_{\mu,1} = \mathcal{H}_{\mu,1}$, and introduce the random variables $\eta_j = Z_{\mu,1}(g_j)$, $j = 1,2,\ldots$. These random variables are independent with standard normal distribution, and by Itô's formula for Wiener–Itô integrals (Theorem 10.3) all products $H_{r_1}(\eta_{j_1})\ldots H_{r_p}(\eta_{j_p})$ with $r_1 + \cdots + r_p = k$ are in the space $\mathcal{G}_k$, where $H_r(\cdot)$ denotes the Hermite polynomial of order r with leading coefficient 1. We also recall the following results from the classical analysis:

(a) Hermite polynomials constitute a complete orthonormal system in the L_2-space on the real line with respect to the standard normal distribution. (This result will be proved in Appendix C in Proposition C2.)
(b) If a random variable ζ is measurable with respect to the σ-algebra generated by some random variables $\eta_1, \eta_2, \ldots$, then there exists a Borel measurable function $f(x_1, x_2, \ldots)$ on the infinite product of the real line $(R^\infty, \mathcal{B}^\infty)$ in such a way that $\zeta = f(\eta_1, \eta_2, \ldots)$.

This means in our case that any random variable ζ measurable with respect to the σ-algebra generated by the random variables $\eta_j = Z_{\mu,1}(g_j)$, $j = 1,2,\ldots$, can be written in the form $\zeta = f(\eta_1, \eta_2, \ldots)$ with the above introduced independent, standard normal random variables $\eta_1, \eta_2, \ldots$. If ζ has finite second moment, then the function f appearing in its representation is a function of finite L_2-norm in the infinite product of the real line with the infinite product of the standard normal distribution on it. Hence some classical results in analysis enable us to expand the function f with respect to products of Hermite polynomials, and this also yields the identity

$$\zeta = \sum c(j_1, r_1, \dots, j_s, r_s) H_{r_1}(\eta_{j_1}) \cdots H_{r_s}(\eta_{j_s})$$

with some coefficients $c(j_1, r_1, \dots, j_s, r_s)$ such that

$$\sum c^2(j_1, r_1, \dots, j_s, r_s) \|H_{r_1}(u)\|^2 \cdots \|H_{r_s}(u)\|^2 < \infty.$$

(Actually it is known that $\|H_k(u)\|^2 = k!$, but here we do not apply this fact.)

The above relations yield the desired representation of a random variable ζ with finite second moment, if it is measurable with respect to the σ-algebra generated by the random variables in $\mathcal{G}_1$. Indeed, the identity $\zeta = \sum_{k=0}^{\infty} \zeta_k$ holds with

$$\zeta_k = \sum_{r_1+\dots+r_s=k} c(j_1, r_1, \dots, j_s, r_s) H_{r_1}(\eta_{j_1}) \cdots H_{r_s}(\eta_{j_s}),$$

and $\zeta_k \in \mathcal{G}_k$ by Itô's formula.

To complete the proof it is enough to remark that the σ-algebra generated by the random variables $\eta_1, \eta_2, \dots$ and $\mu_W(A)$, $A \in \mathcal{X}$, $\mu(A) < \infty$ agree, as it was stated in part (f) of Theorem 10.1. □

The results about Wiener–Itô integrals discussed in this Chapter are useful in the study of non-linear functionals of a white noise. In my Lecture Note [32] similar problems were discussed, but in that work a slightly different version of Wiener–Itô integrals was introduced. The reason for this modification was that the solution of the problems studied in [32] demanded different methods.

In work [32] stationary Gaussian random fields were considered, and I was mainly interested in it in limit theorems for sequences of non-linear functionals on a stationary Gaussian random field. In a stationary Gaussian random field a shift operator can be introduced. This shift operator can be extended in a natural way to all random variables measurable with respect to the underlying stationary Gaussian random field. In [32] we needed a technique which helps in working with this shift operator. In an analogous case, when functions on the real line are considered, the Fourier analysis is a useful tool in the study of the shift operator. In the work [32] we tried to unify the tools of multiple Wiener–Itô integrals and Fourier analysis. This led to the definition of a slightly different version of Wiener–Itô integrals.

In the work [32] we have shown that not only the correlation function of a stationary Gaussian field can be given by means of the Fourier transform of its spectral measure, but also a random spectral measure can be constructed whose Fourier transform expresses the stationary Gaussian process itself. After the introduction of this random spectral measure a version of the multiple Wiener–Itô integral can be defined with respect to it, and all square integrable random variables, measurable with respect to the σ-algebra generated by the underlying Gaussian stationary random field can be expressed as the sum of such integrals. Moreover, such an approach enables us to apply the methods of multiple Wiener–Itô integrals and Fourier analysis simultaneously. The modified Wiener–Itô integral introduced in [32] behaves similarly to the original Wiener–Itô integral, only it has to be taken into account that the random spectral measure behaves not like a white noise, but as its "Fourier transform". I omit the details. They can be found in [32].

The spaces $\mathcal{G}_k$ consisting of all k-fold Wiener–Itô integrals were introduced also in [32], and this was done for a special reason. In that work the Hilbert space of square integrable functions, measurable with respect to an underlying stationary Gaussian field was studied together with the shift operator acting on this Gaussian field. The shift operator could be extended to a unitary operator on this Hilbert space. The introduction of the subspaces $\mathcal{G}_k$ turned out to be useful, because they supplied such a decomposition of this Hilbert space which consists of orthogonal subspaces invariant with respect to the shift operator.

In the present work no shift operator was defined, and limit theorems for non-linear functionals of a Gaussian field were not studied here. The introduction of the spaces $\mathcal{G}_k$ was useful because of a different reason. In the study of our problems we need good estimates on the $2p$-th moment of random variables, measurable with respect to the underlying white noise for large integers p. As it will be shown, the high moments of the random variables in the spaces $\mathcal{G}_k$ with different indices k show an essentially different behaviour. The high moments of a random variable in $\mathcal{G}_k$ behave similarly to those of the k-th power ξ^k of a Gaussian random variable ξ with zero expectation. This statement will be formulated in a more explicit form in Proposition 13.1 or in its consequence, formula (13.2). A partial converse of this result will be presented in Theorem 13.6.

Chapter 11
The Diagram Formula for Products of Degenerate U-Statistics

There is a natural analogue of the diagram formula for the products of Wiener–Itô integrals both for the products of multiple integrals with respect to normalized empirical measures and for the products of degenerate U-statistics. These two results are closely related. They express the product of multiple random integrals or degenerate U-statistics as a sum of multiple random integrals or degenerate U-statistics respectively. The kernel functions of the random integrals or U-statistics appearing in this sum are defined—similarly to the case of Wiener–Itô integrals—by means of diagrams. This is the reason why these results are also called the diagram formula. The main difference between these diagram formulas and their version for Wiener–Itô integrals is that in the present case we have to work with a more general class of diagrams. The diagram formula for multiple integrals with respect to a normalized empirical measure will be discussed only at an informal level, while a complete proof of the analogous result about degenerate U-statistics will be given. The reason for such an approach is that the diagram formula for the product of degenerate U-statistics can be better applied in this work.

We want to prove the estimates about the tail distribution of degenerate U-statistics and multiple integrals with respect to a normalized empirical distribution formulated in Theorems 8.3 and 8.1 with the help of good bounds on the high moments of degenerate U-statistics and multiple random integrals. In the case of degenerate U-statistics the diagram formula yields an explicit formula for these moments. We exploit that this formula expresses the product of degenerate U-statistics as a sum of degenerate U-statistics of different order. Beside this, the expected value of all degenerate U-statistics of order $k \geq 1$ equals zero. Hence the expected value we are interested in equals the sum of the zero order terms appearing in the diagram formula.

The analogous problem about the moments of multiple integrals with respect to a normalized empirical measure is more difficult. The diagram formula enables us to express the moments of multiple random integrals as the sum of the expectation of such integrals of different order also in this case. But the expected value of random

P. Major, *On the Estimation of Multiple Random Integrals and U-Statistics*, Lecture Notes in Mathematics 2079, DOI 10.1007/978-3-642-37617-7_11,

integrals of order $k \geq 1$ with respect to a normalized empirical distribution may be non-zero. Before the proof of Theorem 9.4 we showed this in an example.

First I give an informal description of the diagram formula for the product of two random integrals with respect to a normalized empirical measure. Its analogue, the diagram formula for the product of two Wiener–Itô integrals can be described in an informal way by means of formulas (10.15) and (10.16) together with the 'identity' $(\mu_W(dx))^2 = \mu(dx)$ in their interpretation. The diagram formula for the product of two multiple integrals with respect to a normalized empirical measure has a similar representation. (Observe that in the definition of the random integral $J_{n,k}(\cdot)$ given in formula (4.8) the diagonals are omitted from the domain of integration, similarly to the case of Wiener–Itô integrals.) In this case such a version of formulas (10.15) and (10.16) can be applied, where the random integrals $Z_{\mu,k}$ are replaced by $J_{n,k}$, and the white noise measure μ_W is replaced by the normalized empirical measure $\nu_n = \sqrt{n}(\mu_n - \mu)$. But the analogue of the "identity" $(\mu_W(dx))^2 = \mu(dx)$ needed in the interpretation of these formulas has a different form. It states that $(\nu_n(dx))^2 = \mu(dx) + \frac{1}{\sqrt{n}}\nu_n(dx)$. Let us "prove" this new "identity".

Take a small set Δ, i.e. a set Δ such that $\mu(\Delta)$ is very small, write down the identity $(\nu_n(\Delta))^2 = n(\mu_n(\Delta))^2 + n(\mu(\Delta))^2 - 2n\mu_n(\Delta)\mu(\Delta)$ and observe that only a second order error is committed if the terms $n(\mu(\Delta))^2$ and $2n\mu_n(\Delta)\mu(\Delta)$ are omitted at the right-hand side of this identity. Moreover, also a second order error is committed if $n(\mu_n(\Delta))^2$ is replaced by $\mu_n(\Delta)$, because it has second order small probability that there are at least two sample points in the small set Δ. On the other hand, $n(\mu_n(\Delta))^2 = \mu_n(\Delta)$ if Δ contains only zero or one sample point. The above considerations suggest that $(\nu_n(dx))^2 = \mu_n(dx) = \mu(dx) + \frac{1}{\sqrt{n}}[\sqrt{n}(\mu_n(dx) - \mu(dx))] = \mu(dx) + \frac{1}{\sqrt{n}}\nu_n(dx)$. (This means that in the "identity" expressing the square $(\nu_n(dx))^2$ of a normalized empirical measure a correcting term $\frac{1}{\sqrt{n}}\nu_n(dx)$ appears. If the sample size $n \to \infty$, then the normalized empirical measure tends to a white noise with counting measure μ, and this correcting term disappears.)

The diagram formula for the product of two multiple integrals with respect to a normalized empirical measure was proved in paper [35] with a different notation. Informally speaking, the result in this work states that the identity suggested by the above heuristic argument really holds. We remark that if the form of this identity is found, then it can be proved with the help of some algebraic calculations similarly to the proof of Lemma 9.5. We omit the proof of this result, since we shall not work with it. We shall prove instead a version of it about the product of degenerate U-statistics that we can better apply. This result is similar to the diagram formula for the products of multiple integrals with respect to a normalized empirical distribution. This similarity will be discussed in *Remark* 4 after Theorem 11.1.

In this chapter first I formulate the diagram formula about the product of two degenerate U-statistics in Theorem 11.1, then its generalization about the product of finitely many degenerate U-statistics in Theorem 11.2. Their proofs is postponed to the next chapter. I also present a Corollary of Theorem 11.2 about the expected value of the product of degenerate U-statistics which follows from this result and the observation that the expected value of a U-statistic of order $k \geq 1$ equals zero. This result together with Lemma 11.3 which yields a bound on the L_2-norm of the kernel functions of the degenerate U-statistics appearing in the diagram formula will enable us to prove good estimates on the high moments of degenerate U-statistics. We can prove Theorem 8.3 about the tail distribution of degenerate U-statistics with the help of such estimates. One might try to prove the analogous result, Theorem 8.1 about the tail distribution of multiple integrals with respect to a normalized empirical distribution in a similar way with the help of the diagram formula for multiple random integrals. But that would be much harder, since the diagram formula for multiple integrals with respect to a normalized empirical distribution does not supply such a good formula for the moments of random integrals as the analogous result about degenerate U-statistics.

To describe the results of this chapter we introduce some new notions. In the formulation of the diagram formula for the product of degenerate U-statistics a more general class of diagrams has to be considered than in the case of multiple Wiener–Itô integrals. I shall define them under the name coloured diagrams. The kernel functions of the U-statistics appearing in the diagram formula will be defined with their help. First I introduce the notations needed in the formulation of the diagram formula for the product of two degenerate U-statistics, then I present this result in Theorem 11.1. After this, to understand the notations better I explain with the help of an example how to calculate a general term in this diagram formula.

A class of coloured diagrams $\Gamma(k_1,\dots,k_m)$ will be defined whose vertices will be the pairs (p,r), $1 \le p \le m$, $1 \le r \le k_p$, and the set of vertices (p,r), $1 \le r \le k_p$, with a fixed number p will be called the p-th row of the diagram. To define the coloured diagrams of the class $\Gamma(k_1,\dots,k_m)$ first the notions of chains and coloured chains will be introduced. A sequence $\beta = \{(p_1,r_1),\dots,(p_s,r_s)\}$ with $1 \le p_1 < p_2 < \dots < p_s \le m$ and $1 \le r_u \le k_{p_u}$ for all $1 \le u \le s$ will be called a chain. The number s of vertices (p_u,r_u) in this sequence, denoted by $\ell(\beta)$, will be called the length of the chain β. Chains of length $\ell(\beta)=1$, i.e. chains consisting only of one vertex (p_1,r_1) are also allowed. We shall define a function $c(\beta)=\pm 1$ which will be called the colour of the chain β, and the pair $(\beta,c(\beta))$ will be called a coloured chain. We shall allow arbitrary colouring $c(\beta)=\pm 1$ of a chain with the only restriction that a chain of length 1 can only get the colour -1, i.e. $c(\beta)=-1$ if $\ell(\beta)=1$.

A coloured diagram $\gamma \in \Gamma(k_1,\dots,k_m)$ is a partition of the set of vertices $A(k_1,\dots,k_m)=\{(p,r)\colon 1 \le p \le m,\ 1 \le r \le k_p\}$ to the union of some coloured chains $\beta \in \gamma$, i.e. $\bigcup\limits_{\beta\in\gamma} \beta = A(k_1,\dots,k_m)$, and each vertex $(p,r) \in A(k_1,\dots,k_m)$ is the element of exactly one chain $\beta \in \gamma$. Beside this, each chain $\beta \in \gamma$ has a colour $c_\gamma(\beta)=\pm 1$. The set $\Gamma(k_1,\dots,k_m)$ consists of all coloured diagrams γ with the above properties with the only restriction that for a chain $\beta=\{(p,r)\} \in \gamma$ of length $\ell(\beta)=1$ of a diagram $\gamma \in \Gamma(k_1,\dots,k_m)$ we have $c_\gamma(\beta)=-1$.

Let us define for all coloured diagrams $\gamma \in \Gamma(k_1,\dots,k_m)$ the set of open chains $O(\gamma)=\{\beta\colon \beta \in \gamma,\ c_\gamma(\beta)=-1\}$ and the set of closed chains $C(\gamma)=\{\beta\colon \beta \in \gamma,\ c_\gamma(\beta)=1\}$ of this diagram γ. We shall define for all sets of bounded functions $f_p=f_p(x_1,\dots,x_{k_p}) \in L_2(X^{k_p},\mathcal{X}^{k_p},\mu^{k_p})$, $1 \le p \le m$, and diagrams $\gamma \in \Gamma(k_1,\dots,k_m)$ a bounded function $F_\gamma(f_1,\dots,f_m) = F_\gamma(f_1,\dots,f_m)(x_1,\dots,x_{|O(\gamma)|})$ with $|O(\gamma)|$ variables on the product space $(X^{|O(\gamma)|},\mathscr{X}^{|O(\gamma)|},\mu^{|O(\gamma)|})$, where $|O(\gamma)|$ denotes the number of open chains in the diagram γ. The arguments of the function $F_\gamma(f_1,\dots,f_m)$ will correspond to the open chains of the diagram γ. We will see that the function $F_\gamma(f_1,\dots,f_m)$ is canonical (with respect to the measure μ) if the same relation holds for the functions $f_1,\dots,f_m$. In the diagram formula we shall express the product of normalized degenerate U-statistics $\prod\limits_{p=1}^{m} n^{-k_p/2}k_p!I_{n,k_p}(f_p)$ as a linear combination of the normalized degenerate U-statistics

$$n^{-|O(\gamma)|/2}|O(\gamma)|!I_{n,|O(\gamma)|}(F_\gamma(f_1,\dots,f_m)).$$

To define the above mentioned functions $F_\gamma(f_1,\dots,f_m)$ first we fix for all pairs of positive integers $k_1,k_2 = 1,2,\dots$ and diagrams $\gamma \in \Gamma(k_1,k_2)$ an enumeration of the chains of γ, and beside this we also fix an enumeration of the open chains of all diagrams $\gamma \in \Gamma(k_1,\dots,k_m)$, $m=2,3,\dots$. (For $m \ge 3$ we shall need an enumeration only for the open chains.) For the sake of simpler notation we choose such an enumeration of the chains of a diagram γ for $m=2$ where the chains get the labels $1,2,\dots,|O(\gamma)|+|C(\gamma)|$, and the open chains get the first $|O(\gamma)|$ labels, i.e. $\beta(l)$ is an open chain if $1 \le l \le |O(\gamma)|$, and it is a closed chain if $|O(\gamma)|+1 \le l \le |O(\gamma)|+|C(\gamma)|$. In the case $m \ge 3$ we give an enumeration only of the open chains of a diagram γ, and they will be indexed by the numbers $1 \le l \le |O(\gamma)|$. This means that $\beta(l)$ will be defined for $1 \le l \le |O(\gamma)|$, and it is an open chain of γ.

We shall fix an enumeration of the chains of the diagrams with two rows and of the open chains of the diagrams with at least three rows at the start, and during the application of the diagram formula we shall always apply this enumeration of the chains. The subsequent definition of the functions $F_\gamma(f_1,\dots,f_m)$ will depend on this enumeration, but the results formulated with the help of these functions are valid for an arbitrary (previously fixed) enumeration of the chains. Hence the non-uniqueness in the definition of the functions $F_\gamma(f_1,\dots,f_m)$ will cause no problem.

First we formulate the diagram formula for the product of two degenerate U-statistics, i.e. we consider the case $m=2$. Let us have a measurable space $(X,\mathscr{X})$ with a probability measure μ on it together with two measurable functions $f_1(x_1,\dots,x_{k_1})$ and $f_2(x_1,\dots,x_{k_2})$ of k_1 and k_2 variables on this space which are canonical with respect to the measure μ. Let $\xi_1,\xi_2,\dots$ be a sequence of $(X,\mathscr{X})$ valued, independent and identically distributed random variables with distribution μ. We want to express the product $n^{-k_1/2}k_1!I_{n,k_1}(f_1)n^{-k_2/2}k_2!I_{n,k_2}(f_2)$ of normalized degenerate U-statistics defined with the help of the above random variables and kernel functions f_1 and f_2 as a sum of normalized degenerate U-statistics. For this goal we define some functions $F_\gamma(f_1,f_2)$ for all $\gamma \in \Gamma(k_1,k_2)$.

We shall define the function $F_\gamma(f_1,f_2)$ with the help of the previously fixed enumeration of the chains of the diagram γ. We shall introduce with the help of this enumeration also an enumeration of the vertices $(1,p)$, $(2,q)$, $1\le p\le k_1$, $1\le q\le k_2$, of the diagram γ. We put

$$\alpha_\gamma((p,r))=l \quad \text{if } (p,r)\in\beta(l), \quad p=1,2, \quad 1\le r\le k_p. \tag{11.1}$$

Let us have two functions $f_1(x_1,\dots,x_{k_1})$ and $f_2(x_1,\dots,x_{k_2})$ together with a coloured diagram $\gamma\in\Gamma(k_1,k_2)$. We define the function $F_\gamma(f_1,f_2)$ in two steps. First we define the function

$$\begin{aligned}&(f_1\circ f_2)_\gamma(x_1,\dots,x_{s(\gamma)})\\&\qquad=f_1(x_{\alpha_\gamma((1,1))},\dots,x_{\alpha_\gamma((1,k_1))})f_2(x_{\alpha_\gamma((2,1))},\dots,x_{\alpha_\gamma((2,k_2))}),\end{aligned} \tag{11.2}$$

where $s(\gamma)=|O(\gamma)|+|C(\gamma)|$ is the number of chains in γ, and the indices $\alpha_\gamma(1,j)$ and $\alpha_\gamma(2,j')$ were defined in (11.1). (In formula (11.2) the arguments of both functions f_1 and f_2 have different indices. But two indices $\alpha_\gamma((1,j))$ and $\alpha_\gamma((2,j'))$ may agree in some cases. This happens if the vertices $(1,j)$ and $(2,j')$ belong to the same chain $\beta\in\gamma$ of length 2.) In the second step we define the function

$$\begin{aligned}&F_\gamma(f_1,f_2)(x_1,\dots,x_{|O(\gamma)|})\\&\quad=\left(\prod_{j:\beta(j)\in C(\gamma)}P_j\prod_{j':\beta(j')\in O_2(\gamma)}Q_{j'}\right)(f_1\circ f_2)_\gamma(x_1,\dots,x_{|O(\gamma)|+|C(\gamma)|})\end{aligned} \tag{11.3}$$

with the operators P_j and $Q_{j'}$ defined in formulas (9.1) and (9.2), where $C(\gamma)$ is the set of closed chains of the diagram γ, and $O_2(\gamma)\subset O(\gamma)$ is the set of open chains of γ with length 2, i.e. $O_2(\gamma)=\{\beta\colon c_\gamma(\beta)=-1, \text{ and } \ell(\beta)=2\}$. Let us also remark that the operators P_j and $Q_{j'}$ in formula (11.3) are exchangeable, hence it is not important in what order we apply them.

Let me remark that if we applied a different enumeration of the diagrams $\gamma\in\Gamma(k_1,k_2)$ then we would get a different function $F_\gamma(f_1,f_2)$. This would be a reindexed version of the original function $F_\gamma(f_1,f_2)$. But the value of the U-statistic $I_{n,|O(n)|}(F_\gamma(f_1,f_2))$ does not depend on the indexation of the variables in its kernel function. Hence the identity which will be formulated in formula (11.4) of the subsequent Theorem 11.1 does not depend on the enumeration of the chains of the diagrams $\gamma\in\Gamma(k_1,k_2)$. Now we can formulate the following result.

Theorem 11.1 (The diagram formula for the product of two degenerate U-statistics). *Let a sequence of independent and identically distributed random variables $\xi_1,\xi_2,\dots$ be given with some distribution μ on a measurable space $(X,\mathcal{X})$ together with two bounded, canonical functions $f_1(x_1,\dots,x_{k_1})$ and $f_2(x_1,\dots,x_{k_2})$ with respect to the probability measure μ on the product spaces $(X^{k_1},\mathcal{X}^{k_1})$ and $(X^{k_2},\mathcal{X}^{k_2})$ respectively. Let us take the class of coloured diagrams $\Gamma(k_1,k_2)$ introduced above together with the functions $F_\gamma(f_1,f_2)$ defined in formulas (11.1)–(11.3).*

The functions $F_\gamma(f_1, f_2)$ are bounded and canonical with respect to the measure μ with $|O(\gamma)|$ arguments for all coloured diagrams $\gamma \in \Gamma$, where $O(\gamma)$ and $C(\gamma)$ denote the set of open and closed chains of the diagram γ. The product of the normalized degenerate U-statistics $n^{-k_1/2}k_1!I_{n,k_1}(f_1)$ and $n^{-k_2/2}k_2!I_{n,k_2}(f_2)$, $n \ge \max(k_1, k_2)$, defined in (8.8) can be expressed as

$$n^{-k_1/2}k_1!I_{n,k_1}(f_1) \cdot n^{-k_2/2}k_2!I_{n,k_2}(f_2) = \sum_{\gamma\in\Gamma(k_1,k_2)}{}^{\prime(n)} \prod_{j=1}^{|C(\gamma)|} \left(\frac{n-s(\gamma)+j}{n}\right) n^{-W(\gamma)/2} \cdot n^{-|O(\gamma)|/2}|O(\gamma)|!I_{n,|O(\gamma)|}(F_\gamma(f_1, f_2)) \tag{11.4}$$

with $W(\gamma) = k_1+k_2-|O(\gamma)|-2|C(\gamma)|$ (we explain in Remark 1 after Theorem 11.1 that $W(\gamma) = |O_2(\gamma)|$, i.e. it equals the number of open chains with length 2) and $s(\gamma) = |O(\gamma)| + |C(\gamma)|$ (which equals the number of coloured chains in γ). Here $\sum^{\prime(n)}$ means that summation is taken only for such coloured diagrams $\gamma \in \Gamma(k_1, k_2)$ which satisfy the inequality $s(\gamma) \le n$, and $\prod_{j=1}^{|C(\gamma)|}$ equals 1 in the case $|C(\gamma)| = 0$. The term $I_{n,|O(\gamma)|}(F_\gamma(f_1, f_2))$ can be replaced by $I_{n,|O(\gamma)|}(\mathrm{Sym}F_\gamma(f_1, f_2))$ in formula (11.4).

Consider the L_2-norm of the functions $F_\gamma(f_1, f_2)$

$$\|F_\gamma(f_1, f_2)\|_2^2 = \int F_\gamma(f_1, f_2)^2(x_1 \dots, x_{|O(\gamma)|}) \prod_{p=1}^{|O(\gamma)|} \mu(dx_p).$$

The inequality

$$\|F_\gamma(f_1, f_2)\|_2 \le \|f_1\|_2\|f_2\|_2 \quad \text{if } W(\gamma) = 0 \tag{11.5}$$

holds for this norm. The condition $W(\gamma) = 0$ in formula (11.4) means that the diagram $\gamma \in \Gamma(k_1, k_2)$ has no chains β of length $\ell(\beta) = 2$ with colour $c_\gamma(\beta) = -1$. For a general diagram $\gamma \in \Gamma(k_1, k_2)$ under the condition $\sup|f_2(x_1, \dots, x_{k_2})| \le 1$ the inequality

$$\|F_\gamma(f_1, f_2)\|_2 \le 2^{W(\gamma)}\|f_1\|_2 \tag{11.6}$$

holds. Inequalities (11.5) and (11.6) remain valid also in the case when f_1 and f_2 may be non-canonical functions.

Inequality (11.5) is actually a repetition of estimate (10.13) about the diagrams appearing in the case of Wiener–Itô integrals. Inequality (11.6) yields a weaker bound about the L_2-norm $\|F_\gamma(f_1, f_2)\|_2$ for a general diagram γ. We formulated it in a form where the functions f_1 and f_2 do not play a symmetrical role. This estimate depends on the L_2-norm of the function f_1, and it is assumed in it that the supremum of the function $|f_2|$ is less than 1. We chose such a formulation of this inequality because it can be well generalized to the case when the product of several U-statistics is considered. The appearance of the condition about the supremum of the function $|f_2|$ in the estimate (11.6) is closely related to the fact that in the estimates on the tail distribution of U-statistics—unlike the case of Wiener–Itô integrals—a condition is imposed not only on the L_2-norm of the kernel function f, but also on its L_∞-norm. I return to this question later.

Next I show an example which may help to understand how to apply the diagram formula for the product of two degenerate U-statistics.

Take two normalized degenerate U-statistics $n^{-3/2}3!I_{n,3}(f_1)$ and $n^{-2}4!I_{n,4}(f_2)$ with kernel functions $f_1(x_1,x_2,x_3)$ and $f_2(x_1,x_2,x_3,x_4)$, and let us see how to calculate with the help of formula (11.4) a term of the sum which expresses the product $3!I_{n,3}(f_1)4!I_{n,4}(f_2)$ as a sum of degenerate U-statistics.

Let us first understand which are the coloured diagrams we have to consider in the diagram formula (11.4), and then let us calculate the term corresponding to a coloured diagram at the right-hand side of this formula.

The coloured diagrams we have to consider have two rows with vertices labelled by (1,1), (1,2), (1,3) and (2.1), (2,2), (2,3), (2,4) respectively. The coloured diagrams are such partitions of the vertices whose elements contain from each row at most one element. The elements of these partitions which we call chains contain 1 or 2 elements. (We speak here about chains and not about graphs, because we want to apply such a terminology which also works in the more general case when we consider the diagram formula for the product of several degenerate U-statistics.) We give each chain either the colour $+1$ or -1. Chains consisting of only 1 vertex (chains of length 1) get the colour -1 while chains containing 2 vertices (chains of length 2) can get both colours $+1$ and -1. We take all coloured diagrams satisfying the above properties, and each of them yields a contribution to the sum at the right-hand side of (11.4). Let us look what kind of contribution yields the coloured diagram γ which contains a closed chain $((1,1),(2,2))$ (with colour $+1$) and an open chain $((1,3),(2,4))$ (with colour -1) of length two, and beside this it contains chains of length 1 and colour -1. They are $(1,2)$ from the first row, and $(2,1)$, $(2,3)$ from the second row. (See the picture.)

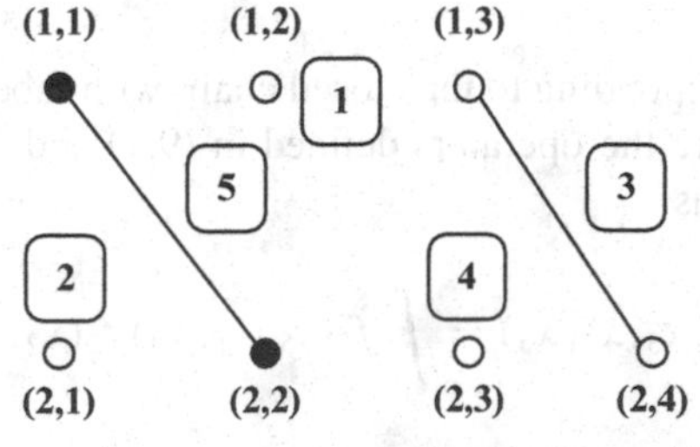

The diagram with the labelling of its chains
(o–o denotes open and •–• denotes closed chains)

We fix a labelling of the chains of the diagram γ, and define with its help a relabelling of the vertices. We label the chains subsequently from 1 to 5 in such a way that the open chains get the smaller labels, 1, 2, 3 and 4. Otherwise, we choose arbitrary labelling. We have the right for it, since although the kernel function of the U-statistic we shall define with the help of the diagram γ will depend on this labelling, but the U-statistic determined by it will not depend on it. Let us give the following labels for the chains: $(1,2)$—label 1, $(2,1)$—label 2, $((1,3),(2,4))$—label 3, $(2,3)$—label 4, $((1,1),(2,2))$—label 5. (This was an arbitrary choice.) Then we relabel the vertices contained in a chain with the label of this chain. (See the picture). (We used such a notation where the labels of the chains are put in a box, like this $\boxed{1}$.)

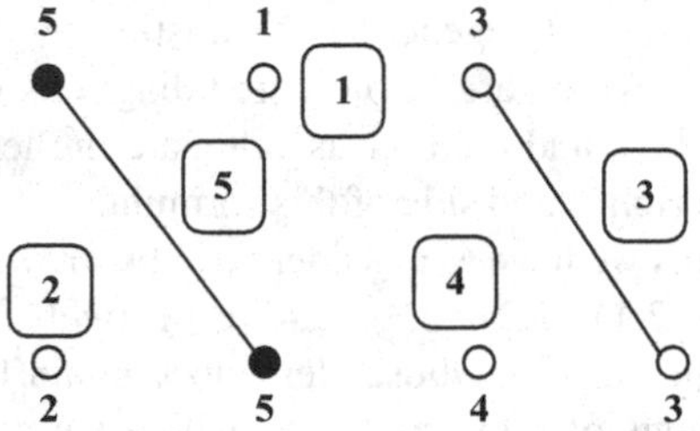

The relabelled version of our diagram

Then we reindex the variables of the functions f_1 and f_2 corresponding to the new labels of the vertices in the first and second row respectively. In the present case we take the reindexed functions $f_1(x_5, x_1, x_3)$ and $f_2(x_2, x_5, x_4, x_3)$. Then we define the product of these reindexed functions

$$(f_1 \circ f_2)_\gamma(x_1, x_2, x_3, x_4, x_5) = f_1(x_5, x_1, x_3) f_2(x_2, x_5, x_4, x_3).$$

Next we define the function $F_\gamma(f_1, f_2)$ introduced in (11.3) as

$$F_\gamma(f_1, f_2)(x_1, x_2, x_3, x_4) = Q_3 P_5 (f_1 \circ f_2)_\gamma(x_1, x_2, x_3, x_4, x_5),$$

where P_5 and Q_3 corresponding to the closed chain with label 5 and open chain of length 2 with label 3 are the operators defined in (9.1) and (9.2) with $j = 5$ and $j = 3$ respectively. Thus

$$P_5(f_1 \circ f_2)_\gamma(x_1, x_2, x_3, x_4, x_5) = \int f_1(x_5, x_1, x_3) f_2(x_2, x_5, x_4, x_3) \mu(dx_5),$$

and

$$
\begin{aligned}
F_\gamma(f_1,f_2)(x_1,x_2,x_3,x_4) &= Q_3P_5(f_1\circ f_2)_\gamma(x_1,x_2,x_3,x_4,x_5) \qquad (11.7)\\
&= \int f_1(x_5,x_1,x_3)f_2(x_2,x_5,x_4,x_3)\mu(dx_5)\\
&\quad -\int f_1(x_5,x_1,x_3)f_2(x_2,x_5,x_4,x_3)\mu(dx_3)\mu(dx_5).
\end{aligned}
$$

The normalized degenerate U-statistic corresponding to the diagram γ is

$$n^{-2}4!I_{n,4}(F_\gamma(f_1,f_2)),$$

and the contribution of the diagram γ to the sum in the diagram formula, i.e. to the sum at the right-hand side of (11.4) is $\frac{n-4}{n}\cdot n^{-1/2}\cdot n^{-2}4!I_{n,4}(F_\gamma(f_1,f_2))$. Here the factor $n^{-1/2}$ is the term $n^{-W(\gamma)/2}$ in (11.4) which is a contraction term which roughly speaking depends on the difference of the diagram γ from the "regular diagrams" appearing also in the diagram formula for Wiener–Itô integrals. The factor $\frac{n-4}{n}$ is a technical term which has no great importance. Its appearance is related to the form of the Hoeffding decomposition. In formula (9.5), expressing this relation a factor of the form $(n-|V|)(n-|V|-1)\cdots(n-k+1)$ appears instead of the "regular term" $n^{k-|V|}$, and this is the reason for the appearance of this factor.

Finally the notation $\sum'^{(n)}$ in formula (11.4) means that the above calculated term corresponding to the diagram γ takes part in the summation only if the sample size n of the U-statistic satisfies the inequality $n\ge 5$. This restriction is related to the fact that a k-fold U-statistic can be defined only if $n\ge k$ for the sample size. The U-statistic with kernel function $F_\gamma(f_1,f_2)$ has order 4. Nevertheless, a slightly stronger restriction is imposed. The reason for it is that, as the proof of Theorem 11.1 will show, the U-statistic we considered here appears as a term in the Hoeffding decomposition of the U-statistic with kernel function $(f_1\circ f_2)_\gamma$. This is a U-statistic of order 5, and the condition $n\ge 5$ comes from here.

Next we make some comments to Theorem 11.1.

Remark 1. The expression $W(\gamma)=k_1+k_2-|O(\gamma)|-2|C(\gamma)|$ appearing in formulas (11.4) and (11.5) equals $|O_2(\gamma)|$, i.e. it is the number of the chains $\beta\in\gamma$ for which $\ell(\beta)=2$, and $c_\gamma(\beta)=-1$. Indeed, if $\bar W(\gamma)$ equals the number of chains $\beta\in\gamma$ for which $\ell(\beta)=1$ (and as a consequence $c_\gamma(\beta)=-1$), then $|O_2(\gamma)|+\bar W(\gamma)=|O(\gamma)|$, and $2C|(\gamma)|+2|O(\gamma)|-\bar W(\gamma)=k_1+k_2$. (In the last identity we calculated the number of vertices in γ in two different ways.) Because of the definition of $W(\gamma)$ the last identity can be rewritten as $W(\gamma)+\bar W(\gamma)=|O(\gamma)|$. These relations imply the statement of this remark.

Remark 2. The term $I_{n,|O(\gamma)|}(F_\gamma(f_1,f_2))$ with some coloured diagram $\gamma\in\Gamma(k_1,k_2)$ appeared in the sum at the right-hand side of (11.4) only if the condition $s(\gamma)\le n$ was satisfied, which means that the sample size n of the U-statistic is sufficiently large. This restriction in the summation had a technical character, which has no great importance in our investigations. It is related to the fact that a U-statistic $I_{n,k}(f)$ was defined only if $n\ge k$. As a consequence, some U-statistics disappear at the right-hand side of (11.4) if the sample size n of the U-statistics is relatively small. The term $I_{n,|O(\gamma)|}(F_\gamma(f_1, f_2))$ appeared in (11.4) through the Hoeffding decomposition of a U-statistic with kernel function $(f_1 \circ f_2)_\gamma$ defined in (11.2). This function has $s(\gamma)$ arguments, and the U-statistic corresponding to it appears in our calculations only if the sample size n is not smaller than the number $s(\gamma)$.

Remark 3. As I earlier mentioned the functions $F_\gamma(f_1, f_2)$ depended on the labelling of the chains $\beta \in \Gamma(k_1, k_2)$. This non-uniqueness in the formulation of identity (11.4) has no importance in its applications. Moreover, we can get rid of this non-uniqueness by working with symmetrical functions f_1 and f_2 (with functions which do not change by a permutation of their variables) and by replacing the functions $F_\gamma(f_1, f_2)$ by their symmetrizations. A similar remark holds for the general version of the diagram formula to be discussed later, where we may consider the product of several degenerate U-statistics.

Remark 4. The diagram formula formulated in Theorem 11.1 is similar to its version about the product of two multiple integrals with respect to a normalized empirical distribution. The latter result was not written up here explicitly, but its form was explained in an informal way at the beginning of this chapter. The kernel functions of the U-statistics and random integrals appearing in these formulas are indexed by the same diagrams. Their definitions are different, because in the U-statistic case we have to work with canonical functions while in the multiple integral case we have no such restriction. As a consequence we define the functions $F_\gamma(f_1, f)$ in this case by means of a modified version of formula (11.3), where the operators $Q_{j'}$ are omitted from the definition. The coefficients of the normalized degenerate U-statistics and random integrals in the two results are slightly different. In the multiple integral case we have to multiple with $n^{-W(\gamma)/2}$ while in the U-statistic case this term is multiplied with a factor between 0 and 1. This is related to the form of the Hoeffding decomposition of U-statistics given in (9.3). The restriction in the summation $\sum'^{(n)}$ is also related to the properties of U-statistics.

Let us turn to the formulation of the general form of the diagram formula for the product of finitely many degenerate U-statistics. After introduction of some notations we present this result in Theorem 11.2. Then we discuss an example to understand its notation better.

This result has a more complicated form than its analogue about Wiener–Itô integrals, because in the present case we cannot define the kernel functions of the U-statistics appearing in the diagram formula in a simple, direct way. We shall define them with the help of an inductive procedure. To do this first we introduce some conventions which will be useful later.

Let us recall the convention introduced after the definition of canonical degenerate U-statistics by which $I_{n,0}(c)$ is a degenerate U-statistic of order zero, and $I_{n,0}(c) = c$ for a constant c. If $\gamma \in \Gamma(k_1, k_2)$ is such a diagram for which $|O(\gamma)| = 0$, i.e. $c_\gamma(\beta) = 1$ for all chains $\beta \in \gamma$, then the expression $F_\gamma(f_1, f_2)$

defined in (11.3) is a constant, and for such a diagram γ we define the term $I_{n,|O(\gamma)|}(F_\gamma(f_1, f_2))$ in relation (11.4) by means of the previous convention.

We introduce another convention (similarly to the discussion of Wiener–Itô integrals in Chap. 10) which enables us to extend the validity of Theorem 11.1 to the case when $k_1 = 0$, and the function $f_{k_1} = c$ with a constant c. In this case $\Gamma(k_1, k_2)$ consists of only one diagram γ containing the chains $\beta_p = \{(2, p)\}$ of length one and with colour $c_\gamma(\{(2, p)\}) = -1$, $1 \le p \le k_2$, and we define $F_\gamma(f_1, f_2) = cf_2(x_1, \dots, x_{k_2})$. Beside this, we have $C(\gamma) = \emptyset$, $O(\gamma) = \{(2,1), \dots, (2, k_2)\}$, hence $W(\gamma) = k_1 + k_2 - |O(\gamma)| - 2|C(\gamma)| = 0$, $|C(\gamma)| = 0$. We also have $s(\gamma) = k_2$, thus the inequality $(\gamma) \le n$ holds under the conditions of Theorem 11.1. Hence formula (11.4) remains valid also in the case $k_1 = 0$. For the sake of completeness we introduce a listing of the (open) chains $\beta \in O(\gamma)$ of the diagram(s) of the set $\Gamma(0, k_2)$. We define $\beta(l) = \{(2, l)\}$, $1 \le l \le k_2$ in this case. We have introduced the above conventions because they are useful in the inductive argument we shall apply in the proof of the diagram formula for the product of degenerate U-statistics in the general case.

To formulate the diagram formula for the product of degenerate U-statistics in the general case first we define a function $F_\gamma(f_1, \dots, f_m) = F_\gamma(f_1, \dots, f_m)(x_1, \dots, x_{|O(\gamma)|})$ for each coloured diagram $\gamma \in \Gamma(k_1, \dots, k_m)$ and collection of canonical functions (canonical with respect to a probability measure μ on a measurable space $(X, \mathcal{X})$) $f_1, \dots, f_m$ with $k_1, \dots, k_m$ variables. The function $F_\gamma(f_1, \dots, f_m)$ we shall define has $|O(\gamma)|$ arguments. It will appear as the kernel function of the degenerate U-statistic corresponding to the diagram γ at the right-hand side of the diagram formula.

The functions $F_\gamma(f_1, \dots, f_m)$ will be defined by induction with respect to the number m of the components in the product of degenerate U-statistics. For $m = 2$ we have already defined them. Let the functions $F_\gamma(f_1, \dots, f_{m-1})$ be defined for each coloured diagram $\gamma \in \Gamma(k_1, \dots, k_{m-1})$. To define $F_\gamma(f_1, \dots, f_m)$ for a coloured diagram $\gamma \in \Gamma(k_1, \dots, k_m)$ first we define the predecessor $\gamma_{pr} = \gamma_{pr}(\gamma) \in \Gamma(k_1, \dots, k_{m-1})$ of γ. It consist of the restrictions of the chains of the diagram γ to the first $m - 1$ rows of this diagram together with an appropriate colouring of these restricted chains. Then we define the function $F_{\gamma_{pr}}(f_1, \dots, f_{m-1})$ with $|O(\gamma_{pr})|$ arguments in our inductive procedure. We shall also define a coloured diagram $\gamma_{cl} \in \Gamma(|O(\gamma_{pr})|, k_m)$ of two rows, which has the heuristic content that it contains the additional information we need to reconstruct the diagram $\gamma \in \Gamma(k_1, \dots, k_m)$ from its predecessor γ_{pr}. We shall define $F_\gamma(f_1, \dots, f_m)$ which will be a canonical function with $|O(\gamma)|$ variables with the help of the diagram γ_{cl} and the pair of functions $F_{\gamma_{pr}}(f_1, \dots, f_{m-1})$ and f_m.

The diagram $\gamma_{pr} \in \Gamma(k_1, \dots, k_{m-1})$ will consist of the chains

$$\beta_{pr} = \beta \setminus \{(m, 1), \dots, (m, k_m)\}, \quad \beta \in \gamma,$$

i.e. we get the chain β_{pr} by dropping from β its vertex contained in the last row $\{(m, 1), \dots, (m, k_m)\}$ of the diagram if it contains such a vertex. If we get an empty set in such a way (this happens if β consists of a single vertex of the form(m, p))

then we disregard it, i.e the empty set will be not taken as a chain of γ_{pr}. We define the colour of β_{pr} as $c_{\gamma_{pr}}(\beta_{pr}) = c_\gamma(\beta)$ if $\beta = \beta_{pr}$, i.e. if $\beta \cap \{(m,1),\dots,(m,k_m)\} = \emptyset$, and $c_{\gamma_{pr}}(\beta_{pr}) = -1$ if β contains a vertex of the form (m,p), $1 \le p \le k_m$. After the definition of the diagrams $\gamma_{pr} \in \Gamma(k_1,\dots,k_{m-1})$ we can define the canonical function $F_{\gamma_{pr}}(f_1,\dots,f_{m-1})$ with arguments $x_1,\dots,x_{|O(\gamma_{pr})|}$ by means of our inductive procedure.

We also define the diagram $\gamma_{cl} \in \Gamma(|O(\gamma_{pr})|,k_m)$ for a diagram $\gamma \in \Gamma(k_1,\dots,k_m)$. We must tell which are the chains $\{(1,p),(2,r)\}$, $1 \le p \le |O(\gamma_{pr})|$, $1 \le r \le k_m$, of length two of the diagram γ_{cl}, and we have to define their colour. The set $\{(1,p),(2,r)\}$ is a chain of length two of the diagram γ_{cl} if and only if the open chain $\beta(p) \in \gamma_{pr}$ (the chain $\beta(p)$ is that open chain of $\gamma_{pr} \in \Gamma(k_1,\dots,k_{m-1})$ which got the label p in the enumeration of the open chains of γ_{pr}) is the restriction β_{pr} of that chain $\beta \in \gamma$ for which $(m,r) \in \beta$. If $\{(1,p),(2,r)\} \in \gamma_{cl}$, then its colour in γ_{cl} is defined as $c_{\gamma_{cl}}(\{(1,p),(2,r)\}) = c_\gamma(\beta)$ with the chain $\beta = \beta(p) \in \gamma$, which is the chain for which $(m,r) \in \beta$. Those vertices $(1,p)$ and $(2,r)$, $1 \le p \le |O(\gamma_{pr})|$, $1 \le r \le k_m$, which are not contained in such a chain of length 2 will be chains of length 1 of γ_{cl} with colour -1.

Given some bounded functions $f_1,\dots,f_m$ of k_p variables, $1 \le p \le m$, and a diagram $\gamma \in \Gamma(k_1,\dots,k_m)$ we shall define the function $F_\gamma(f_1,\dots,f_m)$ with the help of the pair of functions $F_{\gamma_{pr}}(f_1,\dots,f_{m-1})$ and f_m and the diagram $\gamma_{cl} \in \Gamma(|O(\gamma_{pr})|),k_m)$ by the formula

$$
\begin{aligned}
&F_\gamma(f_1,\dots,f_m)(x_1,x_2,\dots,x_{|O(\gamma)|}) \\
&\quad = F_{\gamma_{cl}}(F_{\gamma_{pr}}(f_1,\dots,f_{m-1}),f_m)(x_1,\dots,x_{|O(\gamma_{cl})|})).
\end{aligned} \tag{11.8}
$$

Here we applied formula (11.3) with the choice $\gamma = \gamma_{cl}$ and pair of functions $f_1 = F_{\gamma_{pr}}(f_1,\dots,f_{m-1})$ and $f_2 = f_m$. To justify the correctness of formula (11.8) we still have to show that $|O(\gamma)| = |O(\gamma_{cl})|$.

To prove this identity observe that the number of those open chains of γ_{cl} which contain a vertex from the first row of γ_{cl} equals the number of those open chains of $\beta \in \gamma$ which have a vertex outside of the m-th row of the diagram γ. The remaining open chains of γ_{cl} contain one vertex from the second row of γ_{cl}, and they correspond to those open diagrams of γ which consist of one vertex from the m-th row of the diagram. The above observations imply the desired identity.

To formulate the general form of the diagram formula for the product of degenerate U-statistics we introduce some quantities which are the versions of the quantities $W(\gamma)$, $s(\gamma)$ appearing in the identity (11.4) in Theorem 11.1 in the case $m > 2$. Put

$$
W(\gamma) = \sum_{\beta\in O(\gamma)} (\ell(\beta)-1) + \sum_{\beta\in C(\gamma)} (\ell(\beta)-2), \quad \gamma \in \Gamma(k_1,\dots,k_m), \tag{11.9}
$$

where $\ell(\beta)$ denotes the length of the chain β.

To define the next quantity let us introduce some notations. We consider the chains of the form $\beta = \{(p_1, r_1), \dots, (p_l, r_l)\}$, $1 \le p_1 < p_2 < \dots < p_l \le m$, with elements in the set $A(k_1, \dots, k_m) = \{(p,r)\colon 1 \le p \le m,\ 1 \le r \le k_p\}$, and define their upper level $u(\beta) = p_1$, and deepest level $d(\beta) = p_l$. With the help of these notions we introduce for all diagrams $\gamma \in \Gamma(k_1, \dots, k_m)$ and integers p, $1 \le p \le m$, the following subsets of the diagram γ. Put $\mathscr{B}_1(\gamma, p) = \{\beta\colon \beta \in \gamma,\ c_\gamma(\beta) = 1,\ d(\beta) = p\}$, and $\mathscr{B}_2(\gamma, p) = \{\beta\colon \beta \in \gamma,\ c_\gamma(\beta) = -1,\ d(\beta) \le p\} \cup \{\beta\colon \beta \in \gamma,\ u(\beta) \le p,\ d(\beta) > p\}$. In words, $\mathscr{B}_1(\gamma, p)$ consists of those chains $\beta \in \gamma$ which have colour 1, all their vertices are in the first p rows of the diagram, and contain a vertex in the p-th row. The set $\mathscr{B}_2(\gamma, p)$ consists of those chains $\beta \in \gamma$ which have either colour -1, and all their vertices are in the first p rows of the diagram, or they have (with an arbitrary colour) a vertex both in the first p rows and in the remaining rows of the diagram. Put $B_1(\gamma, p) = |\mathscr{B}_1(\gamma, p)|$ and $B_2(\gamma, p) = |\mathscr{B}_2(\gamma, p)|$. With the help of these numbers we define

$$J_n(\gamma, p) = \begin{cases} \prod\limits_{j=1}^{B_1(\gamma,p)} \left(\frac{n - B_1(\gamma,p) - B_2(\gamma,p) + j}{n} \right) & \text{if } B_1(\gamma, p) \ge 1 \\ 1 \quad \text{if } B_1(\gamma, p) = 0 \end{cases} \tag{11.10}$$

for all $2 \le p \le m$ and diagrams $\gamma \in \Gamma(k_1, \dots, k_m)$.

Theorem 11.2 will be formulated with the help of the above notations.

Theorem 11.2 (The diagram formula for the product of several degenerate U-statistics). *Let a sequence of independent and identically distributed random variables $\xi_1, \xi_2, \dots$ be given with some distribution μ on a measurable space $(X, \mathscr{X})$ together with $m \ge 2$ bounded functions $f_p(x_1, \dots, x_{k_p})$ on the spaces $(X^{k_p}, \mathscr{X}^{k_p})$, $1 \le p \le m$, canonical with respect to the probability measure μ. Let us consider the class of coloured diagrams $\Gamma(k_1, \dots, k_m)$ together with the functions $F_\gamma = F_\gamma(f_1, \dots, f_m)$, $\gamma \in \Gamma(k_1, \dots, k_m)$, defined in formulas (11.8) and the constants $W(\gamma)$ and $J_n(\gamma, p)$, $1 \le p \le m$, given in formulas (11.9) and (11.10).*

The functions $F_\gamma(f_1, \dots, f_m)$ are bounded and canonical with respect to the measure μ with $|O(\gamma)|$ variables, and the product of the normalized degenerate U-statistics $n^{-k_p/2}!k_p I_{n,k_p}(f_p)$, $1 \le p \le m$, $n \ge \max\limits_{1 \le p \le m} k_p$, defined in (8.8) can be written in the form

$$\prod_{p=1}^{m} n^{-k_p/2} k_p! I_{n,k_p}(f_p) = \sum_{\gamma \in \Gamma(k_1, \dots, k_m)}^{\prime(n,m)} \left(\prod_{p=2}^{m} J_n(\gamma, p) \right) n^{-W(\gamma)/2} \cdot n^{-|O(\gamma)|/2} |O(\gamma)|! I_{n,|O(\gamma)|}(F_\gamma(f_1, \dots, f_m)), \tag{11.11}$$

where $\sum^{\prime(n,m)}$ means that summation is taken for those $\gamma \in \Gamma(k_1, \dots, k_m)$ which satisfy the relation $B_1(\gamma, p) + B_2(\gamma, p) \le n$ for all $2 \le p \le m$

with the quantities $B_1(\gamma, p)$ and $B_2(\gamma, p)$ introduced before the definition of $J_n(\gamma, p)$ in (11.10), and the expression $W(\gamma)$ was defined in (11.9). The terms $I_{n,|O(\gamma)|}(F_\gamma(f_1, \dots, f_m))$ at the right-hand side of formula (11.11) can be replaced by $I_{n,|O(\gamma)|}(\text{Sym}\, F_\gamma(f_1, \dots, f_m))$.

To understand better the formulation of Theorem 11.2 let us consider the following example.
Take three normalized degenerate U-statistics $n^{-3/2}3!I_{n,3}(f_1)$, $n^{-2}4!I_{n,4}(f_2)$ and $n^{-3/2}3!I_{n,3}(f_3)$ with canonical kernel functions $f_1(x_1, x_2, x_3)$, $f_2(x_1, x_2, x_3, x_4)$ and $f_3(x_1, x_2, x_3)$, and let us see how to calculate a term from the sum at the right-hand side of formula (11.11) which expresses the product

$$n^{-3/2}3!I_{n,3}(f_1)n^{-2}4!I_{n,4}(f_2)n^{-3/2}3!I_{n,3}(f_3)$$

in the form of a linear combination of degenerate U-statistics.

In this case we have to consider coloured diagrams with rows of vertices (1,1), (1,2), (1,3), then (2,1), (2,2), (2,3), (2,4), and finally (3,1), (3,2), (3,3). We have to consider all coloured diagrams with these rows, and to calculate their contribution to the sum at the right-hand side of (11.11). Let us consider for instance the diagram containing two closed chains (with colour 1) $((1,3),(2,4),(3,3))$ of length 3, $((1,1),(2,2))$ of length 2, an open chain (with colour -1) $((2,1),(3,1))$ of length 2, and the remaining vertices (1,2), (2,3), (3,2) are chains of length 1 which are consequently open (with colour -1). (See picture.)

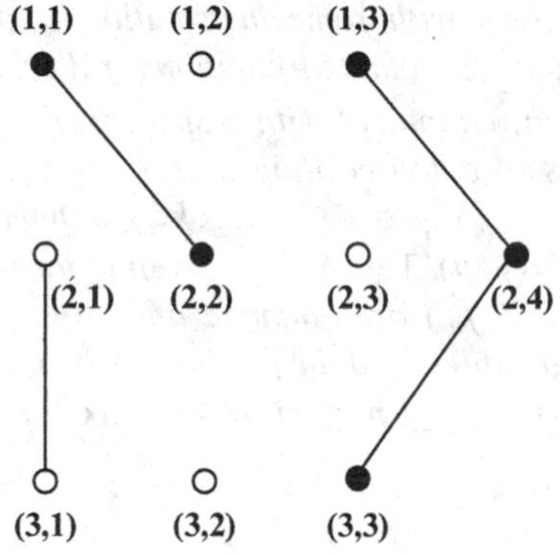

Our diagram γ

We want to calculate $F_\gamma(f_1, f_2, f_3)$. For this goal first we have to determine the coloured diagrams $\gamma_{pr} \in \Gamma(3,4)$ and $\gamma_{cl} \in \Gamma(4,3)$ (here the first parameter 4 in the definition of the class of diagrams where γ_{cl} belongs to is the number of open chains in γ_{pr}, which is, as we will see, equals 4), and the kernel function $F_{\gamma_{pr}}(f_1, f_2)$. (See the picture of the diagram γ_{pr} together with a labelling of its chains and the diagram γ_{cl} to which we also attached a labelling.)

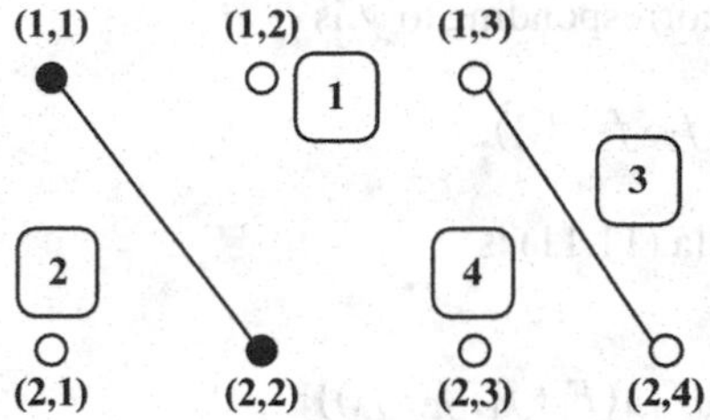

The diagram γ_{pr} corresponding to γ together with the enumeration of its open chains

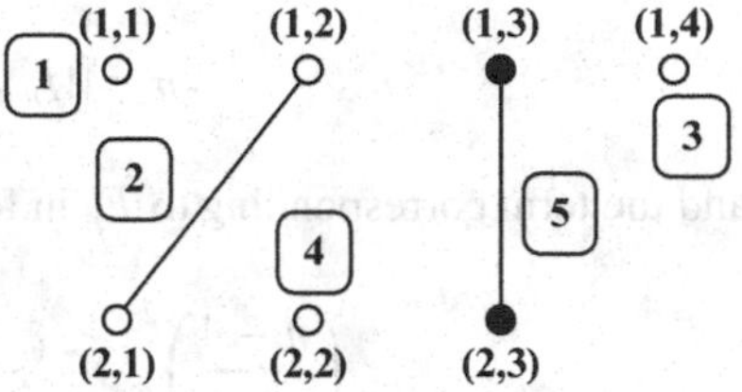

The diagram γ_{cl} constructed with the help of γ_{pr} and of the enumeration of its open chains

In our example γ_{pr} is a diagram with two rows (1,1), (1,2), (1,3) and (2,1), (2,2), (2,3), (2,4). It contains a closed chain $((1,1),(2,2))$ and an open chain $((1,3),(2,4))$ of length 2, (the latter is the restriction of a chain of length 3), and open chains of length 1, which are the vertices (1,2), (2,1), (2,3). This is the same diagram which we considered in the example after Theorem 11.1. In that example we have fixed an enumeration of the chains of this diagram. We also made the convention that the enumeration of the chains of a diagram fixed at the start cannot be modified later. Hence we have the following enumeration of the open chains of this diagram: (1,2)—label 1, (2,1)—label 2, $((1,3),(2,4))$—label 3, and (2,3)—label 4.

We define the coloured diagram γ_{cl} with the help of the diagram γ_{pr} and the enumeration of its open chains. It has two rows. The vertices of the first row $(1,1)$, $(1,2)$, $(1,3)$ and $(1,4)$ correspond to the open chains of the diagram γ_{pr} with labels 1, 2, 3 and 4 respectively. The vertices of the second row, $(2,1)$, $(2,2)$ and $(2,3)$ correspond to the vertices $(3,1)$, $(3,2)$ and $(3,3)$ of the last row of the original diagram γ. The diagram γ_{cl} has an open chain $((1,2),(2,1))$ of length two, (here the open chain (2,1) of γ_{pr} labelled by 2, is connected to the vertex (3,1) with second index 1), a closed chain of length 2 $((1,3),(2,3))$ (here the open chain of γ_{pr} labelled by 3 is connected with the vertex (3,3)), and the remaining open chains of γ_{cl} of length 1 are (1,1), (1,4) (the open chains (1,2) and (2,3) of γ_{pr} with labels 1, and 4), and (2,2).

Actually we have already calculated the function $F_{\gamma_{pr}}(f_1, f_2)$ in formula (11.7). We can calculate similarly the function $F_\gamma(f_1, f_2, f_3) = F_{\gamma_{cl}}(F_{\gamma_{pr}}(f_1, f_2), f_3)$. First we fix a labelling of the chains of the diagram γ_{cl}, say (1,1)—label 1, $((1,2),(2,1))$—label 2, (1,4)—label 3, (2,2)—label 4, and $((1,3),(2,3))$—label 5. (I have denoted this labelling in the corresponding picture.) With such a labelling

$$
\begin{aligned}
F_\gamma(f_1, f_2, f_3)(x_1, x_2, x_3, x_4) &= Q_2 P_5(F_{\gamma_{pr}}(x_1, x_2, x_5, x_3) f_3(x_2, x_4, x_5)) \\
&= \int F_{\gamma_{pr}}(x_1, x_2, x_5, x_3) f_3(x_2, x_4, x_5) \mu(dx_5) \\
&\quad - \int F_{\gamma_{pr}}(x_1, x_2, x_5, x_3) f_3(x_2, x_4, x_5) \mu(dx_2) \mu(dx_5).
\end{aligned}
$$

The normalized degenerate U-statistic corresponding to γ is

$$n^{-2}4!I_{n,4}(F_\gamma(f_1, f_2, f_3)),$$

and the term corresponding to F_γ in formula (11.11) is

$$\left(\frac{n-4}{n}\right)^2 \cdot n^{-1} \cdot n^{-2}4!I_{n,4}(F_\gamma(f_1, f_2, f_3))$$

if $n \geq 5$. In the case $n \leq 4$ this term disappears.

In Theorem 11.2 the product of such degenerate U-statistics were considered whose kernel functions were bounded. This also implies that all functions F_γ appearing at the right-hand side of (11.11) are well-defined (i.e. the integrals appearing in their definition are convergent) and bounded. In the applications of Theorem 11.2 it is useful to have a good bound on the L_2-norm of the functions $F_\gamma(f_1, \dots, f_m)$. Such a result is formulated in the following

Lemma 11.3 (Estimate about the L_2-norm of the kernel functions of the U-statistics appearing in the diagram formula). *Let m functions $f_p(x_1, \dots, x_{k_p})$, $1 \leq p \leq m$, be given on the products $(X^{k_p}, \mathcal{X}^{k_p}, \mu^{k_p})$ of some measure space $(X, \mathcal{X}, \mu)$, $1 \leq p \leq m$, with a probability measure μ, which satisfy inequality (8.1) (if the index k is replaced by the index k_p in formula (8.1)). Let us take a coloured diagram $\gamma \in \Gamma(k_1, \dots, k_m)$, and consider the function $F_\gamma(f_1, \dots, f_m)$ defined inductively by means of formula (11.8). The L_2-norm of the function $F_\gamma(f_1, \dots, f_m)$ (with respect to the product measure $\mu \times \dots \times \mu$ on the space where $F_\gamma(f_1, \dots, f_m)$ is defined) satisfies the inequality*

$$\|F_\gamma(f_1, \dots, f_m)\|_2 \leq 2^{W(\gamma)} \prod_{p \in U(\gamma)} \|f_p\|_2,$$

where $W(\gamma)$ is given in (11.9), and the set $U(\gamma) \subset \{1, \dots, m\}$ is defined as

$$\begin{aligned} U(\gamma) = \{p\colon\ 1 \leq p \leq m, \quad & \text{for all vertices } (p,r),\ 1 \leq r \leq k_p \text{ the chain } \beta \in \gamma \\ & \text{for which } (p,r) \in \beta \text{ has the property that either } u(\beta) = p \\ & \text{or } d(\beta) = p \text{ and } c_\gamma(\beta) = 1\}. \end{aligned} \tag{11.12}$$

(If the point (p,r) is contained in a chain $\beta = \{(p,r)\} \in \gamma$ of length 1, then $u(\beta) = d(\beta) = p$, and $c_\gamma(\beta) = -1$. In this case the vertex (p,r) satisfies that condition which all vertices (p,r), $1 \leq r \leq k_p$, must satisfy to guarantee the property $p \in U(\gamma)$.)

Remark. Let us give a less formal definition of the set $U(\gamma)$ in formula (11.12). It contains the indices of those rows of the diagram γ whose vertices behave in a sense nicely. This nice behaviour means the following. Each vertex is contained in a chain

β of the diagram γ. We say that a vertex behaves nicely if it is either at the highest or the lowest level of the chain $\beta \in \gamma$ containing it. Moreover, if it is at its lower level, then we also demand that β must be closed, i.e. $c(\beta) = 1$. If a vertex is contained in a chain containing no other vertex, then it is both at the higher and lower level of this chain. In this case we say that the vertex behave nicely.

The last result of this chapter is a corollary of Theorem 11.2. In this corollary we give an estimate on the expected value of a product of degenerate U-statistics. To formulate this result we introduce the following terminology. We call a (coloured) diagram $\gamma \in \Gamma(k_1,\dots,k_m)$ closed if $c_\gamma(\beta) = 1$ for all chains $\beta \in \gamma$, and denote the set of all closed diagrams by $\bar\Gamma(k_1,\dots,k_m)$. Observe that $F_\gamma(f_1,\dots,f_m)$ is constant (a function of zero variable) if and only if γ is a closed diagram, i.e. $\gamma \in \bar\Gamma(k_1,\dots,k_m)$, and

$$I_{n,|O(\gamma)|}(F_\gamma(f_1,\dots,f_m)) = I_{n,0}(F_\gamma(f_1,\dots,f_m)) = F_\gamma(f_1,\dots,f_m)$$

in this case. Now we formulate the following result.

Corollary of Theorem 11.2 about the expectation of a product of degenerate U-statistics. *Let a finite sequence of functions $f_p(x_1,\dots,x_{k_p})$, $1 \le p \le m$, be given on the products $(X^{k_p},\mathcal{X}^{k_p})$ of some measurable space $(X,\mathcal{X})$ together with a sequence of independent and identically distributed random variables with value in the space $(X,\mathcal{X})$ and some distribution μ which satisfy the conditions of Theorem 11.2.*

Let us apply the notation of Theorem 11.2 together with the notion of the above introduced class of closed diagrams $\bar\Gamma(k_1,\dots,k_m)$. The identity

$$E\left(\prod_{p=1}^m n^{-k_p/2}k_p!I_{n,k_p}(f_{k_p})\right) = \sum_{\gamma\in\bar\Gamma(k_1,\dots,k_m)}^{\prime(n,m)} \left(\prod_{p=1}^m J_n(\gamma,p)\right) n^{-W(\gamma)/2}\cdot F_\gamma(f_1,\dots,f_m) \tag{11.13}$$

holds. This identity has the consequence

$$\left|E\left(\prod_{p=1}^m n^{-k_p/2}k_p!I_{n,k_p}(f_{k_p})\right)\right| \le \sum_{\gamma\in\bar\Gamma(k_1,\dots,k_m)} n^{-W(\gamma)/2}|F_\gamma(f_1,\dots,f_m)|. \tag{11.14}$$

Beside this, if the functions f_p, $1 \le p \le m$, satisfy conditions (8.1) and (8.2) (with indices k_p instead of k in them), then the numbers $|F_\gamma(f_1,\dots,f_m)|$ at the right-hand side of (11.14) satisfy the inequality

$$|F_\gamma(f_1,\dots,f_m)| \le 2^{W(\gamma)}\sigma^{|U(\gamma)|} \quad \text{for all } \gamma\in\bar\Gamma(k_1,\dots,k_m). \tag{11.15}$$

In formula (11.15) the same number $W(\gamma)$ and set $U(\gamma)$ appear as in Lemma 11.3. The only difference is that in the present case the definition of $U(\gamma)$ becomes a bit simpler, since $c_\gamma(\beta) = 1$ for all chains $\beta \in \gamma$.

Remark. We have applied a different terminology for diagrams in this chapter and in Chap. 10, where the theory of Wiener–Itô integrals was discussed. But there is a simple relation between their terminology. If we take only those diagrams considered in this chapter which contain only chains of length 1 or 2, and beside this the chains of length 1 have colour -1, and the chains of length 2 have colour 1, then we get the diagrams considered in the previous chapter. Moreover, the functions $F_\gamma = F_\gamma(f_1, \dots, f_m)$ are the same in the two cases. Hence formula (10.22) in the Corollary of Theorem 10.2 and formula (11.14) in the Corollary of Theorem 11.2 make possible to compare the moments of Wiener–Itô integrals and degenerate U-statistics.

The main difference between the estimates of this chapter and those given in the Gaussian case is that formula (11.14) contains some additional terms. They are the contributions of those diagrams $\gamma \in \bar{\Gamma}(k_1, \dots, k_m)$ which contain chains $\beta \in \gamma$ with length $\ell(\beta) > 2$. These are those diagrams $\gamma \in \bar{\Gamma}(k_1, \dots, k_m)$ for which $W(\gamma) \ge 1$. The estimate (11.15) given for the terms F_γ corresponding to such diagrams is weaker, than the estimate given for the terms F_γ with $W(\gamma) = 0$, since $|U(\gamma)| < m$ if $W(\gamma) \ge 1$, and $|U(\gamma)| = m$ if $W(\gamma) = 0$. On the other hand, such terms have a coefficient $n^{-W(\gamma)/2}$ at the right-hand side of formula (11.14). A closer study of these formulas may explain the relation between the estimates given for the tail distribution of Wiener–Itô integrals and degenerate U-statistics.

Chapter 12
The Proof of the Diagram Formula for U-Statistics

In this chapter the results of the previous chapter will be proved. First I prove its main result, the diagram formula for the product of two degenerate U-statistics.

Proof of Theorem 11.1. In the first step of the proof the product

$$k_1!I_{n,k_1}(f_1)k_2!I_{n,k_2}(f_2)$$

of two degenerate U-statistics will be rewritten as a sum of not necessarily degenerate U-statistics. In this step a term by term multiplication is carried out for the product $k_1!I_{n,k_1}(f_1)k_2!I_{n,k_2}(f_2)$, and the terms of the sum obtained in such a way are put into different classes indexed by the (non-coloured) diagrams with two rows of length k_1 and k_2. This step is very similar to the heuristic argument leading to formulas (10.15) and (10.16) in our explanation about the diagram formula for Wiener–Itô integrals.

In this step we consider all sets of pairs

$$\{(l_1, l'_1), \dots, (l_r, l'_r)\}, \quad 1 \le r \le \min(k_1, k_2),$$

with the following properties: $1 \le l_1 < l_2 < \dots < l_r \le k_1$, the numbers $l'_1, \dots, l'_r$ are all different, and $1 \le l'_s \le k_2$, for all $1 \le s \le r$.

To a set of pairs $\{(l_1, l'_1), \dots, (l_r, l'_r)\}$ with the above properties let us correspond the following diagram $\bar\gamma((l_1, l'_1), \dots, (l_r, l'_r)) \in \bar\Gamma(k_1, k_2)$, where $\bar\Gamma(k_1, k_2)$ denotes the set of (non-coloured) diagrams with two rows of length k_1 and k_2. The diagram $\bar\gamma((l_1, l'_1), \dots, (l_r, l'_r))$ has two rows, $\{(1,1)\dots,(1,k_1)\}$, and $\{(2,1),\dots,(2,k_2)\}$, its chains of length 2 are the sets $\{(1,l_s),(2,l'_s)\}$, $1 \le s \le r$, and beside this it contains the chains $\{(1,p)\}$, $p \in \{1,\dots,k_1\} \setminus \{l_1,\dots,l_r\}$, and $\{(2,p)\}$, $p \in \{1,\dots,k_2\} \setminus \{l'_1,\dots,l'_r\}$ of length 1. All (non-coloured) diagrams $\bar\gamma \in \bar\Gamma(k_1,k_2)$ can be represented in the form $\bar\gamma = \bar\gamma((l_1, l'_1), \dots, (l_r, l'_r))$ with the help of a set of pairs $\{(l_1, l'_1), \dots, (l_r, l'_r)\}$, $1 \le r \le \min(k_1, k_2)$, with the above properties in a unique way.

P. Major, *On the Estimation of Multiple Random Integrals and U-Statistics*, Lecture Notes in Mathematics 2079, DOI 10.1007/978-3-642-37617-7_12,

To make the notation in the subsequent discussion simpler we introduce, similarly to the notation of Chap. 11, a labelling of the chains of the diagrams $\bar\gamma \in \bar\Gamma(k_1,k_2)$, and then we define the labelling of the vertices of this diagram $\bar\gamma$ with its help.

Let us choose the following natural labelling of the chains of a diagram. Consider the diagram $\bar\gamma = \bar\gamma((l_1,l'_1),\dots,(l_r,l'_r)) \in \bar\Gamma(k_1,k_2)$ which has $s(\bar\gamma) = k_1+k_2-r$ chains. The chain $\beta \in \bar\gamma$ containing the vertex $(1,p)$ gets the label p, i.e. $\{(1,p)\} = \beta(p)$ if $1 \le p \le k_1$, and $p \notin \{l_1,\dots,l_r\}$, and $\{(1,l_s),(2,l'_s)\} = \beta(p)$ if $p = l_s$ with some $1 \le s \le r$. The remaining chains of $\bar\gamma$ have the form $\{(2,p)\}$ with $p \in \{1,\dots,k_2\} \setminus \{l'_1,\dots,l'_r\}$. Let us list the numbers p with this property in an increasing order, i.e. write $\{1,\dots,k_2\}\setminus\{l'_1,\dots,l'_r\} = \{\bar l_1,\dots,\bar l_{k_2-r}\}$ with $1 \le \bar l_1 < \dots < \bar l_{k_2-r}$, and define $\{(2,\bar l_p)\} = \beta(k_1+p)$ for $1 \le p \le k_2-r$. In such a way we have labelled the chains of a diagram $\bar\gamma \in \bar\Gamma(k_1,k_2)$. After this, we label its vertices (p,r) by the formula $\alpha_{\bar\gamma}((p,r)) = l$ with that label l for which $(p,r) \in \beta(l)$. Let us also define the sets $V_1 = V_1(\bar\gamma) = \{1,\dots,k_1+k_2-r\}\setminus\{l_1,\dots,l_r\}$ and $V_2 = V_2(\bar\gamma) = \{l_1,\dots,l_r\}$. These sets yield the labels of the chains of length 1 and length 2 respectively, i.e. $\beta(p)$ is a chain of length 1 if $p \in V_1$, and it is a chain of length 2 if $p \in V_2$.

We have defined a special labelling of the chains of the diagrams $\bar\gamma \in \bar\Gamma(k_1,k_2)$, and we shall work with it during the proof. First we prove a slightly modified version of relation (11.4) with functions $F_\gamma(f_1,f_2)$ defined with the help of the above labelling of the chains, which may not satisfy all conditions we imposed for a labelling of the chains before the formulation of Theorem 11.1. Then we show that identity (11.4) remains valid with the formulation of Theorem 11.1 (i.e. with that labelling of the chains which we considered there).

Let us consider the product $k_1!I_{n,k_1}(f_1)k_2!I_{n,k_2}(f_2)$, and let us rewrite it in the form of the sum we get by carrying out a term by term multiplication in this expression. We put the terms obtained in such a way into disjoint classes indexed by the diagrams $\bar\gamma \in \bar\Gamma(k_1,k_2)$ in the following way: A product

$$f_1(\xi_{j_1},\dots,\xi_{j_{k_1}})f_2(\xi_{j'_1},\dots,\xi_{j'_{k_2}})$$

belongs to the class indexed by the diagram $\bar\gamma((l_1,l'_1),\dots,(l_r,l'_r))$ with the parameters $(l_1,l'_1),\dots,(l_r,l'_r)$, $1 \le r \le \min(k_1,k_2)$, where $1 \le l_1 < l_2 < \dots < l_r \le k_1$, the numbers $l'_1,\dots,l'_r$ are different, and $1 \le l'_s \le k_2$, for all $1 \le s \le r$ if the indices $j_1,\dots,j_{k_1},j'_1,\dots,j'_{k_2}$ in the arguments of the variables in $f_1(\cdot)$ and $f_2(\cdot)$ satisfy the relation $j_{l_s} = j'_{l'_s}$, $1 \le s \le r$, and there is no more coincidence between the indices $j_1,\dots,j_{k_1},j'_1,\dots,j'_{k_2}$.

It is not difficult to see by applying the above partition of the terms in the product $k_1!I_{n,k_1}(f_1)k_2!I_{n,k_2}(f_2)$, and exploiting that each diagram $\bar\gamma \in \bar\Gamma(k_1,k_2)$ can be represented in the form $\bar\gamma((l_1,l'_1),\dots,(l_r,l'_r))$ in a unique way that the identity

$$\begin{aligned} &n^{-k_1/2}k_1!I_{n,k_1}(f_1)k_2!n^{-k_2/2}I_{n,k_2}(f_2)\\ &\quad= \sum_{\bar\gamma\in\bar\Gamma(k_1,k_2)}^{\prime(n)} n^{-(k_1+k_2)/2}s(\bar\gamma)!\,I_{n,s(\bar\gamma)}((f_1\circ f_2)_{\bar\gamma}) \end{aligned} \tag{12.1}$$

holds, where the functions $(f_1 \circ f_2)_{\bar\gamma} = (f_1 \circ f_2)_{\bar\gamma}(x_1,\dots,x_{s(\bar\gamma)})$ are defined in formula (11.2) with the help of the above introduced labelling of the chains of the diagram $\bar\gamma$, and $s(\bar\gamma) = k_1 + k_2 - |V_2(\bar\gamma)|$ denotes the number of chains in $\bar\gamma$. (Observe that with our labelling of the chains the indices of the function $(f_1 \circ f_2)_{\bar\gamma}$ are the numbers $1,\dots,s(\bar\gamma)$.) The notation $\sum'^{(n)}$ in (12.1) means that summation is taken only for such diagrams $\bar\gamma \in \bar\Gamma(k_1,k_2)$ for which $n \ge s(\bar\gamma)$. (Let me remark that although formula (11.2) was defined for coloured diagrams, the colours of the chains played no role in it.)

Relation (12.1) is not appropriate for our purposes, since the functions $(f_1 \circ f_2)_{\bar\gamma}$ in it may be non-canonical. To get the desired formula, Hoeffding's decomposition will be applied for the U-statistics $I_{n,s(\bar\gamma)}((f_1 \circ f_2)_{\bar\gamma})$ appearing at the right-hand side of formula (12.1). This decomposition becomes slightly simpler because of some special properties of the function $(f_1 \circ f_2)_{\bar\gamma}$ which follow from the canonical property of the initial functions f_1 and f_2.

To carry out this procedure let us observe that a function $f(x_1,\dots,x_k)$ is canonical if and only if $(P_j f)(x_s,\ s \in \{1,\dots,k\} \setminus \{j\}) = 0$ with the operator P_j defined in (9.1) for all indices j and $\{x_s\colon 1 \le s \le k,\ s \ne j\}$. Beside this, the condition that the functions f_1 and f_2 are canonical implies the relation $P_v(f_1 \circ f_2)_{\bar\gamma} \equiv 0$ if $v \in V_1(\bar\gamma)$ for all diagrams $\bar\gamma \in \bar\Gamma(k_1,k_2)$. (The set $V_1(\bar\gamma)$ denoted the labels of the chains of length 1 in the diagram $\bar\gamma$.) This relation remains valid if the function $(f_1 \circ f_2)_{\bar\gamma}$ is replaced by such functions which we get by applying the product of some transformations $P_{v'}$ and $Q_{v'}$, $v' \in V_2(\bar\gamma)$, for the function $(f_1 \circ f_2)_{\bar\gamma}$ with the transformations $P_{v'}$ and $Q_{v'}$ defined in formulas (9.1) and (9.2).

Beside this, the transformations P_v or Q_v are exchangeable with the operators $P_{v'}$ or $Q_{v'}$ for any pairs of indices v, v', and $P_v + Q_v = I$, where I denotes the identity operator. Beside this, $P_v Q_v = 0$, since $P_v Q_v = P_v - P_v^2 = 0$. The above relations make possible the following decomposition of the function $(f_1 \circ f_2)_{\bar\gamma}$ to the sum of canonical functions for all $\bar\gamma \in \bar\Gamma(k_1,k_2)$. (In the proof of the Hoeffding decomposition a similar argument was applied.)

$$(f_1 \circ f_2)_{\bar\gamma} = \prod_{v \in V_2(\bar\gamma)} (P_v + Q_v)(f_1 \circ f_2)_{\bar\gamma} \tag{12.2}$$

$$= \sum_{A \subset V_2(\bar\gamma)} \left(\prod_{v \in A} P_v \prod_{v \in V_2 \setminus A} Q_v \right) (f_1 \circ f_2)_{\bar\gamma} = \sum_{\gamma \in \Gamma(\bar\gamma)} \bar F_\gamma(f_1, f_2),$$

where $\Gamma(\bar\gamma)$ denotes the set of those coloured diagrams $\gamma \in \Gamma(k_1,k_2)$ which contain the same chains (with colour 1 or -1) as the non-coloured diagram $\bar\gamma$. Here $\Gamma(\bar\gamma)$ denotes the set of all such coloured diagrams which have the same chains as the diagram $\bar\gamma$, their chains of length 2 may have colour 1 or -1, while the colour of their chains with length 1 is -1. The function $\bar F_\gamma(f_1, f_2)$ is defined for a diagram $\gamma \in \Gamma(\bar\gamma)$ in the following way.

If the colouring of the chains of a coloured diagram $\gamma \in \Gamma(\bar\gamma)$ is defined with the help of a set $A \subset V_2(\bar\gamma)$ by the relations $c_\gamma(\beta(v)) = 1$ if $v \in A$, $c_\gamma(\beta(v)) = -1$ if $v \in V_2(\bar\gamma) \setminus A$, (and for the remaining chains $\beta \in \gamma$ with length 1 $c_\gamma(\beta) = -1$), then

$$\begin{aligned} \bar F_\gamma(f_1, f_2) &= \bar F_\gamma(f_1, f_2)(x_{l_1}, \dots, x_{l_{|O(\gamma)|}}) \\ &= \prod_{v \in A} P_v \prod_{v \in V_2 \setminus A} Q_v (f_1 \circ f_2)_{\bar\gamma}(x_1, \dots, x_{s(\bar\gamma)}). \end{aligned} \tag{12.3}$$

Here the indices $l_1, \dots, l_{|O(\gamma)|}$, $l_1 < \dots < l_{|O(\gamma)|}$, of the variables of the function $\bar F_\gamma(f_1, f_2)$ are the labels of the open chains (chains with colour -1) of the diagram γ, i.e, they are the elements of the set $(V_2(\bar\gamma) \setminus A) \cup V_1(\bar\gamma)$. (Clearly, $s(\gamma) = s(\bar\gamma)$ for the number of chains of γ and $\bar\gamma$ if $\gamma \in \Gamma(\bar\gamma)$.) In such a way we have defined $\bar F_\gamma(f_1, f_2)$ for each $\gamma \in \Gamma(\bar\gamma)$. The definition of this function is very similar to that of $F_\gamma(f_1, f_2)$ in formula (11.3). They differ only in the indexation of their variables. (The variables of the function $\bar F_\gamma(f_1, f_2)$ have indices $l_1, \dots, l_{|O(\gamma)|}$, and the set of these indices may be different of the set $\{1, \dots, |O(\gamma)|\}$. But we have defined the U-statistics with a kernel function also in this case.)

It is not difficult to check relation (12.2). We claim that it implies that a U-statistic with kernel function $(f_1 \circ f_2)_{\bar\gamma}$ satisfies the identity

$$\begin{aligned} &n^{-(k_1+k_2)/2} s(\bar\gamma)! I_{n,\bar s(\bar\gamma)}\left((f_1 \circ f_2)_{\bar\gamma}\right) \\ &\quad = \sum_{\gamma \in \Gamma(\bar\gamma)} n^{-(k_1+k_2)/2} n^{|C(\gamma)|} J_n(\gamma) |O(\gamma)|! I_{n,|O(\gamma)|}\left(\bar F_\gamma(f_1, f_2)\right) \end{aligned} \tag{12.4}$$

with the function $\bar F_\gamma(f_1, f_2)$, where $C(\gamma)$ is the set of closed chains of γ, and $J_n(\gamma)$ is defined as $J_n(\gamma) = 1$ if $|C(\gamma)| = 0$, and

$$J_n(\gamma) = \prod_{j=1}^{|C(\gamma)|} \left(\frac{n - s(\gamma) + j}{n}\right) \quad \text{if } |C(\gamma)| > 0 \tag{12.5}$$

for all $\bar\gamma \in \bar\Gamma(k_1, k_2)$.

Relation (12.4) follows from relation (12.2) in the same way as formula (9.5) follows from formula (9.3) in the proof of the Hoeffding decomposition. Let us understand why the coefficient $n^{|C(\gamma)|} J_n(\gamma)$ appears at the right-hand side of (12.4).

This coefficient can be calculated in the following way. Let us write up the identity

$$\begin{aligned} &n^{-(k_1+k_2)/2} (f_1 \circ f_2)_{\bar\gamma}(\xi_{j_1}, \dots, \xi_{j_{s(\bar\gamma)}}) \\ &\quad = \sum_{\gamma \in \Gamma(\bar\gamma)} n^{-(k_1+k_2)/2} \bar F_\gamma(f_1, f_2)(\xi_{j_{l_1}}, \dots, \xi_{j_{l_{|O(\gamma)|}}}) \end{aligned}$$

with the help of (12.2) for all sequences $\xi_{j_1}, \dots, \xi_{j_{s(\bar\gamma)}}$, and let us sum it up for all such sets of arguments $(j_1, \dots, j_{s(\bar\gamma)})$ for which all indices j_p, $1 \le p \le s(\bar\gamma)$, are different, and $1 \le j_p \le n$. Then we get at the left-hand side of the identity the U-statistic

$$n^{-(k_1+k_2)/2} s(\bar\gamma)! I_{n,\bar s(\bar\gamma)}\left((f_1 \circ f_2)_{\bar\gamma}\right).$$

We still have to check that at the right-hand side of this identity we get a sum, where a term of the form $n^{-(k_1+k_2)/2} \bar F_\gamma(f_1, f_2)(\xi_{j_{l_1}}, \dots, \xi_{j_{l_{|O(\gamma)|}}})$ appears with multiplicity $n^{|C(\gamma)|} J_n(\gamma)$. Indeed, such a term appears for such vectors $(j_1, \dots, j_{s(\bar\gamma)})$ for which the value of $|O(\gamma)|$ arguments are fixed, the remaining arguments can take arbitrary value between 1 and n with the only restriction that all coordinates must be different. (The operators P_ν are applied for these remaining coordinates.) There are $n^{|C(\gamma)|} J_n(\gamma)$ such vectors. The above observations imply identity (12.4).

Let us observe that $k_1 + k_2 - 2|C(\gamma)| = |O(\gamma)| + W(\gamma)$ with the number $W(\gamma)$ introduced in the formulation of Theorem 11.1. Hence

$$n^{-(k_1+k_2)/2} n^{|C(\gamma)|} = n^{-W(\gamma)/2} n^{-|O(\gamma)|/2}.$$

Let us replace the left-hand side of the last identity by its right-hand side in (12.4), and let us sum up the identity we get in such a way for all $\bar\gamma \in \bar\Gamma(k_1, k_2)$ such that $s(\bar\gamma) \le n$. The identity we get in such a way together with formulas (12.1) and (12.5) imply such a version of identity (11.4) where the kernel functions $F_\gamma(f_1, f_2)$ of the U-statistics at the right-hand side of the equation are replaced by the kernel functions $\bar F_\gamma(f_1, f_2)$ defined in (12.3). But we can get the function $F_\gamma(f_1, f_2)$ by reindexing the arguments of the function $\bar F_\gamma(f_1, f_2)$. This can be seen by taking the original indexation of the chains of γ and looking at the indexation of the vertices it implies. On the other hand, we know that the reindexation of the variables of the kernel function does not change the value of the U-statistic. Hence $I_{n,|O(\gamma)|}(F_\gamma(f_1, f_2)) = I_{n,|O(\gamma)|}(\bar F_\gamma(f_1, f_2))$, and identity (11.4) holds in its original form.

Clearly, $I_{n,|O(\gamma)|}(F_\gamma(f_1, f_2)) = I_{n,|O(\gamma)|}(\mathrm{Sym} F_\gamma(f_1, f_2))$, hence $I_{n,|O(\gamma)|}(F_\gamma(f_1, f_2))$ can be replaced by $I_{n,|O(\gamma)|}(\mathrm{Sym}\, F_\gamma(f_1, f_2))$ in formula (11.4). Beside this, we have shown that the functions $F_\gamma(f_1, f_2)$ are canonical, and it can be simply shown that they are bounded, if the functions f_1 and f_2 have this property. We still have to prove inequalities (11.5) and (11.6).

Inequality (11.5), the estimate of the L_2-norm of the function $F_\gamma(f_1, f_2)$ follows from the Schwarz inequality, and actually it agrees with inequality (10.13), proved at the start of Appendix B. Hence its proof is omitted here.

To prove inequality (11.6) let us introduce, similarly to formula (9.2), the operators

$$(\tilde Q_j h)(x_1, \dots, x_r) = h(x_1, \dots, x_r) + \int h(x_1, \dots, x_r)\mu(dx_j), \quad 1 \le j \le r,$$

in the space of functions $h(x_1,\dots,x_r)$ with coordinates in the space $(X,\mathcal{X})$. Observe that both the operators $\tilde{Q}_j$ and the operators P_j defined in (9.1) are positive, i.e. they map a non-negative function to a non-negative function. Beside this, $Q_j \le \tilde{Q}_j$, and the norms of the operators $\frac{\tilde{Q}_j}{2}$ and P_j are bounded by 1 both in the $L_1(\mu)$, the $L_2(\mu)$ and the supremum norm.

Let us define the function

$$\tilde{F}_\gamma(f_1,f_2)(x_1,\dots,x_{|O(\gamma)|}) = \left(\prod_{j:\beta(j)\in C(\gamma)} P_j \prod_{j':\beta(j')\in O_2(\gamma)} \tilde{Q}_{j'}\right)(f_1\circ f_2)_\gamma(x_q,\dots,x_{|O(\gamma)|+|C(\gamma)|})$$

with the notation of Chap. 11. The function $\tilde{F}_\gamma(f_1,f_2)$ was defined similarly to $F_\gamma(f_1,f_2)$ defined in (11.3) with the help of $(f_1\circ f_2)_\gamma$ only the operators Q_j were replaced by $\tilde{Q}_j$ in its definition.

The properties of the operators P_j and $\tilde{Q}_j$ listed above together with the condition $\sup|f_2(x_1,\dots,x_k)|\le 1$ imply that

$$|F_\gamma(f_1,f_2)| \le \tilde{F}_\gamma(|f_1|,|f_2|) \le \tilde{F}_\gamma(|f_1|,1), \tag{12.6}$$

where "$\le$" means that the function at the right-hand side is greater than or equal to the function at the left-hand side in all points, and the term 1 in (12.6) denotes the function which equals identically 1. Because of the relation (12.6) to prove relation (11.6) it is enough to show that

$$\begin{aligned}\|(\tilde{F}_\gamma(|f_1|,1)_\gamma\|_2 &= \left\|\left(\prod_{j:\beta(j)\in C(\gamma)} P_j \prod_{j':\beta(j'),\in O_2(\gamma)} \tilde{Q}_{j'}\right)|f_1(x_{\alpha_\gamma((1,1))},\dots,x_{\alpha_\gamma((1,k_1))})|\right\|_2\\ &\le 2^{|O_2(\gamma)|}\|f_1\|_2 = 2^{W(\gamma)}\|f_1\|_2.\end{aligned} \tag{12.7}$$

But this inequality trivially holds, since the norm of all operators P_j in formula (12.7) is bounded by 1, the norm of all operators $\tilde{Q}_j$ is bounded by 2 in the $L_2(\mu)$ norm, and $|O_2(\gamma)| = W(\gamma)$. □

Proof of Theorem 11.2. Theorem 11.2 will be proved with the help of Theorem 11.1 by induction with respect to the number m of the terms in the product of the degenerate U-statistics $k_p!I_{n,k_p}(f_p)$, $1\le p\le m$. It is not difficult to check with the help of Theorem 11.1 and the recursive definition of the functions F_γ by applying induction with respect to m that the functions $F_\gamma(f_1,\dots,f_m)$ are bounded and canonical if the functions $f_1,\dots,f_m$ satisfy the same properties. We still have to prove the identity (11.11). This will be proved also by induction with respect to m with the help of Theorem 11.1.

For $m = 2$ formula (11.11) follows from Theorem 11.1, since in this case it agrees with relation (11.4). To prove this formula for $m \ge 3$ first we express with the help of our inductive hypothesis the product of the first $m-1$ terms in the product of degenerate U-statistics as a sum of degenerate U-statistics. Then we express the product of each term in this sum with the last U-statistic of the product as a sum of U-statistics with the help of Theorem 11.1, and sum up these identities. In such a way we express the product of m degenerate U-statistics in the form of a sum of degenerate U-statistics. We have to show that in such a way we get formula (11.11). In the proof of this statement we shall exploit that in the calculation of the product of the first $m-1$ U-statistics we have to work with the diagrams γ_{pr} and if we calculate the product of these terms with the m-th the U-statistic, then we calculate with the diagrams γ_{cl}.

To carry out the above program first we observe that a diagram $\gamma \in \Gamma(k_1,\dots,k_m)$ is uniquely determined by the pairs of $(\gamma_{pr},\gamma_{cl})$ defined with the help of γ, i.e. if $\gamma,\gamma' \in \Gamma(k_1,\dots,k_m)$, and $\gamma \ne \gamma'$, then either $\gamma_{pr} \ne \gamma'_{pr}$ or $\gamma_{cl} \ne \gamma'_{cl}$. Hence we can identify each diagram $\gamma \in \Gamma(k_1,\dots,k_m)$ with the pair $(\gamma_{pr},\gamma_{cl})$ we defined with its help. Beside this, the pairs of diagrams $(\gamma_{pr},\gamma_{cl})$ satisfy the relations $\gamma_{pr} \in \Gamma(k_1,\dots,k_{m-1})$ and $\gamma_{cl} \in \Gamma(|O(\gamma_{pr})|,k_m)$. Moreover, the class of pairs of diagrams $(\gamma_{pr},\gamma_{cl})$, $\gamma \in \Gamma(k_1,\dots,k_m)$, have the following characterization. Take all such pairs of diagrams $(\bar\gamma,\hat\gamma)$ for which $\bar\gamma \in \Gamma(k_1,\dots,k_{m-1})$ and $\hat\gamma \in \Gamma(|O(\bar\gamma)|,k_m)$. There is a one to one correspondence between the pairs of diagrams $(\bar\gamma,\hat\gamma)$ with this property and the diagrams $\gamma \in \Gamma(k_1,\dots,k_m)$ in such a way that $\bar\gamma = \gamma_{pr}$ and $\hat\gamma = \gamma_{cl}$. (This correspondence depends on the labelling of the open chains of the diagrams $\bar\gamma \in \Gamma(k_1,\dots,k_{m-1})$ that we have previously fixed.) It is not difficult to check the above statements, and I leave it to the reader.

Because of our inductive hypothesis we can write by applying relation (11.11) of Theorem 11.2 with parameter $m-1$ the identity

$$\prod_{p=1}^{m-1} n^{-k_p/2} k_p! I_{n,k_p}(f_p) = \sum_{\bar\gamma\in\Gamma(k_1,\dots,k_{m-1})}^{\prime(n,m-1)} \left(\prod_{p=2}^{m-1} J_n(\bar\gamma,p)\right) n^{-W(\bar\gamma)/2}\cdot n^{-|O(\bar\gamma)|/2}|O(\bar\gamma)|! I_{n,|O(\bar\gamma)|}(F_{\bar\gamma}(f_1,\dots,f_{m-1})). \tag{12.8}$$

(Here we use the notations of Chap. 11.)

We get by applying the identity (11.4) of Theorem 11.1 for the product

$$n^{-|O(\bar\gamma)|/2}|O(\bar\gamma)|! I_{n,|O(\bar\gamma)|}(F_{\bar\gamma}(f_1,\dots,f_{m-1}))\cdot n^{-k_m/2} k_m! I_{n,k_m}(f_m),$$

and by multiplying it with $\left(\prod_{p=2}^{m-1} J_n(\bar\gamma,p)\right) n^{-W(\bar\gamma)/2}$ that the identity

$$\left(\prod_{p=2}^{m-1} J_n(\bar\gamma, p)\right) n^{-W(\bar\gamma)/2} n^{-|O(\bar\gamma)|/2} O(\bar\gamma)! I_{n,|O(|\bar\gamma|}(F_{\bar\gamma}(f_1,\dots,f_{m-1}))$$
$$\cdot n^{-k_m/2} k_m! I_{n,k_m}(f_m)$$
$$= \left(\prod_{p=2}^{m-1} J_n(\bar\gamma, p)\right) n^{-W(\bar\gamma)/2} \sum_{\hat\gamma\in\Gamma(|O(\bar\gamma|,k_m)}^{\prime(n)} \prod_{j=1}^{|C(\hat\gamma)|}\left(\frac{n-s(\hat\gamma)+j}{n}\right) \tag{12.9}$$
$$n^{-W(\hat\gamma)/2}\cdot n^{-|O(\hat\gamma)|/2} |O(\hat\gamma)|! I_{n,|O(\hat\gamma)|}(F_{\hat\gamma}(F_{\bar\gamma}(f_1,\dots,f_{m-1}), f_m)).$$

holds for all $\bar\gamma \in \Gamma(k_1,\dots,k_{m-1})$, where $\sum\limits_{\hat\gamma\in\Gamma(|O(\bar\gamma|),k_m)}^{\prime(n)}$ means that summation is taken for such diagrams $\hat\gamma \in \Gamma(|O(\bar\gamma)|, k_m)$ for which $s(\hat\gamma) = |O(\hat\gamma)| + |C(\hat\gamma)| \le n$, and $\prod\limits_{j=1}^{|C(\hat\gamma)|}$ equals 1, if $|C(\hat\gamma)| = 0$.

We shall prove relation (11.11) for the parameter m with the help of relations (12.8) and (12.9).

Let us sum up formula (12.9) for all such diagrams $\bar\gamma \in \Gamma(k_1,\dots,k_{m-1})$ for which $B_1(\bar\gamma, p)+B_2(\bar\gamma, p) \le n$ for all $2 \le p \le m-1$. The numbers $B_1(\cdot)$ and $B_2(\cdot)$ in these inequalities are the numbers introduced before formula (11.10), only in this case the diagram γ is replaced by $\bar\gamma$. We imposed those conditions on the terms $\bar\gamma$ in this summation which appear in the conditions of the summation in $\sum^{\prime(n,m-1)}$ at the right-hand side of formula (12.8) when it is applied with parameter $m-1$. Hence formula (12.8) implies that the sum of the terms at the left-hand side of these identities equals $\prod\limits_{p=1}^{m} n^{-k_p/2} k_p! I_{n,k_p}(f_p)$, i.e. the left-hand side of (11.11) for parameter m. To prove formula (11.11) for the parameter m it is enough to show that the sum of the right-hand side terms of the above identities equals the right-hand side of (11.11).

In the proof of this relation we shall apply the properties of the pairs of diagrams $(\gamma_{pr}, \gamma_{cl})$ coming from a diagram $\gamma \in \Gamma(k_1,\dots,k_m)$ mentioned before. Namely, we shall exploit that there is a one to one correspondence between the diagrams $\gamma \in \Gamma(k_1,\dots,k_m)$ and pairs of diagrams $(\bar\gamma, \hat\gamma)$, $\bar\gamma \in \Gamma(k_1,\dots,k_{m-1})$, $\hat\gamma \in \Gamma(|O(\bar\gamma)|, k_m)$ in such a way that γ and the pair $(\bar\gamma, \hat\gamma)$ correspond to each other if and only if $\bar\gamma = \gamma_{pr}$ and $\hat\gamma = \gamma_{cl}$. This correspondence enables us to reformulate the statement we have to prove in the following way. Let us rewrite formula (12.9) by replacing $\bar\gamma$ with γ_{pr} and $\hat\gamma$ by γ_{cl}, with that diagram $\gamma \in \Gamma(k_1,\dots,k_m)$ for which $\bar\gamma = \gamma_{pr}$ and $\hat\gamma = \gamma_{cl}$. It is enough to show that if we take those modified versions of (12.9) which we get by replacing the pairs $(\bar\gamma, \hat\gamma)$ by the pairs $(\gamma_{pr}, \gamma_{cl})$ with some $\gamma \in \Gamma(k_1,\dots,k_m)$ and sum up them for those γ for which $B_1(\gamma_{pr}, p) + B_2(\gamma_{pr}, p) \le n$ for all $2 \le p \le m-1$, then the sum of the right-hand side expressions in these identities equals the right-hand side of (11.11).

We shall prove the above identity with the help of the following statements to be verified later.

For all $\gamma \in \Gamma(k_1,\dots,k_m)$ the identities $W(\gamma_{pr})+W(\gamma_{cl})=W(\gamma)$ and

$$\prod_{p=2}^{m-1} J_n(\gamma_{pr},p)\prod_{j=1}^{|C(\gamma_{cl})|}\left(\frac{n-s(\gamma_{cl})+j}{n}\right)=\prod_{p=2}^{m} J_n(\gamma,p),$$

hold, where $\prod_{j=1}^{|C(\gamma_{cl})|}=1$ if $|C(\gamma_{cl})|=0$. The inequalities $B_1(\gamma,p)+B_2(\gamma,p)\le n$ hold simultaneously for all $2\le p\le m$ for a diagram γ if and only if the inequalities $B_1(\gamma_{pr},p)+B_2(\gamma_{pr},p)\le n$ for all $2\le p\le m-1$ and $s(\gamma_{cl})\le n$ hold simultaneously for this γ.

To prove the identity we claimed to hold with the help of the above relations let us first check that we sum up for the same set of $\gamma\in\Gamma(k_1,\dots,k_m)$ if we take the sum of modified versions of (12.9) for all γ such that $B_1(\gamma_{pr},p)+B_2(\gamma_{pr},p)\le n$ for all $2\le p\le m-1$ and if we take the $\sum^{\gamma(n,m)}$ at the right-hand side of (11.11). Indeed, in the second case we have to take those diagrams γ for which $B_1(\gamma,p)+B_2(\gamma,p)\le n$ for all $2\le p\le m$, while in the first case we take those diagrams γ for which $B_1(\gamma_{pr},p)+B_2(\gamma_{pr},p)\le n$ for all $2\le p\le m-1$, and $s(\gamma_{cl})\le n$. The last condition is contained in a slightly hidden form in the summation $\sum^{\prime(n)}$ of formula (12.9). Hence the above mentioned relations imply that have to sum up for the same diagrams γ in the two cases.

Beside this, it follows from (11.8) that the same U-statistics appear for a diagram $\gamma\in\Gamma(k_1,\dots,k_m)$ in (11.11) and in the modified version of (12.9). We still have to check that they have the same coefficients in the two cases. But this holds, because the previously formulated identities imply that

$$n^{-(W(\gamma_{pr})/2}n^{-W(\gamma_{cl})/2}=n^{-W(\gamma)/2},$$

$$\prod_{p=2}^{m-1} J_n(\gamma_{pr},p)\prod_{j=1}^{|C(\gamma_{cl})|}\left(\frac{n-s(\gamma_{cl})+j}{n}\right)=\prod_{p=2}^{m} J_n(\gamma,p)$$

and $n^{-|O(\gamma_{cl})|/2}|O(\gamma_{cl})|!=n^{-|O(\gamma)|/2}|O(\gamma)|!$, since $|O(\gamma)|=|O(\gamma_{cl})|$, as we have seen before.

To complete the proof of the identity it remained to check the relations we applied in the previous argument. We start with the proof of the identity $W(\gamma_{pr})+W(\gamma_{cl})=W(\gamma)$ for the function $W(\cdot)$ defined in (11.9).

Let us first remark that $W(\gamma_{cl})=|O_2(\gamma_{cl})|$, where $O_2(\gamma_{cl})$ is the set of open chains in γ_{cl} with length 2. Beside this if $\beta\in\gamma$ is such that $\beta\cap\{(m,1),\dots,(m,k)\}=\emptyset$, i.e. if the chain β contains no vertex from the last row of the diagram γ, then $\ell(\beta)=\ell(\beta_{pr})$, and $c_\gamma(\beta)=c_{\gamma_{pr}}(\beta_{pr})$. If $\beta\cap\{(m,1),\dots,(m,k)\}\neq\emptyset$, then either $c_\gamma(\beta)=1$, $\ell(\beta_{pr})=\ell(\beta)-1$, and $c_{\gamma_{pr}}(\beta) = -1$ or $c_\gamma(\beta) = -1$ and one of the following cases appears. Either $\ell(\beta) = 1$, and the chain β_{pr} does not exists, or $\ell(\beta) > 1$, and $\ell(\beta_{pr}) = \ell(\beta) - 1$, $c_{\gamma_{pr}}(\beta_{pr}) = -1$. We get by calculating $W(\gamma)$ with the help of the above relations that $W(\gamma) = W(\gamma_{pr}) + |\mathscr{V}(\gamma)|$, where $\mathscr{V}(\gamma) = \{\beta\colon\ \beta \in \gamma, \beta \cap \{(m,1),\dots,(m,k)\} \neq \emptyset, \ell(\beta) > 1, c_\gamma(\beta) = -1\}$. Since $|\mathscr{V}(\gamma)| = |O_2(\gamma_{cl})|$, the above relations imply the desired identity.

To prove the remaining relations first we observe that for each diagram $\gamma \in \Gamma(k_1,\dots,k_m)$ and number $2 \le p \le m-1$ the identities $B_1(\gamma_{pr}, p) = B_1(\gamma, p)$ and $B_2(\gamma_{pr}, p) = B_2(\gamma, p)$ hold. Beside this, $|C(\gamma_{cl})| = B_1(\gamma, m)$ and $|O(\gamma_{cl})| = B_2(\gamma, m)$. The identity about $|C(\gamma_{cl})|$ simply follows from the definition of γ_{cl} and $B_1(\gamma, m)$. To prove the identity about $|O(\gamma_{cl})|$ observe that $|O(\gamma_{cl})| = |O(\gamma)|$, and $|O(\gamma)| = B_2(\gamma, m)$. (Observe that in the case $p = m$ the definition of the set $\mathscr{B}_2(\gamma, m)$ becomes simpler, because there is no chain $\beta \in \gamma$ for which $d(\beta) > m$.)

The remaining relations can be deduced from these facts. Indeed, they imply that $J_n(\gamma_{pr}, p) = J_n(\gamma, p)$ for all $2 \le p \le m-1$. Beside this, $\prod\limits_{j=1}^{|C(\gamma_{cl})|} \left(\frac{n-s(\gamma_{cl})+j}{n}\right) = J_n(\gamma, m)$ because of the relations $|C(\gamma_{cl})| = B_1(\gamma, m)$ $|O(\gamma_{cl})| = B_2(\gamma, m)$, $s(\gamma_{cl}) = |C(\gamma_{cl})| + |O(|\gamma_{cl})|$ and the definition of $J_n(\gamma, m)$. Hence the identity about the product of the terms $J_n(\gamma, p)$ holds. It can be seen similarly that the relations $B_1(\gamma, p) + B_2(\gamma, p) \le n$ holds for all $2 \le p \le m-1$ if and only if $B_1(\gamma_{pr}, p) + B_2(\gamma_{pr}, p) \le n$ for all $2 \le p \le m-1$, and $B_1(\gamma, m) + B_2(\gamma, m) \le n$ if and only if $s(\gamma_{cl}) \le n$.

Thus we have proved identity (11.11). To complete the proof of Theorem 11.2 we still have to show that under its conditions $F_\gamma(f_1,\dots,f_m)$ is a bounded, canonical function. But this follows from Theorem 11.1 and relation (11.8) by a simple induction argument. □

Proof of Lemma 11.3. Lemma 11.3 will be proved by induction with respect to the number m of the terms in the product of U-statistics with the help of inequalities (11.5) and (11.6). These relations imply the desired inequality for $m = 2$. In the case $m > 2$ we apply the identity (11.8) $F_\gamma(f_1,\dots,f_m) = F_{\gamma_{cl}}(F_{\gamma_{pr}}(f_1,\dots,f_{m-1}), f_m)$. We have seen that $W(\gamma) = W(\gamma_{pr}) + W(\gamma_{cl})$, and it is not difficult to show that $U(\gamma) = U(\gamma_{pr}) + U(\gamma_{cl})$. Hence if $U(\gamma_{cl}) = 0$, i.e. if γ_{cl} contains a chain of length 2 with colour -1, then $U(\gamma) = U(\gamma_{pr})$, and an application of (11.8) and (11.6) for the diagram γ_{cl} implies Lemma 11.3 in this case.

If $U(\gamma_{cl}) = 1$, then $W(\gamma_{cl}) = 0$, $U(\gamma) = U(\gamma_{pr}) + 1$, $W(\gamma) = W(\gamma_{pr})$, and the application of (11.8) and (11.5) for the diagram γ_{cl} implies Lemma 11.3 in this case. □

The corollary of Theorem 11.2 is a simple consequence of Theorem 11.2 and Lemma 11.3.

Proof of the corollary of Theorem 11.2. Observe that F_γ is a function of $|O(\gamma)|$ arguments. Hence a coloured diagram $\gamma \in \Gamma(k_1,\dots,k_m)$ is in the class of closed diagrams, i.e. $\gamma \in \bar\Gamma(k_1,\dots,k_m)$ if and only if $F_\gamma(f_1,\dots,f_m)$ is a constant. Thus formula (11.13) is a simple consequence of relation (11.11) and the observation that $EI_{n,|O(\gamma)|}(F_\gamma(f_1,\dots,f_m)) = 0$ if $|O(\gamma)| \ge 1$, i.e. if $\gamma \notin \bar\Gamma(k_1,\dots,k_m)$, and

$$I_{n,|O(\gamma)|}(F_\gamma(f_1,\dots,f_m)) = I_{n,0}(F_\gamma(f_1,\dots,f_m)) = F_\gamma(f_1,\dots,f_m) \quad \text{if } \gamma \in \bar\Gamma(k_1,\dots,k_m).$$

Relations (11.14) and (11.15) follow from relation (11.13) and Lemma 11.3. □

Chapter 13
The Proof of Theorems 8.3, 8.5 and Example 8.7

In this chapter we prove the estimates on the distribution of a multiple Wiener–Itô integral or degenerate U-statistic formulated in Theorems 8.5 and 8.3, and also present the proof of Example 8.7. Beside this, we prove a multivariate version of Hoeffding's inequality (Theorem 3.4). The latter result is useful in the estimation of the supremum of a class of degenerate U-statistics. The estimate on the distribution of a multiple random integral with respect to a normalized empirical distribution given in Theorem 8.1 is omitted, because, as it was shown in Chap. 9, this result follows from the estimate of Theorem 8.3 on degenerate U-statistics. We finish this chapter with a separate part Chap. 13 B, where the results proved in this chapter are discussed together with the method of their proofs and some recent results. These new results state that in certain cases the estimates on the tail distribution of Wiener–Itô integrals and U-statistics considered in this chapter can be improved if we have some additional information on the kernel function of these Wiener–Itô integrals or U-statistics.

The proof of Theorems 8.5 and 8.3 is based on a good estimate on high moments of Wiener–Itô integrals and degenerate U-statistics. Such estimates can be proved with the help of the corollaries of Theorems 10.2 and 11.2. This approach slightly differs from the classical proof in the one-variate case. The one-variate version of the above problems is an estimate about the tail distribution of a sum of independent random variables. Such an estimate can be obtained with the help of a good bound on the moment generating function of the sum. This method does not work in the multivariate case, because, as later calculations will show, there is no good estimate on the moment-generating function of U-statistics or multiple Wiener–Itô integrals of order $k \geq 3$. Actually, the moment-generating function of a Wiener–Itô integral of order $k \geq 3$ is always divergent, because the tail distribution behaviour of such a random integral is similar to that of the k-th power of a Gaussian random variable. On the other hand, good bounds on the moments EZ^{2M} of a random variable Z for all positive integers M (or at least for a sufficiently rich class of parameters M) together with the application of the Markov inequality for Z^{2M} and an appropriate choice of the parameter M yield a good estimate on the tail distribution of Z.

P. Major, *On the Estimation of Multiple Random Integrals and U-Statistics*, Lecture Notes in Mathematics 2079, DOI 10.1007/978-3-642-37617-7_13,

Propositions 13.1 and 13.2 contain some estimates on the moments of Wiener–Itô integrals and degenerate U-statistics.

Proposition 13.1 (Estimate on the moments of Wiener–Itô integrals). *Let us consider a function $f(x_1,\dots,x_k)$ of k variables on some measurable space $(X,\mathcal{X})$ which satisfies formula (8.15) with some σ-finite non-atomic measure μ. Take the k-fold Wiener–Itô integral $Z_{\mu,k}(f)$ of this function with respect to a white noise μ_W with reference measure μ. The inequality*

$$E\left(|k!Z_{\mu,k}(f)|\right)^{2M} \le 1\cdot 3\cdot 5\cdots(2kM-1)\sigma^{2M} \quad \textit{for all } M=1,2,\dots \tag{13.1}$$

holds.

By Stirling's formula Proposition 13.1 implies that

$$E(|k!Z_{\mu,k}(f)|)^{2M} \le \frac{(2kM)!}{2^{kM}(kM)!}\sigma^{2M} \le A\left(\frac{2}{e}\right)^{kM}(kM)^{kM}\sigma^{2M} \tag{13.2}$$

for any $A>\sqrt{2}$ if $M \ge M_0 = M_0(A)$. Formula (13.2) can be considered as a simpler, better applicable version of Proposition 13.1. It can be better compared with the moment estimate on degenerate U-statistics given in formula (13.3).

Proposition 13.2 provides a similar, but weaker inequality for the moments of normalized degenerate U-statistics.

Proposition 13.2 (Estimate on the moments of degenerate U-statistics). *Let us consider a degenerate U-statistic $I_{n,k}(f)$ of order k with sample size n and with a kernel function f satisfying relations (8.1) and (8.2) with some $0<\sigma^2\le 1$. Fix a positive number $\eta>0$. There exist some universal constants $A<\infty$ and $C<\infty$ such that*

$$E\left(n^{-k/2}k!I_{n,k}(f)\right)^{2M} \le A\left(1+C\sqrt{\eta}\right)^{2kM}\left(\frac{2}{e}\right)^{kM}(kM)^{kM}\sigma^{2M} \quad \textit{for all integers } M \textit{ such that } 0\le kM\le \eta n\sigma^2. \tag{13.3}$$

In formula (13.3) the constant C can be chosen as $C=\sqrt{2}$.

Proposition 13.2 yields a good estimate on $E\left(n^{-k/2}k!I_{n,k}(f)\right)^{2M}$ with a fixed exponent $2M$ with the choice $\eta=\frac{kM}{n\sigma^2}$. With such a choice of the number η formula (13.3) yields an estimate on the moments $E\left(n^{-k/2}k!I_{n,k}(f)\right)^{2M}$ comparable with the estimate on the corresponding Wiener–Itô integral if $M\le n\sigma^2$, while it yields a much weaker estimate if $M\gg n\sigma^2$.

Now I turn to the proof of these propositions.

Proof of Proposition 13.1. Proposition 13.1 can be simply proved by means of the Corollary of Theorem 10.2 with the choice $m = 2M$, and $f_p = f$ for all $1 \le p \le 2M$. Formulas (10.22) and (10.23) yield that

$$E\left(k!Z_{\mu,k}(f)^{2M}\right) \le \left(\int f^2(x_1,\dots,x_k)\mu(dx_1)\dots\mu(dx_k)\right)^M |\Gamma_{2M}(k)|$$
$$\le |\Gamma_{2M}(k)|\sigma^{2M},$$

where $|\Gamma_{2M}(k)|$ denotes the number of closed diagrams γ in the class $\bar\Gamma(\underbrace{k,\dots,k}_{2M \text{ times}})$ introduced in the corollary of Theorem 10.2. Thus to complete the proof of Proposition 13.1 it is enough to show that $|\Gamma_{2M}(k)| \le 1\cdot 3\cdot 5\cdots(2kM-1)$. But this can easily be seen with the help of the following observation. Let $\bar\Gamma_{2M}(k)$ denote the class of all graphs with vertices (l,j), $1 \le l \le 2M$, $1 \le j \le k$, such that from all vertices (l,j) exactly one edge starts, all edges connect different vertices, but edges connecting vertices (l,j) and (l,j') with the same first coordinate l are also allowed. Let $|\bar\Gamma_{2M}(k)|$ denote the number of graphs in $\bar\Gamma_{2M}(k)$. Then clearly $|\Gamma_{2M}(k)| \le |\bar\Gamma_{2M}(k)|$. On the other hand, $|\bar\Gamma_{2M}(k)| = 1\cdot 3\cdot 5\cdots(2kM-1)$. Indeed, let us list the vertices of the graphs from $\bar\Gamma_{2M}(k)$ in an arbitrary way. Then the first vertex can be paired with another vertex in $2kM-1$ way, after this the first vertex from which no edge starts can be paired with $2kM-3$ vertices from which no edge starts. By following this procedure the next edge can be chosen $2kM-5$ ways, and by continuing this calculation we get the desired formula. □

Proof of Proposition 13.2. Relation (13.3) will be proved by means of relations (11.14) and (11.15) in the Corollary of Theorem 11.2 with the choice $m = 2M$ and $f_p = f$ for all $1 \le p \le 2M$. Let us take the class of closed coloured diagrams $\Gamma(k,M) = \bar\Gamma(\underbrace{k,\dots,k}_{2M \text{ times}})$. This will be partitioned into subclasses $\Gamma(k,M,r)$, $1 \le r \le kM$, where $\Gamma(k,M,r)$ contains those closed diagrams $\gamma \in \Gamma(k,M)$ for which $W(\gamma) = 2r$. Let us recall that $W(\gamma)$ was defined in (11.9), and in the case of closed diagrams $W(\gamma) = \sum\limits_{\beta\in\gamma}(\ell(\beta)-2)$. For a diagram $\gamma \in \Gamma(k,M)$, $W(\gamma)$ is an even number, since $W(\gamma)+2s(\gamma) = 2kM$, i.e. $W(\gamma) = 2r$ with $r = kM - s$, where $s = s(\gamma)$ denotes the number of chains in γ.

First we prove an estimate about the cardinality of $\Gamma(k,M,r)$. We claim that there exists a universal constant $A < \infty$ such that

$$|\Gamma(k,M,r)| \le \binom{2kM}{2r} 1\cdot 3\cdot 5\cdots(2kM-2r-1)(kM-r)^{2r} \tag{13.4}$$
$$\le A\left(\frac{2}{e}\right)^{kM}\binom{2kM}{2r}2^{-r}(kM)^{kM+r} \quad \text{for all } 0 \le r \le kM$$

with some universal constant $A < \infty$.

To prove formula (13.4) let us first observe that $|\Gamma(k,M,r)|$ can be bounded from above with the number of such partitions of a set with $2kM$ points which consists of $s = kM - r$ sets containing at least two points. Indeed, for each $\gamma \in \Gamma(k,M,r)$ the chains of the diagram γ yield a partition of the set $\{(p,r)\colon 1 \le p \le 2M,\ 1 \le k \le r\}$ consisting of $2r$ sets such that each of them contains at least two points. Moreover, the partition given in such a way determines the chains of γ, because the vertices of a chain are listed in a prescribed order. Namely, the indices of the rows which contain them follow each other in increasing order. This implies that we can correspond to each diagram $\gamma \in \Gamma(k,M,r)$ a different partition of a set of $2Mk$ elements with the prescribed properties.

The number of the partitions with the above properties can be bounded from above in the following way. Let us calculate the number of possibilities for choosing $s = kM - r$ disjoint subsets of cardinality two from a set of cardinality $2kM$, and multiply this number with the possibility of attaching each of the remaining $2r$ points of the original set to one of these sets of cardinality 2.

We can choose these sets of cardinality 2 in $\binom{2kM}{2r} 1 \cdot 3 \cdot 5 \cdots (2kM-1)$ ways, since we can choose the union of these sets, which consists of $2kM - 2r$ points in $\binom{2kM}{2kM-2r} = \binom{2kM}{2r}$ ways, and then we can choose the pair of the first element in $2kM-2r-1$ ways, then the pair of the first still not chosen element in $2kM-2r-3$ ways, and continuing this procedure we get the above formula for the number of choices for these sets of cardinality 2. Then the remaining $2r$ points of the original set can be put in $(kM-r)^{2r}$ ways in one of these $kM-r$ sets of cardinality 2. The above relations imply the first inequality of formula (13.4).

To get the second inequality observe that by the Stirling formula $1 \cdot 3 \cdot 5 \cdots (2kM - 2r - 1) = \frac{(2kM-2r)!}{2^{kM-r}(kM-r)!} \le A\left(\frac{2}{e}\right)^{kM-r}(kM-r)^{kM-r}$ with some universal constant $A < \infty$. Beside this, we can write $(kM-r)^{kM+r} \le (kM)^r (kM-r)^{kM} = (kM)^{kM+r}(1-\frac{r}{kM})^{kM} \le e^{-r}(kM)^{kM+r}$. These estimates imply the second inequality in (13.4).

We prove the estimate (13.3) with the help of the relations (11.14), (11.15) and (13.4). First we estimate the term $n^{-W(\gamma)/2}|F_\gamma|$ for a diagram $\gamma \in \Gamma(k,M,r)$ under the conditions $kM \le \eta n \sigma^2$ and $\sigma^2 \le 1$ with the help of relation (11.15).

In this case we can write $|U(\gamma)| \ge 2M - W(\gamma) = 2M - 2r$ for the function $U(\gamma)$ defined in (11.12). Hence by relation (11.15)

$$n^{-W(\gamma)/2}|F_\gamma| \le 2^{2r} n^{-r} \sigma^{|U(\gamma)|} \le 2^{2r}\left(n\sigma^2\right)^{-r}\sigma^{2M} \le \eta^r 2^{2r}(kM)^{-r}\sigma^{2M}$$

for $\gamma \in \Gamma(k,M,r)$ because of the conditions $kM \le \eta n \sigma^2$ and $\sigma^2 \le 1$.

This estimate together with relation (11.14) imply that for $kM \le \eta n \sigma^2$

$$\begin{aligned} E\left(n^{-k/2}k!I_{n,k}(f_k)\right)^{2M} &\le \sum_{\gamma \in \Gamma(k,M)} n^{-W(\gamma)/2} \cdot |F_\gamma| \\ &\le \sum_{r=0}^{kM} |\Gamma(k,M,r)| \eta^r 2^{2r} (kM)^{-r} \sigma^{2M}. \end{aligned}$$

Hence by formula (13.4)

$$E\left(n^{-k/2}k!I_{n,k}(f_k)\right)^{2M} \le A\left(\frac{2}{e}\right)^{kM}(kM)^{kM}\sigma^{2M}\sum_{r=0}^{kM}\binom{2kM}{2r}\left(\sqrt{2\eta}\right)^{2r}$$
$$\le A\left(\frac{2}{e}\right)^{kM}(kM)^{kM}\sigma^{2M}\left(1+\sqrt{2}\sqrt{\eta}\right)^{2kM}$$

if $0 \le kM \le \eta n\sigma^2$. Thus we have proved Proposition 13.2 with $C=\sqrt{2}$. □

It is not difficult to prove Theorem 8.5 with the help of Proposition 13.1.

Proof of Theorem 8.5. By formula (13.2) which is a consequence of Proposition 13.1 and the Markov inequality

$$P\left(|k!Z_{\mu,k}(f)|>u\right) \le \frac{E\left(k!Z_{\mu,k}(f)\right)^{2M}}{u^{2M}} \le A\left(\frac{2kM\sigma^{2/k}}{eu^{2/k}}\right)^{kM} \tag{13.5}$$

with some constant $A>\sqrt{2}$ if $M \ge M_0$ with some constant $M_0 = M_0(A)$, and M is an integer.

Put $\bar{M} = \bar{M}(u) = \frac{1}{2k}\left(\frac{u}{\sigma}\right)^{2/k}$, and $M = M(u) = [\bar{M}]$, where $[x]$ denotes the integer part of a real number x. Choose some number u_0 such that $\frac{1}{2k}\left(\frac{u_0}{\sigma}\right)^{2/k} \ge M_0+1$. Then relation (13.5) can be applied with $M=M(u)$ for $u \ge u_0$, and this yields that

$$P\left(|k!Z_{\mu,k}(f)|>u\right) \le A\left(\frac{2kM\sigma^{2/k}}{eu^{2/k}}\right)^{kM} \le e^{-kM} \le Ae^k e^{-k\bar{M}}$$
$$= Ae^k \exp\left\{-\frac{1}{2}\left(\frac{u}{\sigma}\right)^{2/k}\right\} \quad \text{if } u \ge u_0. \tag{13.6}$$

Relation (13.6) means that relation (8.17) holds for $u \ge u_0$ with the pre-exponential coefficient Ae^k. Beside this $u_0 \le$ const. By enlarging this coefficient if it is needed it can be guaranteed that relation (8.17) holds for all $u>0$. Theorem 8.5 is proved. □

Theorem 8.3 can be proved similarly by means of Proposition 13.2. Nevertheless, the proof is technically more complicated, since in this case the optimal choice of the parameter in the Markov inequality cannot be given in such a direct form as in the proof of Theorem 8.5. In this case the Markov inequality is applied with an almost optimal choice of the parameter M.

Proof of Theorem 8.3. The Markov inequality and relation (13.3) with $\eta = \frac{kM}{n\sigma^2}$ imply that

$$P(n^{-k/2}|k!I_{n,k}(f)| > u) \le \frac{E\left(n^{-k/2}k!I_{n,k}(f)\right)^{2M}}{u^{2M}} \le A\left(\frac{1}{e}\cdot 2kM\left(1+C\frac{\sqrt{kM}}{\sqrt{n}\sigma}\right)^2\left(\frac{\sigma}{u}\right)^{2/k}\right)^{kM} \tag{13.7}$$

for all integers $M \ge 0$.

Relation (8.12) will be proved with the help of estimate (13.7) under the condition $0 \le \frac{u}{\sigma} \le n^{k/2}\sigma^k$. To this end let us introduce the number $\bar M$ by means of the formula

$$k\bar M = \frac{1}{2}\left(\frac{u}{\sigma}\right)^{2/k}\frac{1}{1+\frac{B}{\sqrt{n}\sigma}\left(\frac{u}{\sigma}\right)^{1/k}} = \frac{1}{2}\left(\frac{u}{\sigma}\right)^{2/k}\frac{1}{1+B\left(un^{-k/2}\sigma^{-(k+1)}\right)^{1/k}}$$

with a sufficiently large number $B = B(C) > 0$ and $M = [\bar M]$, where $[x]$ means the integer part of the number x.

Observe that $\sqrt{k\bar M} \le \left(\frac{u}{\sigma}\right)^{1/k}$, $\frac{\sqrt{k\bar M}}{\sqrt{n}\sigma} \le \left(un^{-k/2}\sigma^{-(k+1)}\right)^{1/k} \le 1$, and

$$\left(1+C\frac{\sqrt{k\bar M}}{\sqrt{n}\sigma}\right)^2 \le 1+B\frac{\sqrt{k\bar M}}{\sqrt{n}\sigma} \le 1+B\left(un^{-k/2}\sigma^{-(k+1)}\right)^{1/k}$$

with a sufficiently large $B = B(C) > 0$ if $\frac{u}{\sigma} \le n^{k/2}\sigma^k$. Hence

$$\begin{aligned}\frac{1}{e}\cdot 2kM\left(1+C\frac{\sqrt{kM}}{\sqrt{n}\sigma}\right)^2\left(\frac{\sigma}{u}\right)^{2/k} &\le \frac{1}{e}\cdot 2k\bar M\left(1+C\frac{\sqrt{k\bar M}}{\sqrt{n}\sigma}\right)^2\left(\frac{\sigma}{u}\right)^{2/k}\\ &= \frac{1}{e}\cdot\frac{\left(1+C\frac{\sqrt{k\bar M}}{\sqrt{n}\sigma}\right)^2}{1+B\left(un^{-k/2}\sigma^{-(k+1)}\right)^{1/k}} \le \frac{1}{e}\end{aligned} \tag{13.8}$$

if $\frac{u}{\sigma} \le n^{k/2}\sigma^k$. Inequalities (13.7) and (13.8) together yield that

$$P(n^{-k/2}k!|I_{n,k}(f)| > u) \le Ae^{-kM} \le Ae^k e^{-k\bar M}$$

if $0 \le \frac{u}{\sigma} \le n^{k/2}\sigma^k$. Hence the choice of the number $\bar M$ implies that inequality (8.12) holds with the pre-exponential constant Ae^k and the sufficiently large but fixed number $B > 0$. Theorem 8.3 is proved. □

Remark. One would like to understand why the introduction of the quantities $\bar M$ and M in the proof of Theorem 8.3 was a good choice. The natural choice for M

would have been that number where the right-hand side expression in (13.7) takes its minimum. But we cannot calculate this number in a simple way. Hence we chose instead a sufficiently good and simple approximation for it. We get a first order approximation of this quantity if we consider the minimum of the simplified expression we get by dropping the factor $\left(1+C\frac{\sqrt{kM}}{\sqrt{n}\sigma}\right)^2$ from the formula at the right-hand side of (13.7). We get in such a way the approximation $M_0=\frac{1}{2k}(\frac{u}{\sigma})^{2/k}$, but this is not a sufficiently good choice of the number M for our purposes. We get a better approximation by determining the place of minimum of the expression we get by replacing the number M with the number M_0 in the factor we omitted in the previous approximation, i.e. we look for the place of minimum of

$$A\left(\frac{1}{e}\cdot 2kM\left(1+C\frac{\sqrt{kM_0}}{\sqrt{n}\sigma}\right)^2\left(\frac{\sigma}{u}\right)^{2/k}\right)^{kM}$$

$$=A\left(\frac{1}{e}\cdot 2kM\left(1+\frac{C}{\sqrt{2n}\sigma}\left(\frac{u}{\sigma}\right)^{1/k}\right)^2\left(\frac{\sigma}{u}\right)^{2/k}\right)^{kM}.$$

This suggests the approximation $M_1=\frac{1}{2k}\left(\frac{u}{\sigma}\right)^{2/k}\frac{1}{\left(1+\frac{C}{\sqrt{2n}\sigma}(\frac{u}{\sigma})^{1/k}\right)^2}$ for the place of minimum we are looking for. We can choose a similar expression for the parameter M which is almost as good as this number, but it is simpler to work with it. To find it observe that under the conditions of Theorem 8.3 we commit a small error by replacing the term $(1+\frac{C}{\sqrt{2n}\sigma}(\frac{u}{\sigma})^{1/k})^2$ in the denominator of the formula defining M_1 by $1+\frac{2C}{\sqrt{2n}\sigma}(\frac{u}{\sigma})^{1/k}$. To see this observe that the condition $\frac{u}{\sigma}\le n^{k/2}\sigma^k$ of Theorem 8.3 implies that $\frac{1}{\sqrt{n}\sigma}(\frac{u}{\sigma})^{1/k}\le 1$. Moreover, in the really interesting cases this expression is very close to zero. This suggests to expand the above square, and make an approximation by omitting the quadratic term. We can try to choose the number M obtained in such a way in the proof of Theorem 8.3. Moreover, it is useful to replace the parameter C with another number with which we can work better. It turned out that we can work better if the number C is replaced with another large coefficient. This led to the introduction of the quantity $k\bar{M}=\frac{1}{2}\left(\frac{u}{\sigma}\right)^{2/k}\frac{1}{1+\frac{B}{\sqrt{n}\sigma}\left(\frac{u}{\sigma}\right)^{1/k}}$ with a sufficiently large (but fixed) number B in the proof of Theorem 8.3.

Example 8.7 is a relatively simple consequence of Itô's formula for multiple Wiener–Itô integrals.

Proof of Example 8.7. We may restrict our attention to the case $k\ge 2$. Itô's formula for multiple Wiener–Itô integrals, more explicitly relation (10.25), implies that the random variable $k!Z_{\mu,k}(f)$ can be expressed as $k!Z_{\mu,k}(f)=\sigma H_k\left(\int f_0(x)\mu_W(dx)\right)=\sigma H_k(\eta)$, where $H_k(x)$ is the k-th Hermite polynomial with leading coefficient 1, and $\eta=\int f_0(x)\mu_W(dx)$ is a standard normal random variable. Hence we get by exploiting that the coefficient of x^{k-1} in the polynomial $H_k(x)$ is zero that $P(k!|Z_{\mu,k}(f)| > u) = P(|H_k(\eta)| \ge \frac{u}{\sigma}) \ge P\left(|\eta^k| - D|\eta^{k-2}| > \frac{u}{\sigma}\right)$ with a sufficiently large constant $D > 0$ if $\frac{u}{\sigma} > 1$. There exist such positive constants A and B for which

$$P\left(|\eta^k| - D|\eta^{k-2}| > \frac{u}{\sigma}\right) \ge P\left(|\eta^k| > \frac{u}{\sigma} + A\left(\frac{u}{\sigma}\right)^{(k-2)/k}\right) \quad \text{if } \frac{u}{\sigma} > B.$$

Hence

$$\begin{aligned} P(k!|Z_{\mu,k}(f)| > u) &\ge P\left(|\eta| > \left(\frac{u}{\sigma}\right)^{1/k}\left(1 + A\left(\frac{u}{\sigma}\right)^{-2/k}\right)\right) \\ &\ge \frac{\bar{C}\exp\left\{-\frac{1}{2}\left(\frac{u}{\sigma}\right)^{2/k}\right\}}{\left(\frac{u}{\sigma}\right)^{1/k} + 1} \end{aligned}$$

with an appropriate $\bar{C} > 0$ if $\frac{u}{\sigma} > B$. Since $P(k!|Z_{\mu,k}(f)| > 0) > 0$, the above inequality also holds for $0 \le \frac{u}{\sigma} \le B$ if the constant $\bar{C} > 0$ is chosen sufficiently small. This means that relation (8.19) holds. □

Next we prove a multivariate version of Hoeffding's inequality. Before its formulation some notations will be introduced.

Let us fix two positive integers k and n and some real numbers $a(j_1, \dots, j_k)$ for all sequences of arguments $\{j_1, \dots, j_k\}$ such that $1 \le j_l \le n$, $1 \le l \le k$, and $j_l \ne j_{l'}$ if $l \ne l'$.

With the help of the above real numbers $a(\cdot)$ and a sequence of independent random variables $\varepsilon_1, \dots, \varepsilon_n$, $P(\varepsilon_j = 1) = P(\varepsilon_j = -1) = \frac{1}{2}$, $1 \le j \le n$, the random variable

$$V = \sum_{\substack{(j_1,\dots,j_k)\colon 1 \le j_l \le n \text{ for all } 1 \le l \le k, \\ j_l \ne j_{l'} \text{ if } l \ne l'}} a(j_1, \dots, j_k)\varepsilon_{j_1} \cdots \varepsilon_{j_k} \tag{13.9}$$

and number

$$S^2 = \sum_{\substack{(j_1,\dots,j_k)\colon 1 \le j_l \le n \text{ for all } 1 \le l \le k, \\ j_l \ne j_{l'} \text{ if } l \ne l'}} a^2(j_1, \dots, j_k). \tag{13.10}$$

will be introduced.

With the help of the above notations the following result can be formulated.

Theorem 13.3 (The multivariate version of Hoeffding's inequality). *The random variable V defined in formula (13.9) satisfies the inequality*

$$P(|V| > u) \le C\exp\left\{-\frac{1}{2}\left(\frac{u}{S}\right)^{2/k}\right\} \quad \textit{for all } u \ge 0 \tag{13.11}$$

with the constant S defined in (13.10) and some constants $C > 0$ depending only on the parameter k in the expression V.

Theorem 13.3 will be proved by means of two simple lemmas. Before their formulation the random variable

$$Z = \sum_{\substack{(j_1,\dots,j_k):\, 1\le j_l\le n \text{ for all } 1\le l\le k,\\ j_l\ne j_{l'} \text{ if } l\ne l'}} |a(j_1,\dots,j_k)|\eta_{j_1}\cdots\eta_{j_k} \tag{13.12}$$

will be introduced, where $\eta_1,\dots,\eta_n$ are independent random variables with standard normal distribution, and the numbers $a(j_1,\dots,j_k)$ agree with those in formula (13.9). The following lemmas will be proved.

Lemma 13.4. *The random variables V and Z introduced in (13.9) and (13.12) satisfy the inequality*

$$EV^{2M} \le EZ^{2M} \quad \text{for all } M = 1,2,\dots.$$

Lemma 13.5. *The random variable Z defined in formula (13.12) satisfies the inequality*

$$EZ^{2M} \le 1\cdot 3\cdot 5\cdots(2kM-1)S^{2M} \quad \text{for all } M = 1,2,\dots \tag{13.13}$$

with the constant S defined in formula (13.10).

Proof of Lemma 13.4. We can write, by carrying out the multiplications in the expressions EV^{2M} and EZ^{2M}, by exploiting the additive and multiplicative properties of the expectation for sums and products of independent random variables together with the identities $E\varepsilon_j^{2k+1}=0$ and $E\eta_j^{2k+1}=0$ for all $k=0,1,\dots$ that

$$EV^{2M} = \sum_{\substack{(j_1,\dots,j_l,\,m_1,\dots,m_l):\\ 1\le j_s\le n,\ m_s\ge 1,\ 1\le s\le l,\ m_1+\dots+m_l=kM}} A(j_1,\dots,j_l,m_1,\dots,m_l)E\varepsilon_{j_1}^{2m_1}\cdots E\varepsilon_{j_l}^{2m_l} \tag{13.14}$$

and

$$EZ^{2M} = \sum_{\substack{(j_1,\dots,j_l,\,m_1,\dots,m_l):\\ 1\le j_s\le n,\ m_s\ge 1,\ 1\le s\le l,\ m_1+\dots+m_l=kM}} B(j_1,\dots,j_l,m_1,\dots,m_l)E\eta_{j_1}^{2m_1}\cdots E\eta_{j_l}^{2m_l} \tag{13.15}$$

with some coefficients $A(j_1,\dots,j_l,m_1,\dots,m_l)$ and $B(j_1,\dots,j_l,m_1,\dots,m_l)$ such that

$$|A(j_1,\dots,j_l,m_1,\dots,m_l)| \le B(j_1,\dots,j_l,m_1,\dots,m_l). \tag{13.16}$$

The coefficients $A(\cdot,\cdot,\cdot)$ and $B(\cdot,\cdot,\cdot)$ could be expressed explicitly, but we do not need such a formula. What is important for us is that $A(\cdot,\cdot,\cdot)$ can be expressed as the sum of certain terms, and $B(\cdot,\cdot,\cdot)$ as the sum of the absolute value of the same terms. Hence relation (13.16) holds. Since $E\varepsilon_j^{2m} \le E\eta_j^{2m}$ for all parameters j and m formulas (13.14)–(13.16) imply Lemma 13.4. □

Proof of Lemma 13.5. Let us consider a white noise $W(\cdot)$ on the unit interval $[0,1]$ with the Lebesgue measure λ on $[0,1]$ as its reference measure, i.e. let us take a set of Gaussian random variables $W(A)$ indexed by the measurable sets $A \subset [0,1]$ such that $EW(A) = 0$, $EW(A)W(B) = \lambda(A \cap B)$ with the Lebesgue measure λ for all measurable subsets of the interval $[0,1]$. Let us introduce n orthonormal functions $\varphi_1(x),\dots,\varphi_n(x)$ with respect to the Lebesgue measure on the interval $[0,1]$, and define the random variables $\eta_j = \int \varphi_j(x)W(dx)$, $0 \le j \le n$. Then $\eta_1,\dots,\eta_n$ are independent random variables with standard normal distribution, hence we may assume that they appear in the definition of the random variable Z in formula (13.12). Beside this, the identity $\eta_{j_1}\cdots\eta_{j_k} = \int \varphi_{j_1}(x_1)\cdots\varphi_{j_k}(x_k)W(dx_1)\dots W(dx_k)$ holds for all k-tuples $(j_1,\dots,j_k)$ such that $1 \le j_s \le n$ for all $1 \le s \le k$, and the indices $j_1,\dots,j_s$ are different. This identity follows from Itô's formula for multiple Wiener–Itô integrals formulated in formula (10.24) of Theorem 10.3.

Hence the random variable Z defined in (13.12) can be written in the form

$$Z = \int f(x_1,\dots,x_k)W(dx_1)\dots W(dx_k)$$

with the function

$$f(x_1,\dots,x_k) = \sum_{\substack{(j_1,\dots,j_k)\colon 1\le j_l\le n \text{ for all } 1\le l\le k,\\ j_l\ne j_{l'} \text{ if } l\ne l'}} |a(j_1,\dots,j_k)|\varphi_{j_1}(x_1)\cdots\varphi_{j_k}(x_k).$$

Because of the orthogonality of the functions $\varphi_j(x)$

$$S^2 = \int_{[0,1]^k} f^2(x_1,\dots,x_k)\,dx_1\dots dx_k.$$

Lemma 13.5 is a straightforward consequence of the above relations and formula (13.1) in Proposition 13.1. □

Proof of Theorem 13.3. The proof of Theorem 13.3 with the help of Lemmas 13.4 and 13.5 is an almost word for word repetition of the proof of Theorem 8.5. By Lemma 13.4 inequality (13.13) remains valid if the random variable Z is replaced by the random variable V at its left-hand side. Hence the Stirling formula yields that

$$EV^{2M} \le EZ^{2M} \le \frac{(2kM)!}{2^{kM}(kM)!}S^{2M} \le C\left(\frac{2}{e}\right)^{kM}(kM)^{kM}S^{2M}$$

for any $C \ge \sqrt{2}$ if $M \ge M_0(A)$. As a consequence, by the Markov inequality the estimate

$$P(|V| > u) \le \frac{EV^{2M}}{u^{2M}} \le C\left(\frac{2kM}{e}\left(\frac{S}{u}\right)^{2/k}\right)^{kM} \tag{13.17}$$

holds for all $C > \sqrt{2}$ if $M \ge M_0(C)$. Put $k\bar{M} = k\bar{M}(u) = \frac{1}{2}\left(\frac{u}{S}\right)^{2/k}$ and $M = M(u) = [\bar{M}]$, where $[x]$ denotes the integer part of the number x. Let us choose a threshold number u_0 by the identity $\frac{1}{2k}\left(\frac{u_0}{S}\right)^{2/k} = M_0(C) + 1$. Formula (13.17) can be applied with $M = M(u)$ for $u \ge u_0$, and it yields that

$$P(|V| > u) \le Ce^{-kM} \le Ce^k e^{-k\bar{M}} = Ce^k \exp\left\{-\frac{1}{2}\left(\frac{u}{S}\right)^{2/k}\right\} \quad \text{if } u \ge u_0.$$

The last inequality means that relation (13.11) holds for $u \ge u_0$ if the constant C is replaced by Ce^k in it. With the choice of a sufficiently large constant C relation (13.11) holds for all $u \ge 0$. Theorem 13.3 is proved. □

13. B) A Short Discussion About the Methods and Results

A comparison of Theorem 8.5 and Example 8.7 shows that the estimate (8.17) is sharp. At least no essential improvement of this estimate is possible which holds for *all* Wiener–Itô integrals with a kernel function f satisfying the conditions of Theorem 8.5. This fact also indicates that the bounds (13.1) and (13.2) on high moments of Wiener–Itô integrals are sharp. It is worth while comparing formula (13.2) with the estimate of Proposition 13.2 on moments of degenerate U-statistics.

Let us consider a normalized k-fold degenerate U-statistic $n^{-k/2}k!I_{n,k}(f)$ with some kernel function f and a μ-distributed sample of size n. Let us compare its moments with those of a k-fold Wiener–Itô integral $k!Z_{\mu,k}(f)$ with the same kernel function f with respect to a white noise μ_W with reference measure μ. Let σ denote the L_2-norm of the kernel function f. If $M \le \varepsilon n\sigma^2$ with a small number $\varepsilon > 0$, then Proposition 13.2 (with an appropriate choice of the parameter η which is small in this case) provides an almost as good bound on the $2M$-th moment of the normalized U-statistic as Proposition 13.1 does on the $2M$-th moment of the corresponding Wiener–Itô integral. In the case $M \le Cn\sigma^2$ with some fixed (not necessarily small) number $C > 0$ the $2M$-th moment of the normalized U-statistic can be bounded by $C(k)^M$ times the natural estimate on the $2M$-th moment of the Wiener–Itô integral with some constant $C(k) > 0$ depending only on the number C. This can be so interpreted that in this case the estimate on the moments of the normalized U-statistic is weaker than the estimate on the moments of the Wiener–Itô integral, but they are still comparable. Finally, in the case $M \gg n\sigma^2$ the estimate on the $2M$-th moment of the normalized U-statistic is much worse than the estimate on the $2M$-th moment of the Wiener–Itô integral.

A similar picture arises if the distribution of the normalized degenerate U-statistic

$$F_n(u) = P(n^{-k/2}|k!I_{n,k}(f)| > u)$$

is compared to the distribution of the Wiener–Itô integral

$$G(u) = P(|k!Z_{\mu,k}(f)| > u).$$

In the case $0 \le u \le \varepsilon n^{k/2}\sigma^{k+1}$ with a small $\varepsilon > 0$ Theorem 8.3 yields an almost as good estimate for the probability $F_n(u)$ as Theorem 8.5 yields for $G(u)$. In the case $0 \le u \le n^{k/2}\sigma^{k+1}$ these results yield similar bound for $F_n(u)$ and $G(u)$, only in the exponent of the estimate on $F_n(u)$ in formula (8.12) a worse constant appears. Finally, if $u \gg n^{k/2}\sigma^{k+1}$, then—as Example 8.8 shows, at least in the case $k = 2$—the (tail) distribution function $F_n(u)$ satisfies a much worse estimate than the function $G(u)$.

A similar picture arose in the one-variate version of this problem discussed in Chap. 3, where the normalized sums of independent random variables were investigated, and their tail-distributions were compared to that of a normally distributed random variable. To understand this similarity better it is useful to recall Theorem 10.4, i.e. the limit theorem about normalized degenerate U-statistics. Theorems 8.3 and 8.5 enable us to compare the tail behaviour of normalized degenerate U-statistics with their limit presented in the form of multiple Wiener–Itô integrals, while the one-variate versions of these results compare the distribution of sums of independent random variables with their Gaussian limit.

The proofs of the above results show that good bounds on the moments of degenerate U-statistics and multiple Wiener–Itô provide a good estimate on their distribution. To understand the behaviour of high moments of degenerate U-statistics better it is useful to have a closer look at the simplest case $k = 1$, when the moments of sums of independent random variables with expectation zero are considered.

Let us consider a sequence of independent and identically distributed random variables $\xi_1, \dots, \xi_n$ with expectation zero, take their sum $S_n = \sum\limits_{j=1}^{n} \xi_j$, and let us try to give a good estimate on the moments ES_n^{2M} for all $M = 1, 2, \dots$. Because of the independence of the random variables ξ_j and the condition $E\xi_j = 0$ the identity

$$ES_n^{2M} = \sum_{\substack{(j_1,\dots,j_s,l_1,\dots,l_s) \\ j_1+\dots+j_s=2M,\ j_u\ge 2, \text{ for all } 1\le u\le s \\ 1\le l_1<l_2<\dots<l_s\le n}} \frac{(2M)!}{j_1!\cdots j_s!} E\xi_{l_1}^{j_1}\cdots E\xi_{l_s}^{j_s} \tag{13.18}$$

holds. Simple combinatorial considerations suggest that the main contribution to the right-hand side of (13.18) is given by such vectors $(j_1,\dots,j_M; l_1,\dots,l_M)$ for which $j_u = 2$ for all $1 \le u \le M$. Their contribution is $\binom{n}{M}\frac{(2M)!}{2^M}(E\xi_1^2)^M \sim$

$n^M \frac{(2M)!}{2^M M!}(E\xi_1^2)^M$. The last asymptotic relation holds if the number n of terms in the random sum S_n is sufficiently large. The above considerations suggest that under not too restrictive conditions $ES_n^{2M} \sim (n\sigma^2)^M \frac{(2M)!}{2^M M!} = E\eta_{n\sigma^2}^{2M}$, where $\sigma^2 = E\xi^2$ is the variance of the terms in the sum S_n, and η_u denotes a random variable with normal distribution with expectation zero and variance u. The question arises when the above heuristic argument gives a valid estimate.

For the sake of simplicity let us restrict our attention to the case when the absolute value of the random variables ξ_j is bounded by 1. Let us observe that even in this case the above heuristic argument holds only under the condition that the variance σ^2 of the random variables ξ_j is not too small. Indeed, let us consider such random variables ξ_j, for which $P(\xi_j = 1) = P(\xi_j = -1) = \frac{\sigma^2}{2}$, $P(\xi_j = 0) = 1 - \sigma^2$. Then these random variables ξ_j have variance σ^2, and the contribution of the terms $E\xi_j^{2M}$, $1 \le j \le n$, to the sum in (13.18) equals $n\sigma^2$. If σ^2 is very small, then it may happen that $n\sigma^2 \gg (n\sigma^2)^M \frac{(2M)!}{2^M M!}$, and the approximation given for ES_n^{2M} in the previous paragraph does not hold any longer. Hence the asymptotic relation for a very high moment ES_n^{2M} suggested by the above heuristic argument may only hold if the variance σ^2 of the summands satisfies an appropriate lower bound.

In the proof of Proposition 13.2 a similar picture appears in a hidden way. In the calculation of the moments of a degenerate U-statistic the contribution of certain (closed) diagrams, more precisely of some integrals defined with their help, has to be estimated. Some of these diagrams (those in which all chains have length 2) appear also in the calculation of the moments of multiple Wiener–Itô integrals. In the calculation of the moments of sums of independent random variables the terms consisting of products of second moments play a similar role in the sum in formula (13.18) as the "nice" diagrams consisting of chains of length 2 play in the calculation of the moments of degenerate U-statistics in formula (11.14). In nice cases the remaining diagrams (multiplied with their small coefficients in formula (11.14)) do not give a greater contribution to the moments of degenerate U-statistics than these "nice" diagrams, and we get an almost as good bound for the moments of a normalized degenerate U-statistic as for the moments of the corresponding multiple Wiener–Itô integral. The proof of Proposition 13.2 shows that such a situation appears under very general conditions.

Let me also remark that there is an essential difference between the tail behaviour of Wiener–Itô integrals and normalized degenerate U-statistics. A good estimate can be given on the tail distribution of Wiener–Itô integrals which depends only on the L_2-norm of the kernel function, while in the case of normalized degenerate U-statistics the corresponding estimate depends not only on the L_2-norm but also on the L_∞ norm of the kernel function. In Theorem 8.3 such an estimate is proved.

For $k \ge 2$ the distribution of k-fold Wiener–Itô integrals are not determined by the L_2-norm of their kernel functions. This is an essential difference between Wiener–Itô integrals of order $k \ge 2$ and $k = 1$. In the case $k = 1$ a Wiener–Itô integral is a Gaussian random variable with expectation zero, and its variance equals the square of the L_2-norm of its kernel function. Hence its distribution is completely determined by the L_2-norm of its kernel function. On the other hand, the distribution of a Wiener–Itô integral of order $k \geq 2$ is not determined by its variance. Theorem 8.5 yields a "worst case" estimate on the distribution of Wiener–Itô integrals if we have a bound on their variance. In the statistical problems which were the main motivation for this work we need such estimates, but it may be interesting to know what kind of estimates are known about the distribution of a multiple Wiener–Itô integral or degenerate U-statistic if we have some additional information about its kernel function. Some results will be mentioned in this direction, but most technical details will be omitted from our discussion.

H.P. Mc. Kean proved the following lower bound on the distribution of multiple Wiener–Itô integrals. (See [32] or [44].)

Theorem 13.6 (Lower bound on the tail distribution of Wiener–Itô integrals). *All k-fold Wiener–Itô integrals $Z_{\mu,k}(f)$ satisfy the inequality*

$$P(|Z_{\mu,k}(f)| > u) > Ke^{-Au^{2/k}} \tag{13.19}$$

with some numbers $K = K(f,\mu) > 0$ and $A = A(f,\mu) > 0$.

The constant A in the exponent $Au^{2/k}$ of formula (13.19) is always finite, but Mc. Kean's proof yields no explicit upper bound on it. The following example shows that in certain cases if we fix the constant K in relation (13.19), then this inequality holds only with a very large constant $A > 0$ even if the variance of the Wiener–Itô integral equals 1.

Take a probability measure μ and a white noise μ_W with reference measure μ on a measurable space $(X, \mathcal{X})$, and let $\varphi_1, \varphi_2, \dots$ be a sequence of orthonormal functions on $(X, \mathcal{X})$ with respect to this measure μ. Define for all $L = 1, 2, \dots,$ the function

$$f(x_1,\dots,x_k) = f_L(x_1,\dots,x_k) = (k!)^{1/2} L^{-1/2} \sum_{j=1}^{L} \varphi_j(x_1)\cdots\varphi_j(x_k) \tag{13.20}$$

and the Wiener–Itô integral

$$Z_{\mu,k}(f) = Z_{\mu,k}(f_L) = \frac{1}{k!} \int f_L(x_1,\dots,x_k)\mu_W(dx_1)\dots\mu_W(dx_k).$$

Then $EZ^2_{\mu,k}(f) = 1$, and the high moments of $Z_{\mu,k}(f)$ can be well estimated. For a large parameter L these moments are much smaller, than the bound given in Proposition 13.1. (The calculation leading to the estimation of the moments of $Z_{\mu,k}(f)$ will be omitted.) These moment estimates also imply that if the parameter L is large, then for not too large numbers u the probability $P(|Z_{\mu,k}(f)| > u)$ has a much better estimate than that given in Theorem 8.5. As a consequence, for a large number L and fixed number K relation (13.19) may hold only with a very big number $A > 0$.

We can expect that if we take a Gaussian random polynomial $P(\xi_1,\dots,\xi_n)$ whose arguments are Gaussian random variables $\xi_1,\dots,\xi_n$, and which is the sum of many small almost independent terms with expectation zero, then a similar picture arises as in the case of a Wiener–Itô integral with kernel function (13.20) with a large parameter L. Such a random polynomial has an almost Gaussian distribution by the central limit theorem, and we can also expect that its not too high moments behave so as the corresponding moments of a Gaussian random variable with expectation zero and the same variance as the Gaussian random polynomial we consider. Such a bound on the moments has the consequence that the estimate on the probability of the event $\{\omega\colon\ P(\xi_1(\omega),\dots,\xi_n(\omega))>u\}$ given in Theorem 8.5 can be improved if the number u is not too large. A similar picture arises if we consider Wiener–Itô integrals whose kernel function satisfies some "almost independence" properties. The problem is to find the right properties under which we can get a good estimate that exploits the almost independence property of a Gaussian random polynomial or of a Wiener–Itô integral. The main result of R. Latała's paper [29] can be considered as a response to this question. This paper has some precedents, see [20] and [23], or paper [19] where such a result was applied. I describe the result of this paper below.

To formulate Latała's result some new notions have to be introduced. Given a finite set A let $\mathcal{P}(A)$ denote the set of all its partitions. If a partition $P=\{B_1,\dots,B_s\}\in\mathcal{P}(A)$ consists of s elements then we say that this partition has order s, and write $|P|=s$. In the special case $A=\{1,\dots,k\}$ the notation $\mathcal{P}(A)=\mathcal{P}_k$ will be used. Given a measurable space $(X,\mathcal{X})$ with a probability measure μ on it together with a finite set $B=\{b_1,\dots,b_j\}$ let us introduce the following notations. Take j different copies $(X_{b_r},\mathcal{X}_{b_r})$ and μ_{b_r}, $1\le r\le j$, of this measurable space and probability measure indexed by the elements of the set B, and define their product $(X^{(B)},\mathcal{X}^{(B)},\mu^{(B)})=\left(\prod\limits_{r=1}^{j}X_{b_r},\prod\limits_{r=1}^{j}\mathcal{X}_{b_r},\prod\limits_{r=1}^{j}\mu_{b_r}\right)$. The points $(x_{b_1},\dots,x_{b_j})\in X^{(B)}$ will be denoted by $x^{(B)}\in X^{(B)}$ in the sequel. With the help of the above notations I introduce the quantities needed in the formulation of the following Theorem 13.7.

Let $f=f(x_1,\dots,x_k)$ be a function on the k-fold product $(X^k,\mathcal{X}^k,\mu^k)$ of a measure space $(X,\mathcal{X},\mu)$ with a probability measure μ. For all partitions $P=\{B_1,\dots,B_s\}\in\mathcal{P}_k$ of the set $\{1,\dots,k\}$ consider the functions $g_r\left(x^{(B_r)}\right)$ on the space $X^{(B_r)}$, $1\le r\le s$, and define with their help the quantities

$$\begin{aligned}\alpha(P)&=\alpha(P,f,\mu)\\&=\sup_{g_1,\dots,g_s}\int f(x_1,\dots,x_k)g_1\left(x^{(B_1)}\right)\cdots g_s\left(x^{(B_s)}\right)\mu(dx_1)\dots\mu(dx_k);\end{aligned}$$

where supremum is taken for such functions

$$g_1,\dots,g_s,\quad g_r\colon X^{B_r}\to R^1 \text{ for which}$$

$$\int g_r^2\left(x^{(B_r)}\right)\mu^{(B_r)}\left(dx^{(B_r)}\right)\le 1\quad\text{for all } 1\le r\le s, \tag{13.21}$$

and put

$$\alpha_s = \max_{P \in \mathscr{P}_k, |P|=s} \alpha(P), \quad 1 \le s \le k. \tag{13.22}$$

In Latała's estimation of Wiener–Itô integrals of order k the quantities α_s, $1 \le s \le k$, play a similar role as the number σ^2 in Theorem 8.5. Observe that in the case $|P| = 1$, i.e. if $P = \{1,\dots,k\}$ the identity $\alpha^2(P) = \int f^2(x_1,\dots,x_k)\mu(dx_1)\dots\mu(dx_k)$ holds, which means that $\alpha_1 = \sigma$. The following estimate is valid for Wiener–Itô integrals of general order.

Theorem 13.7 (Latała's estimate about the tail-distribution of Wiener–Itô integrals). *Let a k-fold Wiener–Itô integral $Z_{\mu,k}(f)$, $k \ge 1$, be defined with the help of a white noise μ_W with a non-atomic reference measure μ and a kernel function f of k variables such that*

$$\int f^2(x_1,\dots,x_k)\mu(dx_1)\dots\mu(dx_k) < \infty.$$

There is some universal constant $C(k) < \infty$ depending only on the order k of the random integral such that the inequalities

$$E(Z_{\mu,k}(f))^{2M} \le \left(C(k) \max_{1\le s\le k} (M^{s/2}\alpha_s)\right)^{2M}, \tag{13.23}$$

and

$$P(|Z_{\mu,k}(f)| > u) \le C(k) \exp\left\{-\frac{1}{C(k)} \min_{1\le s\le k} \left(\frac{u}{\alpha_s}\right)^{2/s}\right\} \tag{13.24}$$

hold for all $M = 1,2,\dots$ and $u > 0$ with the quantities α_s, defined in formulas (13.21) and (13.22).

Inequality (13.24) is a simple consequence of (13.23). In the special case when $\alpha_s \le M^{-(s-1)/2}$ for all $1 \le s \le k$, t inequality (13.23) yields such an estimate on the moment $EZ_{\mu,k}(f)^{2M}$ which has the same magnitude as the $2M$-th moment of a standard Gaussian random variable multiplied by a constant, and (13.24) yields a good estimate on the probability $P(|Z_{\mu,k}(f)| > u)$. Actually the result of Theorem 13.7 can be reduced to the special case when $\alpha_s \le M^{-(s-1)/2}$ for all $1 \le s \le k$. Thus it can be interpreted so that if the quantities α_s of a k-fold Wiener–Itô integral are sufficiently small, then these "almost independence" conditions imply that the $2M$-th moment of this integral behaves similarly to a onefold Wiener–Itô integral with the same variance.

Actually Latała formulated his result in a different form, and he proved a slightly weaker result. He considered Gaussian polynomials of the following form:

$$
\begin{aligned}
&P(\xi_j^{(s)},\, 1 \le j \le n,\, 1 \le s \le k) \\
&\quad = \frac{1}{k!} \sum_{(j_1,\dots,j_k)\colon 1 \le j_s \le n,\, 1 \le s \le k} a(j_1,\dots,j_k)\xi_{j_1}^{(1)} \cdots \xi_{j_k}^{(k)}, \qquad (13.25)
\end{aligned}
$$

where $\xi_j^{(s)}$, $1 \le j \le n$ and $1 \le s \le k$, are independent standard normal random variables. Latała gave an estimate about the moments and tail-distribution of such random polynomials.

The problem about the behaviour of such random polynomials can be reformulated as a problem about the behaviour of Wiener–Itô integrals in the following way: Take a measurable space $(X, \mathcal{X})$ with a non-atomic measure μ on it. Let Z_μ be a white noise with reference measure μ, let us choose a set of orthogonal functions $h_j^{(s)}(x)$, $1 \le j \le n$, $1 \le s \le k$, on the space $(X, \mathcal{X})$ with respect to the measure μ, and define the function

$$
f(x_1,\dots,x_k) = \frac{1}{k!} \sum_{(j_1,\dots,j_k)\colon 1 \le j_s \le n,\, 1 \le s \le k} a(j_1,\dots,j_k) h_{j_1}^{(1)}(x_1) \cdots h_{j_k}^{(k)}(x_k) \qquad (13.26)
$$

together with the Wiener–Itô integral $Z_{\mu,k}(f)$. Since the random integrals $\xi_j^{(s)} = \int h_j^{(s)}(x) Z_\mu(dx)$, $1 \le j \le n$, $1 \le s \le k$, are independent, standard Gaussian random variables, it is not difficult to see with the help of Itô's formula (Theorem 10.3 in this work) that the distributions of the random polynomial $P(\xi_j^{(s)},\, 1 \le j \le n,\, 1 \le s \le k)$ and $Z_{\mu,k}(f)$ agree. Here we reformulated Latała's estimates about random polynomials of the form (13.25) to estimates about Wiener–Itô integrals with kernel function of the form (13.26).

These estimates are equivalent to Latała's result if we restrict our attention to the special class of Wiener–Itô integrals with kernel functions of the form (13.26). But we have formulated our result for Wiener–Itô integrals with a general kernel function. Latała's proof heavily exploits the special structure of the random polynomials given in (13.25), the independence of the random variables $\xi_j^{(s)}$ for different parameters s in it. (It would be interesting to find a proof which does not exploit this property.) On the other hand, this result can be generalized to the case discussed in Theorem 13.7. This generalization can be proved by exploiting the theorem of de la Peña and Montgomery–Smith about the comparison of U-statistics and decoupled U-statistics (formulated in Theorem 14.3 of this work) and the properties of the Wiener–Itô integrals. I omit the details of the proof.

Latała also proved a converse estimate in [29] about random polynomials of Gaussian random polynomials which shows that the estimates of Theorem 13.7 are sharp. We formulate it in its original form, i.e. we restrict our attention to the case of Wiener–Itô integrals with kernel functions of the form (13.26).

Theorem 13.8 (A lower bound about the tail distribution of Wiener–Itô integrals). *A random integral $Z_{\mu,k}(f)$ with a kernel function of the form (13.26) satisfies the inequalities*

$$E(Z_{\mu,k}(f))^{2M} \ge \left(C(k) \max_{1\le s\le k} (M^{s/2}\alpha_s)\right)^{2M},$$

and

$$P(|Z_{\mu,k}(f)| > u) \ge \frac{1}{C(k)} \exp\left\{-C(k) \min_{1\le s\le k} \left(\frac{u}{\alpha_s}\right)^{2/s}\right\}$$

for all $M = 1, 2, \dots$ and $u > 0$ with some universal constant $C(k) > 0$ depending only on the order k of the integral and the quantities α_s, defined in formula (13.21) and (13.22).

Let me finally remark that there is a counterpart of Theorem 13.7 about degenerate U-statistics. Adamczak's paper [1] contains such a result. Here we do not discuss it, because this result is far from the main topic of this work. We only remark that some new quantities have to be introduced to formulate it. The appearance of these conditions is related to the fact that in an estimate about the tail-behaviour of a degenerate U-statistic we need a bound not only on the L_2-norm but also on the supremum norm of the kernel function. In a sharp estimate the bound about the supremum of the kernel function has to be replaced by a more complex system of conditions, just as the condition about the L_2-norm of the kernel function was replaced by a condition about the quantities α_s, $1 \le s \le k$, defined in formulas (13.21) and (13.22) in Theorem 13.7.

Chapter 14
Reduction of the Main Result in This Work

The main result of this work is Theorem 8.4 or its multiple integral version Theorem 8.2. It was shown in Chap. 9 that Theorem 8.2 follows from Theorems 8.4. Hence it is enough to prove Theorem 8.4. It may be useful to study this problem together with its multiple Wiener–Itô integral version, Theorem 8.6.

Theorems 8.6 and 8.4 will be proved similarly to their one-variate versions, Theorems 4.2 and 4.1. Theorem 8.6 will be proved with the help of Theorem 8.5 about the estimation of the tail distribution of multiple Wiener–Itô integrals. A natural modification of the chaining argument applied in the proof of Theorem 4.2 works also in this case. No new difficulties arise. On the other hand, in the proof of Theorem 8.4 several new difficulties have to be overcome. I start with the proof of Theorem 8.6.

Proof of Theorem 8.6. Fix a number $0 < \varepsilon < 1$, and let us list the elements of the countable set $\mathcal{F}$ as $f_1, f_2, \dots$. For all $p = 0, 1, 2, \dots$ let us choose by exploiting the conditions of Theorem 8.6 a set of functions $\mathcal{F}_p = \{f_{a(1,p)}, \dots, f_{a(m_p,p)}\} \subset \mathcal{F}$ with $m_p \le 2D\,2^{(2p+4)L}\varepsilon^{-L}\sigma^{-L}$ elements in such a way that $\inf\limits_{1\le j\le m_p} \int (f - f_{a(j,p)})^2\,d\mu \le 2^{-4p-8}\varepsilon^2\sigma^2$ for all $f \in \mathcal{F}$, and beside this let $f_p \in \mathcal{F}_p$. For all indices $a(j,p)$, $p = 1, 2, \dots$, $1 \le j \le m_p$, choose a predecessor $a(j', p-1)$, $j' = j'(j,p)$, $1 \le j' \le m_{p-1}$, in such a way that the functions $f_{a(j,p)}$ and $f_{a(j',p-1)}$ satisfy the relation $\int |f_{a(j,p)} - f_{a(j',p-1)}|^2\,d\mu \le \varepsilon^2\sigma^2 2^{-4(p+1)}$. Theorem 8.5 with the choice $\bar u = \bar u(p) = 2^{-(p+1)}\varepsilon u$ and $\bar\sigma = \bar\sigma(p) = 2^{-2p-2}\varepsilon\sigma$ yields the estimates

$$
\begin{aligned}
P(A(j,p)) &= P\left(|k!Z_{\mu,k}(f_{a(j,p)} - f_{a(j',p-1)})| \ge 2^{-(1+p)}\varepsilon u\right) \\
&\le C\exp\left\{-\frac{1}{2}\left(\frac{2^{p+1}u}{\sigma}\right)^{2/k}\right\}, \qquad 1 \le j \le m_p, \qquad (14.1)
\end{aligned}
$$

P. Major, *On the Estimation of Multiple Random Integrals and U-Statistics*, Lecture Notes in Mathematics 2079, DOI 10.1007/978-3-642-37617-7_14,

for all $p = 1, 2, \dots$, and

$$P(B(s)) = P\left(|k! Z_{\mu,k}(f_{a(0,s)})| \ge \left(1 - \frac{\varepsilon}{2}\right) u\right)$$
$$\le C \exp\left\{-\frac{1}{2}\left(\frac{\left(1-\frac{\varepsilon}{2}\right)u}{\sigma}\right)^{2/k}\right\}, \quad 1 \le s \le m_0. \qquad (14.2)$$

Since each $f \in \mathcal{F}$ is the element of at least one set $\mathcal{F}_p$, $p = 0, 1, 2, \dots$, (We made a construction, where $f_p \in \mathcal{F}_p$), the definition of the predecessor of an index $a(j, p)$ and of the events $A(j, p)$ and $B(s)$ in formulas (14.1) and (14.2) together with the previous estimates imply that

$$P\left(\sup_{f \in \mathcal{F}} |k! Z_{\mu,k}(f)| \ge u\right) \le P\left(\bigcup_{p=1}^{\infty} \bigcup_{j=1}^{m_p} A(j, p) \cup \bigcup_{s=1}^{m_0} B(s)\right)$$
$$\le \sum_{p=1}^{\infty} \sum_{j=1}^{m_p} P(A(j, p)) + \sum_{s=1}^{m_0} P(B(s))$$
$$\le \sum_{p=1}^{\infty} 2CD 2^{(2p+4)L} \varepsilon^{-L} \sigma^{-L} \exp\left\{-\frac{1}{2}\left(\frac{2^{p+1}u}{\sigma}\right)^{2/k}\right\}$$
$$+ 2^{1+4L} CD \varepsilon^{-L} \sigma^{-L} \exp\left\{-\frac{1}{2}\left(\frac{\left(1-\frac{\varepsilon}{2}\right)u}{\sigma}\right)^{2/k}\right\}. \qquad (14.3)$$

Some calculations show that if $u \ge M L^{k/2} \sigma \frac{1}{\varepsilon} (\log^{k/2} \frac{2}{\varepsilon} + \log^{k/2} \frac{2}{\sigma})$ with a sufficiently large constant $M = M(k)$, then the inequalities

$$2^{(2p+4)L} \varepsilon^{-L} \sigma^{-L} \exp\left\{-\frac{1}{2}\left(\frac{2^{p+1}u}{\sigma}\right)^{2/k}\right\} \le 2^{-p} \left\{-\frac{1}{2}\left(\frac{(1-\varepsilon)u}{\sigma}\right)^{2/k}\right\}$$

hold for all $p = 1, 2 \dots$, and

$$2^{4L} \varepsilon^{-L} \sigma^{-L} \exp\left\{-\frac{1}{2}\left(\frac{\left(1-\frac{\varepsilon}{2}\right)u}{\sigma}\right)^{2/k}\right\} \le \exp\left\{-\frac{1}{2}\left(\frac{(1-\varepsilon)u}{\sigma}\right)^{2/k}\right\}.$$

These inequalities together with relation (14.3) imply relation (8.18). Theorem 8.6 is proved. □

The proof of Theorem 8.4 is harder. In this case the chaining argument in itself does not supply the proof, since Theorem 8.3 gives a good estimate about the distribution of a degenerate U-statistic only if it has a not too small variance. The same difficulty appeared in the proof of Theorem 4.1, and the method applied in that case will be adapted to the present situation.

A multivariate version of Proposition 6.1 will be proved in Proposition 14.1, and another result which can be considered as a multivariate version of Proposition 6.2 will be formulated in Proposition 14.2. It will be shown that Theorem 8.4 follows from Propositions 14.1 and 14.2. Most steps of these proofs can be considered as a simple repetition of the corresponding arguments in the proof of the results in Chap. 6. Nevertheless, I wrote down them for the sake of completeness.

The result formulated in Proposition 14.1 can be proved in almost the same way as its one-variate version, Proposition 6.1. The only essential difference is that now we apply a multivariate version of Bernstein's inequality given in the Corollary of Theorem 8.3. In the calculations of the proof of Proposition 14.1 the term $(\frac{u}{\sigma})^{2/k}$ shows a behaviour similar to the term $(\frac{u}{\sigma})^2$ in Proposition 6.1. Proposition 14.1 contains the information we can get by applying Theorem 8.3 together with the chaining argument. Its main content, inequality (14.4), yields a good estimate on the supremum of degenerate U-statistics if it is taken for an appropriate finite subclass $\mathcal{F}_{\bar\sigma}$ of the original class of kernel functions $\mathcal{F}$. The class of kernel functions $\mathcal{F}_{\bar\sigma}$ is a relatively dense subclass of $\mathcal{F}$ in the L_2 norm. Proposition 14.1 also provides some useful estimates on the value of the parameter $\bar\sigma$ which describes how dense the class of functions $\mathcal{F}_{\bar\sigma}$ is in $\mathcal{F}$.

Proposition 14.1. *Let the k-fold power $(X^k, \mathcal{X}^k)$ of a measurable space $(X, \mathcal{X})$ be given together with some probability measure μ on $(X, \mathcal{X})$ and a countable, L_2-dense class $\mathcal{F}$ of functions $f(x_1, \dots, x_k)$ of k variables with some exponent $L \ge 1$ and parameter $D \ge 1$ with respect to the measure μ^k on the product space $(X^k, \mathcal{X}^k)$ which also has the following properties. All functions $f \in \mathcal{F}$ are canonical with respect to the measure μ, and they satisfy conditions (8.5) and (8.6) with some real number $0 < \sigma \le 1$. Take a sequence of independent, μ-distributed random variables $\xi_1, \dots, \xi_n$, $n \ge \max(k, 2)$, and consider the (degenerate) U-statistics $I_{n,k}(f)$, $f \in \mathcal{F}$, defined in formula (8.8), and fix some number $\bar A = \bar A_k \ge 2^k$.*

There is a number $M = M(\bar A, k)$ such that for all numbers $u > 0$ for which the inequality $n\sigma^2 \ge (\frac{u}{\sigma})^{2/k} \ge M(L \log \frac{2}{\sigma} + \log D)$ holds, a number $\bar\sigma = \bar\sigma(u)$, $0 \le \bar\sigma \le \sigma \le 1$, and a collection of functions $\mathcal{F}_{\bar\sigma} = \mathcal{F}_{\bar\sigma(u)} = \{f_1, \dots, f_m\} \subset \mathcal{F}$ with $m \le D\bar\sigma^{-L}$ elements can be chosen in such a way that the union of the sets $\mathcal{D}_j = \{f\colon f \in \mathcal{F}, \int |f - f_j|^2\, d\mu \le \bar\sigma^2\}$, $1 \le j \le m$ cover the set $\mathcal{F}$. i.e. $\mathcal{F} = \bigcup\limits_{j=1}^{m} \mathcal{D}_j$, and the (degenerate) U-statistics $I_{n,k}(f)$, $f \in \mathcal{F}_{\bar\sigma(u)}$, satisfy the inequality

$$P\left(\sup_{f\in\mathcal{F}_{\bar\sigma(u)}} n^{-k/2}|k!I_{n,k}(f)|\ge\frac{u}{\bar A}\right)\le 2C\exp\left\{-\alpha\left(\frac{u}{10\bar A\sigma}\right)^{2/k}\right\}$$
$$\text{if}\quad n\sigma^2\ge\left(\frac{u}{\sigma}\right)^{2/k}\ge M\left(L\log\frac{2}{\sigma}+\log D\right) \tag{14.4}$$

with the constants $\alpha=\alpha(k)$, $C=C(k)$ appearing in formula (8.13) of the Corollary of Theorem 8.3 and the exponent L and parameter D of the L_2-dense class $\mathcal{F}$. Beside this, also the inequality $4\left(\frac{u}{\bar A\bar\sigma}\right)^{2/k}\ge n\bar\sigma^2\ge\frac{1}{64}\left(\frac{u}{\bar A\sigma}\right)^{2/k}$ holds for this number $\bar\sigma=\bar\sigma(u)$. If the number u satisfies also the inequality

$$n\sigma^2\ge\left(\frac{u}{\sigma}\right)^{2/k}\ge M(L^{3/2}\log\frac{2}{\sigma}+(\log D)^{3/2}) \tag{14.5}$$

with a sufficiently large number $M=M(\bar A,k)$, then the relation $n\bar\sigma^2\ge L\log n+\log D$ holds, too.

Proof of Proposition 14.1. Let us list the elements of the countable set $\mathcal{F}$ as $f_1,f_2,\dots$. For all $p=0,1,2,\dots$ let us choose, by exploiting the L_2-density property of the class $\mathcal{F}$ with respect to the product measure μ^k a set $\mathcal{F}_p=\{f_{a(1,p)},\dots,f_{a(m_p,p)}\}\subset\mathcal{F}$ with $m_p\le D\,2^{2pL}\sigma^{-L}$ elements in such a way that $\inf\limits_{1\le j\le m_p}\int(f-f_{a(j,p)})^2\,d\mu\le 2^{-4p}\sigma^2$ for all $f\in\mathcal{F}$. For all indices $a(j,p)$, $p=1,2,\dots$, $1\le j\le m_p$, choose a predecessor $a(j',p-1)$, $j'=j'(j,p)$, $1\le j'\le m_{p-1}$, in such a way that the functions $f_{a(j,p)}$ and $f_{a(j',p-1)}$ satisfy the relation $\int|f_{a(j,p)}-f_{a(j',p-1)}|^2\,d\mu\le\sigma^2 2^{-4(p-1)}$. Then the inequalities $\int\left(\frac{f_{a(j,p)}-f_{a(j',p-1)}}{2}\right)^2d\mu\le 4\sigma^2 2^{-4p}$ and $\sup\limits_{x_j\in X,\,1\le j\le k}\left|\frac{f_{a(j,p)}(x_1,\dots,x_k)-f_{a(j',p-1)}(x_1,\dots,x_k)}{2}\right|\le 1$ hold. The Corollary of Theorem 8.3 yields that

$$\begin{aligned}P(A(j,p))&=P\left(n^{-k/2}|k!I_{n,k}(f_{a(j,p)}-f_{a(j',p-1)})|\ge\frac{2^{-(1+p)}u}{\bar A}\right)\\&\le C\exp\left\{-\alpha\left(\frac{2^pu}{8\bar A\sigma}\right)^{2/k}\right\}\quad\text{if}\quad 4n\sigma^2 2^{-4p}\ge\left(\frac{2^pu}{8\bar A\sigma}\right)^{2/k},\\&\qquad 1\le j\le m_p,\ p=1,2,\dots,\end{aligned} \tag{14.6}$$

and

$$P(B(s)) = P\left(n^{-k/2}|k!I_{n,k}(f_{0,s})| \ge \frac{u}{2\bar{A}}\right) \le C\exp\left\{-\alpha\left(\frac{u}{2\bar{A}\sigma}\right)^{2/k}\right\},$$

$$1 \le s \le m_0, \quad \text{if } n\sigma^2 \ge \left(\frac{u}{2\bar{A}\sigma}\right)^{2/k}. \tag{14.7}$$

Introduce an integer $R = R(u)$, $R > 0$, which satisfies the relations

$$2^{(4+2/k)(R+1)}\left(\frac{u}{\bar{A}\sigma}\right)^{2/k} \ge 2^{2+6/k}n\sigma^2 \ge 2^{(4+2/k)R}\left(\frac{u}{\bar{A}\sigma}\right)^{2/k},$$

and define $\bar{\sigma}^2 = 2^{-4R}\sigma^2$ and $\mathcal{F}_{\bar{\sigma}} = \mathcal{F}_R$ (this is the class of functions $\mathcal{F}_p$ introduced at the start of the proof with $p = R$). We defined the number R, analogously to the proof of Proposition 6.1, as the largest number p for which the condition formulated in (14.6) holds. As $n\sigma^2 \ge \left(\frac{u}{\sigma}\right)^{2/k}$, and $\bar{A} \ge 2^k$ by our conditions, there exists such a positive integer R.) The cardinality m of the set $\mathcal{F}_{\bar{\sigma}}$ is clearly not greater than $D\bar{\sigma}^{-L}$, and $\bigcup_{j=1}^{m} \mathcal{D}_j = \mathcal{F}$. Beside this, the number R was chosen in such a way that the inequalities (14.6) and (14.7) hold for $1 \le p \le R$. Hence the definition of the predecessor of an index $a(j, p)$ implies that

$$\begin{aligned}
P\left(\sup_{f\in\mathcal{F}_{\bar{\sigma}}} n^{-k/2}k|k!I_{n,k}(f)| \ge \frac{u}{\bar{A}}\right) &\le P\left(\bigcup_{p=1}^{R}\bigcup_{j=1}^{m_p} A(j,p) \cup \bigcup_{s=1}^{m_0} B(s)\right) \\
&\le \sum_{p=1}^{R}\sum_{j=1}^{m_p} P(A(j,p)) + \sum_{s=1}^{m_0} P(B(s)) \\
&\le \sum_{p=1}^{\infty} CD\,2^{2pL}\sigma^{-L}\exp\left\{-\alpha\left(\frac{2^p u}{8\bar{A}\sigma}\right)^{2/k}\right\} \\
&\quad + CD\sigma^{-L}\exp\left\{-\alpha\left(\frac{u}{2\bar{A}\sigma}\right)^{2/k}\right\}.
\end{aligned}$$

If the condition $\left(\frac{u}{\sigma}\right)^{2/k} \ge M(L\log\frac{2}{\sigma} + \log D)$ holds with a sufficiently large constant M (depending on $\bar{A}$), then the inequalities

$$D2^{2pL}\sigma^{-L}\exp\left\{-\alpha\left(\frac{2^p u}{8\bar{A}\sigma}\right)^{2/k}\right\} \le 2^{-p}\exp\left\{-\alpha\left(\frac{2^p u}{10\bar{A}\sigma}\right)^{2/k}\right\}$$

hold for all $p = 1, 2, \dots,$ and

$$D\sigma^{-L}\exp\left\{-\alpha\left(\frac{u}{2\bar{A}\sigma}\right)^{2/k}\right\}\le\exp\left\{-\alpha\left(\frac{u}{10\bar{A}\sigma}\right)^{2/k}\right\}.$$

Hence the previous estimate implies that

$$\begin{aligned}P\left(\sup_{f\in\mathcal{F}_{\bar{\sigma}}} n^{-k/2}|k!I_{n,k}(f)|\ge\frac{u}{\bar{A}}\right)&\le\sum_{p=1}^{\infty}C2^{-p}\exp\left\{-\alpha\left(\frac{2^pu}{10\bar{A}\sigma}\right)^{2/k}\right\}\\&\quad+C\exp\left\{-\alpha\left(\frac{u}{10\bar{A}\sigma}\right)^{2/k}\right\}\\&\le 2C\exp\left\{-\alpha\left(\frac{u}{10\bar{A}\sigma}\right)^{2/k}\right\},\end{aligned}$$

and relation (14.4) holds.

The estimates

$$\begin{aligned}\frac{1}{64}\left(\frac{u}{\bar{A}\sigma}\right)^{2/k}&\le 2^{-2-6/k}2^{2R/k}\left(\frac{u}{\bar{A}\sigma}\right)^{2/k}=2^{-4R}\cdot 2^{(4+2/k)R-2-6/k}\left(\frac{u}{\bar{A}\sigma}\right)^{2/k}\\&\le n\bar{\sigma}^2=2^{-4R}n\sigma^2\le 2^{-4R}\cdot 2^{(4+2/k)(R+1)-2-6/k}\left(\frac{u}{\bar{A}\sigma}\right)^{2/k}\\&=2^{2-4/k}\cdot 2^{2R/k}\left(\frac{u}{\bar{A}\sigma}\right)^{2/k}=2^{2-4/k}\cdot 2^{-2R/k}\left(\frac{u}{\bar{A}\bar{\sigma}}\right)^{2/k}\le 4\left(\frac{u}{\bar{A}\bar{\sigma}}\right)^{2/k}\end{aligned}$$

hold because of the relation $R\ge 1$. This means that $n\bar{\sigma}^2$ has the upper and lower bound formulated in Proposition 14.1. It remained to show that $n\bar{\sigma}^2\ge L\log n+D$ if relation (14.5) holds.

This inequality clearly holds under the conditions of Proposition 14.1 if $\sigma\le n^{-1/3}$, since in this case $\log\frac{2}{\sigma}\ge\frac{\log n}{3}$, and

$$\begin{aligned}n\bar{\sigma}^2&\ge\frac{1}{64}\left(\frac{u}{\bar{A}\sigma}\right)^{2/k}\ge\frac{1}{64}\bar{A}^{-2/k}M\left(L\log\frac{2}{\sigma}+\log D)\right)\\&\ge\frac{1}{192}\bar{A}^{-2/k}M(L\log n+\log D)\ge L\log n+\log D\end{aligned}$$

if $M=M(\bar{A},k)$ is sufficiently large.

If $\sigma\ge n^{-1/3}$, then the inequality $2^{(4+2/k)R}\left(\frac{u}{\bar{A}\sigma}\right)^{2/k}\le 2^{2+6/k}n\sigma^2$ can be applied. This implies that $2^{-4R}\ge 2^{-4(2+6/k))/(4+2/k)}\left[\frac{\left(\frac{u}{\bar{A}\sigma}\right)^{2/k}}{n\sigma^2}\right]^{4/(4+2/k)}$, and

$$n\bar\sigma^2 = 2^{-4R}n\sigma^2 \ge \frac{2^{-16/3}}{\bar A^{4/3}}(n\sigma^2)^{1-\gamma}\left[\left(\frac{u}{\sigma}\right)^{2/k}\right]^{\gamma} \quad \text{with } \gamma = \frac{4}{4+\frac{2}{k}} \ge \frac{2}{3}.$$

The inequalities $n\sigma^2 \ge n^{1/3}$ and $n\sigma^2 \ge (\frac{u}{\sigma})^{2/k} \ge M(L^{3/2}\log\frac{2}{\sigma} + (\log D)^{3/2}) \ge \frac{M}{2}(L^{3/2} + (\log D)^{3/2})$ hold, (since $\log\frac{2}{\sigma} \ge \frac{1}{2}$). They yield that for sufficiently large $M = M(\bar A, k)$

$$(n\sigma^2)^{1-\gamma}\left[\left(\frac{u}{\sigma}\right)^{2/k}\right]^{\gamma} \ge (n\sigma^2)^{1-\gamma}\left[\left(\frac{u}{\sigma}\right)^{2/k}\right]^{2/3} = (n\sigma^2)^{1/(2k+1)}\left[\left(\frac{u}{\sigma}\right)^{2/k}\right]^{2/3},$$

and

$$n\bar\sigma^2 \ge \frac{\bar A^{-4/3}}{50}(n\sigma^2)^{1/(2k+1)}\left[\left(\frac{u}{\sigma}\right)^{2/k}\right]^{2/3}$$

$$\ge \frac{\bar A^{-4/3}}{50}n^{1/3(2k+1)}\left(\frac{M}{2}\right)^{2/3}(L^{3/2} + (\log D)^{3/2})^{2/3} \ge L\log n + \log D.$$

□

A multivariate analogue of Proposition 6.2 is formulated in Proposition 14.2, and it will be shown that Propositions 14.1 and 14.2 imply Theorem 8.4.

Proposition 14.2. *Let a probability measure μ be given on a measurable space $(X, \mathcal{X})$ together with a sequence of independent and μ distributed random variables $\xi_1, \dots, \xi_n$ and a countable L_2-dense class $\mathcal{F}$ of canonical (with respect to the measure μ) kernel functions $f = f(x_1, \dots, x_k)$ with some parameter $D \ge 1$ and exponent $L \ge 1$ on the product space $(X^k, \mathcal{X}^k)$. Let all functions $f \in \mathcal{F}$ satisfy conditions (8.1) and (8.2) with some $0 < \sigma \le 1$ such that $n\sigma^2 > L\log n + D$. Let us consider the (degenerate) U-statistics $I_{n,k}(f)$ with the random sequence $\xi_1, \dots, \xi_n$, $n \ge \max(2, k)$, and kernel functions $f \in \mathcal{F}$. There exists a threshold index $A_0 = A_0(k) > 0$ and two numbers $\bar C = \bar C(k) > 0$ and $\gamma = \gamma(k) > 0$ depending only on the order k of the U-statistics such that the degenerate U-statistics $I_{n,k}(f)$, $f \in \mathcal{F}$, satisfy the inequality*

$$P\left(\sup_{f\in\mathcal{F}} n^{-k/2}|k!I_{n,k}(f)| \ge An^{k/2}\sigma^{k+1}\right) \le \bar C e^{-\gamma A^{1/2k}n\sigma^2} \quad \text{if } A \ge A_0. \tag{14.8}$$

Proposition 14.2 yields an estimate for the tail distribution of the supremum of degenerate U-statistics at level $u \ge A_0 n^{k/2}\sigma^{k+1}$, i.e. in the case when Theorem 8.3 does not give a good estimate on the tail-distribution of the single degenerate U-statistics taking part in the supremum at the left-hand side of (14.8).

Formula (8.14) will be proved by means of Proposition 14.1 with an appropriate choice of the parameter $\bar A$ in it and Proposition 14.2 with the choice $\sigma = \bar\sigma = \bar\sigma(u)$ and the classes of functions $\mathcal{F}_j = \left\{\frac{g-f_j}{2}\colon g \in \mathcal{D}_j\right\}$ with the number $\bar\sigma$, functions f_j and sets of functions $\mathcal{D}_j$, $1 \le j \le m$, introduced in Proposition 14.1. Clearly,

$$P\left(\sup_{f\in\mathcal{F}} n^{-k/2}|k!I_{n,k}(f)|\ge u\right)\le P\left(\sup_{f\in\mathcal{F}_{\bar\sigma}} n^{-k/2}|k!I_{n,k}(f)|\ge \frac{u}{\bar A}\right)$$
$$+\sum_{j=1}^{m} P\left(\sup_{g\in\mathcal{D}_j} n^{-k/2}\left|k!I_{n,k}\left(\frac{f_j-g}{2}\right)\right|\ge\left(\frac12-\frac{1}{2\bar A}\right)u\right), \tag{14.9}$$

where m is the cardinality of the set of functions $\mathcal{F}_{\bar\sigma}$ appearing in Proposition 14.1. We shall estimate the two terms of the sum at the right-hand side of (14.9) by means of Propositions 14.1 and 14.2 with a good choice of the parameters $\bar A$ and the corresponding $M = M(\bar A)$ in Proposition 14.1 together with a parameter $A \ge A_0$ in Proposition 14.2.

We shall choose the parameter $A \ge A_0$ in the application of Proposition 14.2 so that it satisfies also the relation $\gamma A^{1/2k} \ge 2$ with the number γ appearing in relation (14.8), hence we put $A = \max(A_0, (\frac{2}{\gamma})^{2k})$. After this choice we want to define the parameter $\bar A$ in Proposition 14.1 in such a way that the numbers u satisfying the conditions of Proposition 14.1 also satisfy the relation $(\frac12-\frac{1}{2\bar A})u \ge An^{k/2}\bar\sigma^{k+1}$ with the already fixed number A and the number $\bar\sigma = \bar\sigma(u)$ defined in the proof of Proposition 14.1. This inequality can be rewritten in the form $A^{-2/k}(\frac12-\frac{1}{2\bar A})^{2/k}(\frac{u}{\bar\sigma})^{2/k} \ge n\bar\sigma^2$. On the other hand, under the conditions of Proposition 14.1 the inequality $4(\frac{u}{\bar A\bar\sigma})^{2/k} \ge n\bar\sigma^2$ holds. Hence the desired inequality holds if $A^{-2/k}(\frac12-\frac{1}{2\bar A})^{2/k} \ge 4\bar A^{-2/k}$. Thus the number $\bar A = 2^{k+1}A+1$ is an appropriate choice.

With such a choice of $\bar A$ (together with the corresponding $M = M(\bar A, k)$) and A we can write

$$P\left(\sup_{g\in\mathcal{D}_j} n^{-k/2}\left|k!I_{n,k}\left(\frac{f_j-g}{2}\right)\right|\ge\left(\frac12-\frac{1}{2\bar A}\right)u\right)$$
$$\le P\left(\sup_{g\in\mathcal{D}_j} n^{-k/2}\left|k!I_{n,k}\left(\frac{f_j-g}{2}\right)\right|\ge An^{k/2}\bar\sigma^{k+1}\right)\le \bar C e^{-\gamma A^{1/2k}n\bar\sigma^2}$$

for all $1\le j\le m$. (Observe that the set of functions $\frac{f_j-g}{2}$, $g\in\mathcal{D}_j$, is an L_2-dense class with parameter D and exponent L.) Hence Proposition 14.1 (relation (14.4) together with the inequality $m \le D\bar\sigma^{-L}$) and formula (14.8) with our $A \ge A_0$ and relation (14.9) imply that

$$P\left(\sup_{f\in\mathcal{F}} n^{-k/2}|k!I_{n,k}(f)|\ge u\right)$$
$$\le 2C\exp\left\{-\alpha\left(\frac{u}{10\bar A\sigma}\right)^{2/k}\right\}+\bar C D\bar\sigma^{-L}e^{-\gamma A^{1/2k}n\bar\sigma^2}. \tag{14.10}$$

We show by repeating an argument given in Chap. 6 that $D\bar\sigma^{-L} \le e^{n\bar\sigma^2}$. Indeed, we have to show that $\log D + L\log\frac{1}{\bar\sigma} \le n\bar\sigma^2$. But, as we have seen, the relation $n\bar\sigma^2 \ge L\log n + \log D$ with $L \ge 1$ and $D \ge 1$ implies that $n\bar\sigma^2 \ge \log n$, hence $\log\frac{1}{\bar\sigma} \le \log n$, and $\log D + L\log\frac{1}{\bar\sigma} \le \log D + L\log n \le n\bar\sigma^2$. On the other hand, $\gamma A^{1/2k} \ge 2$ by the definition of the number A, and by the estimates of Proposition 14.1 $n\bar\sigma^2 \ge \frac{1}{64}\left(\frac{u}{\bar A\sigma}\right)^{2/k}$. The above relations imply that $D\bar\sigma^{-L}e^{-\gamma A^{1/2k}n\bar\sigma^2} \le e^{-\gamma A^{1/2k}n\bar\sigma^2/2} \le \exp\left\{-\frac{\gamma}{128}A^{1/2k}\bar A^{-2/k}\left(\frac{u}{\sigma}\right)^{2/k}\right\}$. Hence relation (14.10) yields that

$$
\begin{aligned}
&P\left(\sup_{f\in\mathcal{F}} n^{-k/2}|k!I_{n,k}(f)|\ge u\right)\\
&\qquad\le 2C\exp\left\{-\frac{\alpha}{(10\bar A)^2}\left(\frac{u}{\sigma}\right)^{2/k}\right\}+\bar C\exp\left\{-\frac{\gamma}{128}A^{1/2k}\bar A^{-2/k}\left(\frac{u}{\sigma}\right)^{2/k}\right\},
\end{aligned}
$$

and this estimate implies Theorem 8.4. □

To complete the proof of Theorem 8.4 we have to prove Proposition 14.2. It will be proved, similarly to its one-variate version Proposition 6.2, by means of a symmetrization argument. We want to find its right formulation. It would be natural to formulate it as a result about the supremum of degenerate U-statistics. However, we shall choose a slightly different approach. There is a notion, called decoupled U-statistic. Decoupled U-statistics behave similarly to U-statistics, but it is simpler to work with them, because they have more independence properties. It turned out to be useful to introduce them and to apply a result of de la Peña and Montgomery–Smith which enables us to reduce the estimation of U-statistics to the estimation of decoupled U-statistics, and to work out the symmetrization argument for decoupled U-statistics.

Next we introduce the notion of decoupled U-statistics together with their randomized version. We also formulate a result of de la Peña and Montgomery–Smith in Theorem 14.3 which enables us to reduce Proposition 14.2 to a version of it, presented in Proposition 14.2′. It states a result similar to Proposition 14.2 about decoupled U-statistics. The proof of Proposition 14.2′ is the hardest part of the problem. In Chaps. 15–17 we deal essentially with this problem. The result of de la Peña and Montgomery–Smith will be proved in Appendix D.

Definition of decoupled and randomized decoupled U-statistics. *Let us have k independent copies $\xi_1^{(j)},\dots,\xi_n^{(j)}$, $1\le j\le k$, of a sequence $\xi_1,\dots,\xi_n$ of independent and identically distributed random variables taking their values in a measurable space $(X,\mathcal{X})$ together with a measurable function $f(x_1,\dots,x_k)$ on the product space $(X^k,\mathcal{X}^k)$ with values in a separable Banach space. The decoupled U-statistic $\bar I_{n,k}(f)$ determined by the random sequences $\xi_1^{(j)},\dots,\xi_n^{(j)}$, $1\le j\le k$, and kernel function f is defined by the formula*

$$\bar{I}_{n,k}(f) = \frac{1}{k!} \sum_{\substack{(l_1,\dots,l_k):\, 1 \le l_j \le n,\, j=1,\dots,k, \\ l_j \ne l_{j'} \text{ if } j \ne j'}} f\left(\xi_{l_1}^{(1)}, \dots, \xi_{l_k}^{(k)}\right). \tag{14.11}$$

Let us have beside the sequences of random variables $\xi_1^{(j)}, \dots, \xi_n^{(j)}$, $1 \le j \le k$, and function $f(x_1, \dots, x_k)$ a sequence of independent random variables $\varepsilon = (\varepsilon_1, \dots, \varepsilon_n)$, $P(\varepsilon_l = 1) = P(\varepsilon_l = -1) = \frac{1}{2}$, $1 \le l \le n$, which is independent also of the sequences of random variables $\xi_1^{(j)}, \dots, \xi_n^{(j)}$, $1 \le j \le k$. The randomized decoupled U-statistic $\bar{I}_{n,k}(f, \varepsilon)$ (depending on the random sequences $\xi_1^{(j)}, \dots, \xi_n^{(j)}$, $1 \le j \le k$, the kernel function f and the randomizing sequence $\varepsilon_1, \dots, \varepsilon_n$) is defined by the formula

$$\bar{I}_{n,k}^{\varepsilon}(f) = \frac{1}{k!} \sum_{\substack{(l_1,\dots,l_k):\, 1 \le l_j \le n,\, j=1,\dots,k, \\ l_j \ne l_{j'} \text{ if } j \ne j'}} \varepsilon_{l_1} \cdots \varepsilon_{l_k} f\left(\xi_{l_1}^{(1)}, \dots, \xi_{l_k}^{(k)}\right). \tag{14.12}$$

A decoupled or randomized decoupled U-statistics (with a real valued kernel function) will be called degenerate if its kernel function is canonical. This terminology is in full accordance with the definition of (usual) degenerate U-statistics.

A result of de la Peña and Montgomery–Smith will be formulated below. It gives an upper bound for the tail distribution of a U-statistic by means of the tail distribution of an appropriate decoupled U-statistic. It also has a generalization, where the supremum of U-statistics is bounded by the supremum of decoupled U-statistics. It enables us to reduce Proposition 14.2 to a version of it formulated Proposition 14.2′, which gives a bound on the tail distribution of the supremum of decoupled U-statistics. It is simpler to prove this result than the original one.

Before the formulation of the theorem of de la Peña and Montgomery–Smith I make some remark about it. In this result we consider more general U-statistics with kernel functions taking values in a separable Banach space, and we compare the norm of Banach space valued U-statistics and decoupled U-statistics. (Decoupled U-statistics were defined with general Banach space valued kernel functions, and the definition of U-statistics can also be generalized to separable Banach space valued kernel functions in a natural way.) This result was formulated in such a general form for a special reason. This helped us to derive formula (14.14) of the subsequent theorem from formula (14.13). It can be exploited in the proof of formula (14.14) that the constants in the estimate (14.13) do not depend on the Banach space where the kernel function f takes its values.

Theorem 14.3 (Theorem of de la Peña and Montgomery–Smith about the comparison of U-statistics and decoupled U-statistics). *Let us consider a sequence of independent and identically distributed random variables $\xi_1, \dots, \xi_n$ with values in a measurable space $(X, \mathcal{X})$ together with k independent copies $\xi_1^{(j)}, \dots, \xi_n^{(j)}$, $1 \le j \le k$, of this sequence. Let us also have a function $f(x_1, \dots, x_k)$ on the k-fold product space $(X^k, \mathcal{X}^k)$ which takes its values in a separable Banach space B.*

Let us take the U-statistic and decoupled U-statistic $I_{n,k}(f)$ and $\bar{I}_{n,k}(f)$ with the help of the above random sequences $\xi_1,\dots,\xi_n$, $\xi_1^{(j)},\dots,\xi_n^{(j)}$, $1\le j\le k$, and kernel function f. There exist some constants $\bar{C}=\bar{C}(k)>0$ and $\gamma=\gamma(k)>0$ depending only on the order k of the U-statistic such that

$$P\left(\|k!I_{n,k}(f)\|>u\right)\le \bar{C}P\left(\|k!\bar{I}_{n,k}(f)\|>\gamma u\right) \tag{14.13}$$

for all $u>0$. Here $\|\cdot\|$ denotes the norm in the Banach space B where the function f takes its values.

More generally, if we have a countable sequence of functions f_s, $s=1,2,\dots$, taking their values in the same separable Banach-space, then

$$P\left(\sup_{1\le s<\infty}\|k!I_{n,k}(f_s)\|>u\right)\le \bar{C}P\left(\sup_{1\le s<\infty}\|k!\bar{I}_{n,k}(f_s)\|>\gamma u\right). \tag{14.14}$$

Now I formulate the following version of Proposition 14.2.

Proposition 14.2′. *Let a probability measure μ be given on a measurable space $(X,\mathcal{X})$ together with a sequence of independent and μ distributed random variables $\xi_1,\dots,\xi_n$, $n\ge\max(k,2)$, and a countable L_2-dense class $\mathcal{F}$ of canonical (with respect to the measure μ) kernel functions $f=f(x_1,\dots,x_k)$ with some parameter $D\ge 1$ and exponent $L\ge 1$ on the product space $(X^k,\mathcal{X}^k)$. Let all functions $f\in\mathcal{F}$ satisfy conditions (8.1) and (8.2) with some $0<\sigma\le 1$ such that $n\sigma^2>L\log n+\log D$. Let us take k independent copies $\xi_1^{(j)},\dots,\xi_n^{(j)}$, $1\le j\le k$, of the random sequence $\xi_1,\dots,\xi_n$, and consider the decoupled U-statistics $\bar{I}_{n,k}(f)$, $f\in\mathcal{F}$, defined with their help in formula (14.11).*

There exists a threshold index $A_0=A_0(k)>0$ depending only on the order k of the decoupled U-statistics $I_{n,k}(f)$, $f\in\mathcal{F}$, such that the (degenerate) decoupled U-statistics $\bar{I}_{n,k}(f)$, $f\in\mathcal{F}$, satisfy the following version of inequality (14.8):

$$P\left(\sup_{f\in\mathcal{F}} n^{-k/2}|k!\bar{I}_{n,k}(f)|\ge An^{k/2}\sigma^{k+1}\right)\le e^{-2^{-(1/2+1/2k)}A^{1/2k}n\sigma^2} \quad \text{if } A\ge A_0. \tag{14.15}$$

It is clear that Proposition 14.2′ and Theorem 14.3, more explicitly formula (14.14) in it, imply Proposition 14.2. Hence the proof of Theorem 8.4 was reduced to Proposition 14.2′ in this chapter. The proof of Proposition 14.2′ is based on a symmetrization argument. Its main ideas will be explained in the next chapter.

Chapter 15
The Strategy of the Proof for the Main Result of This Work

In the previous chapter the proof of Theorem 8.4 was reduced to that of Proposition 14.2′. Proposition 14.2′ is a multivariate version of Proposition 6.2, and its proof is based on similar ideas. An important step in the proof of Proposition 6.2 was a symmetrization argument in which we reduced the estimation of the probability $P\left(\sup\limits_{f\in\mathscr{F}}\sum\limits_{j=1}^{n} f(\xi_j)>u\right)$ to that of the probability $P\left(\sup\limits_{f\in\mathscr{F}}\sum\limits_{j=1}^{n} \varepsilon_j f(\xi_j)>\frac{u}{3}\right)$, where $\xi_1,\dots,\xi_n$ is a sequence of independent and identically distributed random variables, and ε_j, $1\le j\le n$, is a sequence of independent random variables with distribution $P(\varepsilon_j=1)=P(\varepsilon_j=-1)=\frac{1}{2}$, independent of the sequence ξ_j. We want to prove a similar symmetrization argument which helps to prove Proposition 14.2′.

The symmetrization argument applied in the proof of Proposition 6.2 was carried out in two steps. We took a copy $\xi_1',\dots,\xi_n'$ of the sequence $\xi_1,\dots,\xi_n$, i.e. a sequence of independent random variables which is independent also of the original sequence $\xi_1,\dots,\xi_n$, and has the same distribution. In the first step we compared the tail distribution of the expression $\sup\limits_{f\in\mathscr{F}}\sum\limits_{j=1}^{n}[f(\xi_j)-f(\xi_j')]$ with that of $\sup\limits_{f\in\mathscr{F}}\sum\limits_{j=1}^{n} f(\xi_j)$ with the help of Lemma 7.1. In the second step, in the proof of Lemma 7.2, we applied a "randomization argument" which stated that the distribution of the random fields $\sum\limits_{j=1}^{n}[f(\xi_j)-f(\xi_j')]$ and $\sum\limits_{j=1}^{n}\varepsilon_j[f(\xi_j)-f(\xi_j')]$, $f\in\mathscr{F}$, agree. The symmetrization argument was proved with the help of these two observations.

In the proof of Proposition 14.2′ we would like to reduce the estimation of the tail distribution of the supremum of decoupled U-statistics $\sup\limits_{f\in\mathscr{F}}\bar{I}_{n,k}(f)$ defined in formula (14.11) to the estimation of the tail distribution of the supremum of the randomized decoupled U-statistics $\sup\limits_{f\in\mathscr{F}}\bar{I}^{\varepsilon}_{n,k}(f)$ defined in formula (14.12) in a similar way. To do this we have to find the multivariate version of the "randomization argument" in the proof of Lemma 7.2. This will be done in

P. Major, *On the Estimation of Multiple Random Integrals and U-Statistics*, Lecture Notes in Mathematics 2079, DOI 10.1007/978-3-642-37617-7_15,

the subsequent Lemma 15.1. In Lemma 7.2 this randomization argument was formulated with the help of some random variables introduced in formulas (7.4) and (7.5). We shall define their multivariate versions in formulas (15.1) and (15.2), and they will play a similar role in the formulation of Lemma 15.1.

The adaptation of the first step of the symmetrization argument of the proof of Proposition 6.2 is much harder. The proof of Proposition 6.2 was based on a symmetrization lemma formulated in Lemma 7.1. This result does not work in the present case. Hence we shall generalize it in Lemma 15.2. The proof of the symmetrization argument needed in the proof of Proposition 14.2′ is difficult even with the help of this result. The hardest part of our problem appears at this point. I return to it after the formulation of Lemma 15.2.

To formulate Lemma 15.1 we introduce the following notations.

Let $\mathcal{V}_k = \{(v(1),\dots,v(k))\colon\ v(j) = \pm 1,\ \text{for all } 1 \le j \le k\}$ denote the set of all ± 1 sequences of length k. Let $m(v)$ denote the number of -1 digits in a sequence $v = (v(1),\dots,v(k)) \in \mathcal{V}_k$. Let a (real valued) function $f(x_1,\dots,x_k)$ of k variables be given on a measurable space $(X,\mathcal{X})$ together with a sequence of independent and identically distributed random variables $\xi_1,\dots,\xi_n$ with values in the space $(X,\mathcal{X})$. Take $2k$ independent copies $\xi_1^{(j,1)},\dots,\xi_n^{(j,1)}$ and $\xi_1^{(j,-1)},\dots,\xi_n^{(j,-1)}$, $1 \le j \le k$, of the sequence $\xi_1,\dots,\xi_n$. Let us have beside them another sequence $\varepsilon = (\varepsilon_1,\dots,\varepsilon_n)$, $P(\varepsilon_j = 1) = P(\varepsilon_j = -1) = \frac{1}{2}$, of independent random variables, also independent of all previously introduced random variables. With the help of the above quantities we introduce the random variables

$$\tilde{I}_{n,k}(f) = \frac{1}{k!}\sum_{v\in\mathcal{V}_k}(-1)^{m(v)} \sum_{\substack{(l_1,\dots,l_k)\colon 1\le l_r\le n,\ r=1,\dots,k,\\ l_r\ne l_{r'} \text{ if } r\ne r'}} f\left(\xi_{l_1}^{(1,v(1))},\dots,\xi_{l_k}^{(k,v(k))}\right) \tag{15.1}$$

and

$$\tilde{I}^{\varepsilon}_{n,k}(f) = \frac{1}{k!}\sum_{v\in\mathcal{V}_k}(-1)^{m(v)} \sum_{\substack{(l_1,\dots,l_k)\colon 1\le l_r\le n,\ r=1,\dots,k,\\ l_r\ne l_{r'} \text{ if } r\ne r'}} \varepsilon_{l_1}\cdots\varepsilon_{l_k} f\left(\xi_{l_1}^{(1,v(1))},\dots,\xi_{l_k}^{(k,v(k))}\right) \tag{15.2}$$

The number $m(v)$ in the above formulas denotes the number of the digits -1 in the ± 1 sequence v of length k, hence it counts how many random variables $\xi_{l_j}^{(j,1)}$, $1 \le j \le k$, were replaced by the "secondary copy" $\xi_{l_j}^{(j,-1)}$ for a $v \in \mathcal{V}_k$ in the inner sum in formulas (15.1) or (15.2).

Remark. The definition of the linear combination of decoupled U-statistics $\tilde{I}^{\varepsilon}_{n,k}(f)$ defined in (15.2) shows some similarity to the definition of a Stieltjes measure determined by a function $f(x_1,\dots,x_k)$. One can argue that there is a deeper cause of these resemblance.

The following result holds.

Lemma 15.1. *Let us consider a (non-empty) class of functions $\mathcal{F}$ of k variables $f(x_1,\dots,x_k)$ on the space $(X^k,\mathcal{X}^k)$ together with the random variables $\tilde I_{n,k}(f)$ and $\tilde I^{\varepsilon}_{n,k}(f)$ defined in formulas (15.1) and (15.2) for all $f\in\mathcal{F}$. The distributions of the random fields $\tilde I_{n,k}(f)$, $f\in\mathcal{F}$, and $\tilde I^{\varepsilon}_{n,k}(f)$, $f\in\mathcal{F}$, agree.*

Let me recall that we say that the distribution of two random fields $X(f)$, $f\in\mathcal{F}$, and $Y(f)$, $f\in\mathcal{F}$, agree if for any finite sets $\{f_1,\dots,f_p\}\in\mathcal{F}$ the distribution of the random vectors $(X(f_1),\dots,X(f_p))$ and $(Y(f_1),\dots,Y(f_p))$ agree.

Proof of Lemma 15.1. I even claim that for any fixed sequence

$$u=(u(1),\dots,u(n)),\quad u(l)=\pm 1,\quad 1\le l\le n,$$

of length n the conditional distribution of the field $\tilde I^{\varepsilon}_{n,k}(f)$, $f\in\mathcal{F}$, under the condition $(\varepsilon_1,\dots,\varepsilon_n)=u=(u(1),\dots,u(n))$ agrees with the distribution of the field of $\tilde I_{n,k}(f)$, $f\in\mathcal{F}$.

Indeed, the random variables $\tilde I_{n,k}(f)$, $f\in\mathcal{F}$, defined in (15.1) are functions of a random vector with coordinates $(\xi_l^{(j)},\bar\xi_l^{(j)})=(\xi_l^{(j,1)},\xi_l^{(j,-1)})$, $1\le l\le n$, $1\le j\le k$, and the distribution of this random vector remains the same if the coordinates $(\xi_l^{(j)},\bar\xi_l^{(j)})=(\xi_l^{(j,1)},\xi_l^{(j,-1)})$ are replaced by $(\bar\xi_l^{(j)},\xi_l^{(j)})=(\xi_l^{(j,-1)},\xi_l^{(j,1)})$ for such pairs of indices (l,j) for which $u(l)=-1$ (and the index j is arbitrary), and the coordinates $(\xi_l^{(j)},\bar\xi_l^{(j)})$ with such pairs of indices (l,j) for which $u(l)=1$ are not modified. As a consequence, the distribution of the random field $\tilde I_{n,k}(f|u)$, $f\in\mathcal{F}$, we get by replacing the original vector $(\xi_l^{(j)},\bar\xi_l^{(j)})$, $1\le l\le n$, $1\le j\le k$, in the definition of the expression $\tilde I_{n,k}(f)$ in (15.1) for all $f\in\mathcal{F}$ by this modified vector depending on u has the same distribution as the random field $\tilde I_{n,k}(f)$, $f\in\mathcal{F}$. On the other hand, I claim that the distribution of the random field $\tilde I_{n,k}(f|u)$, $f\in\mathcal{F}$, agrees with the conditional distribution of the random field $\tilde I^{\varepsilon}_{n,k}(f)$, $f\in\mathcal{F}$, defined in (15.2) under the condition that $(\varepsilon_1,\dots,\varepsilon_n)=u$ with $u=(u(1),\dots,u(n))$.

To prove the last statement let us observe that the conditional distribution of the random field $\tilde I^{\varepsilon}_{n,k}(f)$, $f\in\mathcal{F}$, under the condition $(\varepsilon_1,\dots,\varepsilon_n)=u$ is the same as the distribution of the random field we obtain by putting $u(l)=\varepsilon_l$, $1\le l\le n$, in all coordinates ε_l of the random variables $\tilde I^{\varepsilon}_{n,k}(f)$. On the other hand, the random variables we get in such a way agree with the random variables appearing in the sum defining $\tilde I_{n,k}(f|u)$, only the terms in this sum are listed in a different order. Lemma 15.1 is proved. □

Next I prove the following generalized version of Lemma 7.1.

Lemma 15.2 (Generalized version of the Symmetrization Lemma). *Let Z_p and $\bar Z_p$, $p=1,2,\dots$, be two sequences of random variables on a probability space $(\Omega,\mathcal{A},P)$. Let a σ-algebra $\mathcal{B}\subset\mathcal{A}$ be given on the probability space $(\Omega,\mathcal{A},P)$*

together with a $\mathscr{B}$-measurable set B and two numbers $\alpha > 0$ and $\beta > 0$ such that the random variables Z_p, $p = 1, 2, \dots$, are $\mathscr{B}$ measurable, and the inequality

$$P(|\bar{Z}_p| \le \alpha | \mathscr{B})(\omega) \ge \beta \quad \text{for all } p = 1, 2, \dots \text{ if } \omega \in B \tag{15.3}$$

holds. Then

$$P\left(\sup_{1 \le p < \infty} |Z_p| > \alpha + u\right) \le \frac{1}{\beta} P\left(\sup_{1 \le p < \infty} |Z_p - \bar{Z}_p| > u\right) + (1 - P(B)) \tag{15.4}$$

for all $u > 0$.

Proof of Lemma 15.2. Put $\tau = \min\{p\colon |Z_p| > \alpha + u)$ if there exists such an index $p \ge 1$, and put $\tau = 0$ otherwise. Then we have, as $\{\tau = p\} \cap B \in \mathscr{B}$

$$\begin{aligned} P(\{\tau = p\} \cap B) &= \int_{\{\tau = p\} \cap B} 1 \cdot dP \le \int_{\{\tau = p\} \cap B} \frac{1}{\beta} P(|\bar{Z}_p| \le \alpha | \mathscr{B})\, dP \\ &= \frac{1}{\beta} P(\{\tau = p\} \cap \{|\bar{Z}_p| \le \alpha\} \cap B) \\ &\le \frac{1}{\beta} P(\{\tau = p\} \cap \{|Z_p - \bar{Z}_p| > u\}) \quad \text{for all } p = 1, 2, \dots. \end{aligned}$$

Hence

$$\begin{aligned} P\left(\sup_{1 \le p < \infty} |Z_p| > \alpha + u\right) - (1 - P(B)) &\le P\left(\left\{\sup_{1 \le p < \infty} |Z_p| > \alpha + u\right\} \cap B\right) \\ &= \sum_{p=1}^{\infty} P(\{\tau = p\} \cap B) \\ &\le \frac{1}{\beta} \sum_{p=1}^{\infty} P(\{\tau = p\} \cap \{|Z_p - \bar{Z}_p| > u\}) \\ &\le \frac{1}{\beta} P\left(\sup_{1 \le p < \infty} |Z_p - \bar{Z}_p| > u\right). \end{aligned}$$

Lemma 15.2 is proved. □

Next I give a short explanation about the difficulties we meet in the proof of Proposition 14.2′ and the approach applied in this work to overcome them with the help of some symmetrization type arguments.

To find a symmetrization argument useful in the proof of Proposition 14.2′ we want to bound the probability $P\left(n^{-k/2}\sup\limits_{f\in\mathcal{F}}|k!\bar I_{n,k}(f)|>u\right)$ by

$$C\cdot P\left(n^{-k/2}\sup_{f\in\mathcal{F}}|k!\tilde I_{n,k}(f)|>cu\right)+\text{a negligible error term}$$

with some appropriate numbers $C<\infty$ and $0<c<1$. The random variables $\bar I_{n,k}(f)$ and $\tilde I_{n,k}(f)$ appearing in these formulas were defined in (14.11) and (15.1). (Actually we work with a slightly modified version of formula (14.11) where the random variables $\xi_l^{(j)}$ are replaced by the random variables $\xi_l^{(j,1)}$.) We shall prove such an estimate with the help of Lemma 15.2. To find the random variables Z_p and $\bar Z_p$ we want to work with in Lemma 15.2 let us list the elements of the class of functions $\mathcal{F}$ as $\mathcal{F}=\{f_1,f_2,\dots\}$. We shall apply Lemma 15.2 with the choice $Z_p=n^{-k/2}k!\bar I_{n,k}(f_p)$ and $\bar Z_p=n^{-k/2}k![\bar I_{n,k}(f_p)-\tilde I_{n,k}(f_p)]$, $p=1,2,\dots$, together with the σ-algebra $\mathcal{B}=\mathcal{B}(\xi_l^{(j,1)},\ 1\le l\le n,\ 1\le j\le k)$.

Let us observe that Z_p is a decoupled U-statistic depending on the random variables $\xi_l^{(j,1)}$, $1\le j\le k$, $1\le l\le n$, while $\bar Z_p$ is a linear combination of decoupled U-statistics whose arguments may contain not only the random variables of the form $\xi_l^{(j,-1)}$, but also the random variables of the form $\xi_l^{(j,1)}$. As a consequence, the random variables Z_p and $\bar Z_p$ are not independent. This is the reason why we cannot apply Lemma 7.2 in the proof of Proposition 14.2′.

We shall show that Lemma 15.2 with the choice of the above defined random variables Z_p and $\bar Z_p$ and the σ-algebra $\mathcal{B}$ may help us to prove the estimates we need in our considerations. To apply this lemma we have to show that condition (15.3) holds with an appropriate pair of numbers (α,β) and a $\mathcal{B}$ measurable set B of probability almost 1. To check this condition is a hard but solvable problem.

In Lemma 7.2 condition (7.1) played a role similar to condition (15.3) in Lemma 15.2. In that case we could check this condition by estimating the second moments $E\bar Z_n^2$ for all indices n. In the present case we shall estimate the supremum $\sup\limits_{f_p\in\mathcal{F}}E(\bar Z_p^2|\mathcal{B})$ of conditional second moments. In this formula $\bar Z_p$ is a (complicated) random variable depending on the function $f_p\in\mathcal{F}$. The estimation of the supremum of the conditional second moments we want to work with is a hard problem, and the main difficulties of our proof appear at this point.

The conditional second moments whose supremum we want to estimate can be expressed as the integral of a random function that can be written down explicitly. In such a way we get a problem similar to the original one about the estimation of $\sup\limits_{f\in\mathcal{F}}n^{-k/2}k!\bar I_{n,k}(f)$. It turned out that these two problems can be handled similarly. We can work out a symmetrization argument with the help of Lemma 15.2 in both cases, and an inductive argument similar to Proposition 7.3 can be formulated and proved which supplies the results we want to prove.

We shall prove Proposition 14.2′ as a consequence of two inductive propositions formulated in Propositions 15.3 and 15.4. Here we apply an approach similar to the proof of Proposition 6.2 which was done with the help of an inductive proposition formulated in Proposition 7.3. But now we have to prove two inductive propositions simultaneously, because we also have to bound the supremum of some conditional variances, and this demands special attention. To formulate the new inductive propositions first we introduce the notions of *good tail behaviour for a class of decoupled U-statistics* and *good tail behaviour for a class of integrals of decoupled U-statistics.*

Definition of good tail behaviour for a class of decoupled U-statistics. *Let some measurable space $(X, \mathscr{X})$ be given together with a probability measure μ on it. Let us consider a countable class $\mathscr{F}$ of functions $f(x_1,\dots,x_k)$ on the k-fold product $(X^k, \mathscr{X}^k)$ of the space $(X, \mathscr{X})$. Fix some positive integer $n \ge k$ and a positive number $0 < \sigma \le 1$, and take k independent copies $\xi_1^{(j)},\dots,\xi_n^{(j)}$, $1 \le j \le k$, of a sequence of independent, μ-distributed random variables $\xi_1,\dots,\xi_n$. Let us introduce with the help of these random variables the decoupled U-statistics $\bar I_{n,k}(f)$, $f \in \mathscr{F}$, defined in formula (14.11). Given some real number $T > 0$ we say that the set of decoupled U-statistics determined by the class of functions $\mathscr{F}$ has a good tail behaviour at level T (with parameters n and σ^2 which are fixed in the sequel) if*

$$P\left(\sup_{f\in\mathscr{F}} |n^{-k/2}k!\bar I_{n,k}(f)| \ge An^{k/2}\sigma^{k+1}\right) \le \exp\left\{-A^{1/2k}n\sigma^2\right\} \quad \textit{for all } A > T. \tag{15.5}$$

Definition of good tail behaviour for a class of integrals of decoupled U-statistics. *Let us have a product space $(X^k \times Y, \mathscr{X}^k \times \mathscr{Y})$ with some product measure $\mu^k \times \rho$, where $(X^k, \mathscr{X}^k, \mu^k)$ is the k-fold product of some measurable space $(X, \mathscr{X}, \mu)$ with a probability measure μ, and $(Y, \mathscr{Y}, \rho)$ is some other measurable space with a probability measure ρ. Fix some positive integer $n \ge k$ and a positive number $0 < \sigma \le 1$, and consider a countable class $\mathscr{F}$ of functions $f(x_1,\dots,x_k,y)$ on the product space $(X^k \times Y, \mathscr{X}^k \times \mathscr{Y}, \mu^k \times \rho)$. Take k independent copies $\xi_1^{(j)},\dots,\xi_n^{(j)}$, $1 \le j \le k$, of a sequence of independent, μ-distributed random variables $\xi_1,\dots,\xi_n$. For all $f \in \mathscr{F}$ and $y \in Y$ let us define the decoupled U-statistics $\bar I_{n,k}(f,y) = \bar I_{n,k}(f_y)$ by means of these random variables $\xi_1^{(j)},\dots,\xi_n^{(j)}$, $1 \le j \le k$, the kernel function $f_y(x_1,\dots,x_k) = f(x_1,\dots,x_k,y)$ and formula (14.11). Define with the help of these U-statistics $\bar I_{n,k}(f,y)$ the random integrals*

$$H_{n,k}(f) = \int [k!\bar I_{n,k}(f,y)]^2 \rho(dy), \quad f \in \mathscr{F}. \tag{15.6}$$

Choose some real number $T > 0$. We say that the set of random integrals $H_{n,k}(f)$, $f \in \mathcal{F}$, has a good tail behaviour at level T (with parameters n and σ^2 which we fix in the sequel) if

$$P\left(\sup_{f\in\mathcal{F}} n^{-k} H_{n,k}(f) \ge A^2 n^k \sigma^{2k+2}\right) \le \exp\{-A^{1/(2k+1)} n\sigma^2\}$$
$$\text{for all } A > T. \qquad (15.7)$$

Propositions 15.3 and 15.4 will be formulated with the help of the above notions.

Proposition 15.3. *Let us fix a positive integer $n \ge \max(k,2)$, a real number $0 < \sigma \le 2^{-(k+1)}$, a probability measure μ on a measurable space $(X,\mathcal{X})$ together with two real numbers $L \ge 1$ and $D \ge 1$ such that $n\sigma^2 \ge L\log n + \log D$. Let us consider those countable L_2-dense classes $\mathcal{F}$ of canonical kernel functions $f = f(x_1,\dots,x_k)$ (with respect to the measure μ) on the k-fold product space $(X^k,\mathcal{X}^k)$ with exponent L and parameter D for which all functions $f \in \mathcal{F}$ satisfy the inequalities $\sup\limits_{x_j\in X, 1\le j\le k} |f(x_1,\dots,x_k)| \le 2^{-(k+1)}$ and $\int f^2(x_1,\dots,x_k)\mu(\,dx_1)\dots\mu(\,dx_k) \le \sigma^2$.*

There is a real number $A_0 = A_0(k) > 1$ such that if for all classes of functions $\mathcal{F}$ which satisfy the above conditions the sets of decoupled U-statistics $\bar{I}_{n,k}(f)$, $f \in \mathcal{F}$, have a good tail behaviour at level $T^{4/3}$ for some $T \ge A_0$, then they also have a good tail behaviour at level T.

Proposition 15.4. *Fix a positive integer $n \ge \max(k,2)$, a real number $0 < \sigma \le 2^{-(k+1)}$, a product space $(X^k \times Y, \mathcal{X}^k \times \mathcal{Y})$ with some product measure $\mu^k \times \rho$, where $(X^k,\mathcal{X}^k,\mu^k)$ is the k-fold product of some probability space $(X,\mathcal{X},\mu)$, and $(Y,\mathcal{Y},\rho)$ is some other probability space together with two real numbers $L \ge 1$ and $D \ge 1$ such that the inequality $n\sigma^2 > L\log n + \log D$ holds.*

Let us consider those countable L_2-dense classes $\mathcal{F}$ consisting of canonical functions $f(x_1,\dots,x_k,y)$ on the product space $(X^k \times Y, \mathcal{X}^k \times \mathcal{Y})$ with exponent $L \ge 1$ and parameter $D \ge 1$ whose elements $f \in \mathcal{F}$ satisfy the inequalities

$$\sup_{x_j\in X, 1\le j\le k, y\in Y} |f(x_1,\dots,x_k,y)| \le 2^{-(k+1)} \qquad (15.8)$$

and

$$\int f^2(x_1,\dots,x_k,y)\mu(\,dx_1)\dots\mu(\,dx_k)\rho(\,dy) \le \sigma^2 \quad \text{for all } f \in \mathcal{F}. \qquad (15.9)$$

There exists some number $A_0 = A_0(k) > 1$ such that if for all classes of functions $\mathcal{F}$ which satisfy the above conditions the random integrals $H_{n,k}(f)$, $f \in \mathcal{F}$, defined in (15.6) have a good tail behaviour at level $T^{(2k+1)/2k}$ with some $T \ge A_0$, then they also have a good tail behaviour at level T.

Remark. To complete the formulation of Proposition 15.4 we still have to clarify when we call a function $f(x_1,\dots,x_k,y)$ defined on the product space $(X^k \times Y, \mathcal{X}^k \times \mathcal{Y}, \mu^k \times \rho)$ canonical. Here we apply a definition which slightly differs from that given in formula (8.10).

We say that a function $f(x_1,\dots,x_k,y)$ on the product space $(X^k \times Y, \mathcal{X}^k \times \mathcal{Y}, \mu^k \times \rho)$ is canonical if

$$\int f(x_1,\dots,x_{j-1},u,x_{j+1},\dots,x_k,y)\mu(du) = 0$$
$$\text{for all } 1 \le j \le k,\ x_s \in X,\ s \ne j \text{ and } y \in Y.$$

In this definition we do not require the analogous identity if we integrate with respect to the variable Y with fixed arguments $x_j \in X$, $1 \le j \le k$.

Let me also remark that the estimate (15.7) we have formulated in the definition of the property "good tail behaviour for a class of integrals of U-statistics" is fairly natural. We have applied the natural normalization, and with such a normalization it is natural to expect that the tail behaviour of the distribution of $\sup\limits_{f\in\mathcal{F}} n^{-k} H_{n,k}(f)$ is similar to that of const. $\left(\sigma\eta^k\right)^2$, where η is a standard normal random variable. Formula (15.7) expresses such a behaviour, only the power of the number A in the exponent at the right-hand side was chosen in a non-optimal way. Formula (15.5) in the formulation of the property "good tail behaviour for a class of decoupled U-statistics" has a similar interpretation. It says that $\sup\limits_{f\in\mathcal{F}} |n^{-k/2}k!I_{n,k}(f)|$ behaves similarly to const. $\sigma|\eta^k|$ with a standard normal random variable η.

We wanted to prove the property of good tail behaviour for a class of integrals of decoupled U-statistics under appropriate, not too restrictive conditions. Let me remark that in Proposition 15.4 we have imposed beside formula (15.8) a fairly weak condition (15.9) about the L_2-norm of the function f. Most difficulties appear in the proof, because we did not want to impose more restrictive conditions.

It is not difficult to derive Proposition 14.2′ from Proposition 15.3. Indeed, let us observe that the set of decoupled U-statistics determined by a class of functions $\mathcal{F}$ satisfying the conditions of Proposition 15.3 has a good tail-behaviour at level $T_0 = \sigma^{-(k+1)}$, since under the conditions of this Proposition the probability at the left-hand side of (15.5) equals zero for $A > \sigma^{-(k+1)}$. Then we get from Proposition 15.3 by induction with respect to the number j, that this set of decoupled U-statistics has a good tail-behaviour also for all $T = T_j = T_0^{(3/4)^j} = \sigma^{-(k+1)(3/4)^j}$, $j = 0,1,2,\dots$, with such indices j for which $T_j = \sigma^{-(k+1)(3/4)^j} \ge A_0$. This implies that if a class of functions $\mathcal{F}$ satisfies the conditions of Proposition 15.3, then the set of decoupled U-statistics determined by this class of functions has a good tail-behaviour at level $T = A_0^{4/3}$, i.e. at a level which depends only on the order k of the decoupled U-statistics. This result implies Proposition 14.2′, only it has to be applied for the class of function $\mathcal{F}' = \{2^{-(k+1)}f,\ f \in \mathcal{F}\}$ instead of the original class of functions $\mathcal{F}$ which appears in Proposition 14.2′ with the same parameters σ, L and D.

Similarly to the above argument an inductive procedure yields a corollary of Proposition 15.4 formulated below. Actually, we shall need this corollary of Proposition 15.4.

Corollary of Proposition 15.4. *If the class of functions $\mathcal{F}$ satisfies the conditions of Proposition 15.4, then there exists a constant $\bar{A}_0 = \bar{A}_0(k) > 0$ depending only on k such that the class of integrals $H_{n,k}(f)$, $f \in \mathcal{F}$, defined in formula (15.6) have a good tail behaviour at level $\bar{A}_0$.*

Proposition 15.3 will be proved by means of a symmetrization argument which applies Lemma 15.2. The main difficulty arises when we want to check condition (15.3) with the quantities we are working with in Proposition 15.3. This difficulty can be overcome by means of Proposition 15.4, more precisely by means of its corollary. It helps us to estimate the conditional variances of the decoupled U-statistics we have to handle in the proof of Proposition 15.3. The proof of Propositions 15.3 and 15.4 apply similar arguments, and they will be proved simultaneously. The following inductive procedure will be applied in their proof. First Proposition 15.3 and then Proposition 15.4 will be proved for $k = 1$. If Propositions 15.3 and 15.4 are already proved for all $k' < k$ for some number k, then first we prove Proposition 15.3 and then Proposition 15.4 for this number k.

The symmetrization arguments needed in the proof of Propositions 15.3 and 15.4 will be proved in Chap. 16. Then Propositions 15.3 and 15.4 will be proved in Chap. 17 with their help. These results imply Proposition 14.2′, hence also Theorem 8.4.

Chapter 16
A Symmetrization Argument

The proof of Propositions 15.3 and 15.4 applies some ideas similar to the argument in the proof of Proposition 7.3. But here some additional technical difficulties have to be overcome. As a first step, two results formulated in Lemmas 16.1A and 16.1B will be proved. They can be considered as a randomization argument with the help of Rademacher functions. They are analogous to Lemma 7.2 which was applied in the proof of Propositions 7.3. Lemma 16.1A will be applied in the proof of Proposition 15.3 and Lemma 16.1B in the proof of Proposition 15.4. In this chapter these lemmas will be proved. Their proofs will be based on some additional lemmas formulated in Lemmas 16.2A, 16.2B, 16.3A and 16.3B. By exploiting the structure of Propositions 15.3 and 15.4 we may assume when proving them for parameter k that they hold (together with their consequences) for all parameters $k' < k$.

Lemma 16.1A is a natural multivariate version of Lemma 7.2. Lemma 7.2 enabled us to replace the estimation of the supremum of a class of sums of independent random variables with the estimation of the supremum of the randomized version of these sums. Lemma 16.1A will enable us to reduce the proof of Proposition 15.3 to the estimation of the tail-distribution of the supremum of an appropriately defined class of randomized decoupled, degenerate U-statistics. This supremum will be estimated by means of the multi-dimensional version of Hoeffding's inequality given in Theorem 13.3. Lemma 16.B plays a similar role in the proof of Proposition 15.4. But its application is more difficult. In this result the probability investigated in Proposition 15.4 is bounded by means of an expression depending on the supremum of some random variables $\bar{W}(f)$, $f \in \mathcal{F}$, which will be defined in formula (16.7). The expressions $\bar{W}(f)$, $f \in \mathcal{F}$, are rather complicated, and they are worth studying more closely. This will be done in the proof of Corollary of Lemma 16.1B which yields a more appropriate bound for the probability we want to estimate in Proposition 15.4. In the proof of Proposition 15.4 the Corollary of Lemma 16.1B will be applied instead of the original Lemma 16.1B.

The proof of Lemmas 16.1A and 16.1B is similar to that of Lemma 7.2. First we introduce k additional independent copies $\bar{\xi}_1^{(j)}, \dots, \bar{\xi}_n^{(j)}$ beside the k (independent and identically distributed) copies $\xi_1^{(j)}, \dots, \xi_n^{(j)}$, $1 \le j \le k$, of the

P. Major, *On the Estimation of Multiple Random Integrals and U-Statistics*, Lecture Notes in Mathematics 2079, DOI 10.1007/978-3-642-37617-7_16,

sequence $\xi_1, \dots, \xi_n$ applied in the definition of the decoupled U-statistics $\bar{I}_{n,k}(f)$, and construct with their help some appropriate random sums. We shall prove in Lemmas 16.2A and 16.2B that the original random sums we want to estimate have the same distribution as their randomized versions we shall work with in the proof of Lemmas 16.1A and 16.1B. These Lemmas formulate a natural multivariate version of an important argument in the proof of Lemma 7.2. In the proof of Lemma 7.2 we have exploited that the random sums defined in (7.4) have the same joint distribution as their randomized versions defined in (7.5). Lemmas 16.2A and 16.2B are natural multivariate versions of this statement. They enable us (similarly to the corresponding argument in the proof of Lemma 7.2) to reduce the proof of Propositions 16.1A and 16.1B to the study of some simpler questions. This will be done with the help of Lemmas 16.3A and 16.3B. In Lemma 16.3A the supremum of some conditional variances is estimated under appropriate conditions. This lemma plays a similar role in the proof of Lemma 16.1A as condition (7.1) plays in the proof of Lemma 7.1. Its result together with Lemma 15.2, which is a generalized form of the symmetrization lemma, Lemma 7.1, enable us to prove Lemma 16.1A. Lemma 16.1B will be proved similarly, but here the conditional distribution of a more complicated expression has to be estimated. This can be done with the help of Lemma 16.3B. In Lemma 16.3B the supremum of the conditional expectation of some expressions is bounded.

The main results of this chapter are the following two lemmas.

Lemma 16.1A (Randomization argument in the proof of Proposition 15.3). *Let $\mathcal{F}$ be a class of functions on the space $(X^k, \mathcal{X}^k)$ which satisfies the conditions of Proposition 15.3 with some probability measure μ. Let us have k independent copies $\xi_1^{(j)}, \dots, \xi_n^{(j)}$, $1 \le j \le k$, of a sequence of independent μ distributed random variables $\xi_1, \dots, \xi_n$ and a sequence of independent random variables $\varepsilon = (\varepsilon_1, \dots, \varepsilon_n)$, $P(\varepsilon_l = 1) = P(\varepsilon_l = -1) = \frac{1}{2}$, $1 \le l \le n$, which is independent also of the random sequences $\xi_1^{(j)}, \dots, \xi_n^{(j)}$, $1 \le j \le k$. Consider the decoupled U-statistics $\bar{I}_{n,k}(f)$, $f \in \mathcal{F}$, defined with the help of these random variables by formula (14.11) together with their randomized version $\bar{I}^{\varepsilon}_{n,k}(f)$ defined in formula (14.12).*

There exist some constants $A_0 = A_0(k) > 0$ and $\gamma = \gamma_k > 0$ such that the inequality

$$P\left(\sup_{f\in\mathcal{F}} n^{-k/2}\left|k!\bar{I}_{n,k}(f)\right| > An^{k/2}\sigma^{k+1}\right)$$
$$< 2^{k+1}P\left(\sup_{f\in\mathcal{F}}\left|k!\bar{I}^{\varepsilon}_{n,k}(f)\right| > 2^{-(k+1)}An^k\sigma^{k+1}\right)$$
$$+2^k n^{k-1} e^{-\gamma_k A^{1/(2k-1)} n\sigma^2/k} \tag{16.1}$$

holds for all $A \ge A_0$.

It may be worth remarking that the second term at the right-hand side of formula (16.1) yields a small contribution to the upper bound in this relation because of the condition $n\sigma^2 \ge L\log n + \log D$.

To formulate Lemma 16.1B first some new quantities have to be introduced. Some of them will be used somewhat later. The quantities $\bar I^V_{n,k}(f,y)$ introduced in the subsequent formula (16.2) depend on the sets $V\subset\{1,\dots,k\}$, and they are the natural modifications of the inner sum terms in formula (15.1). Such expressions are needed in the formulation of the symmetrization result applied in the proof of Proposition 15.4. Their randomized versions $\bar I^{(V,\varepsilon)}_{n,k}(f,y)$, introduced in formula (16.5), correspond to the inner sum terms in formula (15.2). The integrals of these expressions will be also introduced in formulas (16.3) and (16.6).

Let us consider a class $\mathcal{F}$ of functions $f(x_1,\dots,x_k,y)\in\mathcal{F}$ on a space $(X^k\times Y,\mathcal{X}^k\times\mathcal{Y},\mu^k\times\rho)$ which satisfies the conditions of Proposition 15.4. Let us take $2k$ independent copies $\xi^{(j)}_1,\dots,\xi^{(j)}_n$, $\bar\xi^{(j)}_1,\dots,\bar\xi^{(j)}_n$, $1\le j\le k$, of a sequence of independent μ distributed random variables $\xi_1,\dots,\xi_n$ together with a sequence of independent random variables $(\varepsilon_1,\dots,\varepsilon_n)$, $P(\varepsilon_l=1)=P(\varepsilon_l=-1)=\frac12$, $1\le l\le n$, which is also independent of the previous random sequences. Let us introduce the notation $\xi^{(j,1)}_l=\xi^{(j)}_l$ and $\xi^{(j,-1)}_l=\bar\xi^{(j)}_l$, $1\le l\le n$, $1\le j\le k$. For all subsets $V\subset\{1,\dots,k\}$ of the set $\{1,\dots,k\}$ let $|V|$ denote the cardinality of this set, and define for all functions $f(x_1,\dots,x_k,y)\in\mathcal{F}$ and sets $V\subset\{1,\dots,k\}$ the decoupled U-statistics

$$\bar I^V_{n,k}(f,y)=\frac{1}{k!}\sum_{\substack{(l_1,\dots,l_k):\,1\le l_j\le n,\ j=1,\dots,k\\ l_j\ne l_{j'}\text{ if } j\ne j'}} f\left(\xi^{(1,\delta_1(V))}_{l_1},\dots,\xi^{(k,\delta_k(V))}_{l_k},y\right), \tag{16.2}$$

where $\delta_j(V)=\pm1$, $1\le j\le k$, is defined as $\delta_j(V)=1$ if $j\in V$, and $\delta_j(V)=-1$ if $j\notin V$, together with the random variables

$$H^V_{n,k}(f)=\int[k!\bar I^V_{n,k}(f,y)]^2\rho(dy),\quad f\in\mathcal{F}. \tag{16.3}$$

We shall consider $\bar I^V_{n,k}(f,y)$ defined in (16.2) as a random variable with values in the space $L_2(Y,\mathcal{Y},\rho)$.

Put

$$\bar I_{n,k}(f,y)=\bar I^{\{1,\dots,k\}}_{n,k}(f,y),\quad H_{n,k}(f)=H^{\{1,\dots,k\}}_{n,k}(f), \tag{16.4}$$

i.e. $\bar I_{n,k}(f,y)$ and $H_{n,k}(f)$ are the random variables $\bar I^V_{n,k}(f,y)$ and $H^V_{n,k}(f)$ with $V=\{1,\dots,k\}$, which means that these expressions are defined with the help of the random variables $\xi^{(j)}_l=\xi^{(j,1)}_l$, $1\le j\le k$, $1\le l\le n$.

Let us also define the 'randomized version' of the random variables $\bar I^V_{n,k}(f,y)$ and $H^V_{n,k}(f)$ as

$$\bar{I}_{n,k}^{(V,\varepsilon)}(f,y)=\frac{1}{k!}\sum_{\substack{(l_1,\dots,l_k):\,1\le l_j\le n,\ j=1,\dots,k\\ l_j\ne l_{j'} \text{ if } j\ne j'}}\varepsilon_{l_1}\cdots\varepsilon_{l_k} f\left(\xi_{l_1}^{(1,\delta_1(V))},\dots,\xi_{l_k}^{(k,\delta_k(V))},y\right),$$

$$\text{if } f\in\mathcal{F}, \tag{16.5}$$

and

$$H_{n,k}^{(V,\varepsilon)}(f)=\int[k!\bar{I}_{n,k}^{(V,\varepsilon)}(f,y)]^2\rho(dy),\quad f\in\mathcal{F}, \tag{16.6}$$

where $\delta_j(V)=1$ if $j\in V$, and $\delta_j(V)=-1$ if $j\in\{1,\dots,k\}\setminus V$. Similarly to formula (16.2), we shall consider $\bar{I}_{n,k}^{V,\varepsilon}(f,y)$ defined in (16.5) as a random variable with values in the space $L_2(Y,\mathcal{Y},\rho)$.

Let us also introduce the random variables

$$\bar{W}(f)=\int\left[\sum_{V\subset\{1,\dots,k\}}(-1)^{(k-|V|)}k!\bar{I}_{n,k}^{(V,\varepsilon)}(f,y)\right]^2\rho(dy),\quad f\in\mathcal{F}. \tag{16.7}$$

With the help of the above notations Lemma 16.1B can be formulated in the following way.

Lemma 16.1B (Randomization argument in the proof of Proposition 15.4). *Let $\mathcal{F}$ be a set of functions on $(X^k\times Y,\mathcal{X}^k\times\mathcal{Y})$ which satisfies the conditions of Proposition 15.4 with some probability measure $\mu^k\times\rho$. Let us have $2k$ independent copies $\xi_1^{(j,\pm1)},\dots,\xi_n^{(j,\pm1)}$, $1\le j\le k$, of a sequence of independent μ distributed random variables $\xi_1,\dots,\xi_n$ together with a sequence of independent random variables $\varepsilon_1,\dots,\varepsilon_n$, $P(\varepsilon_j=1)=P(\varepsilon_j=-1)=\frac{1}{2}$, $1\le j\le n$, which is independent also of the previously considered random sequences.*

Then there exist some constants $A_0=A_0(k)>0$ and $\gamma=\gamma_k$ such that if the integrals $H_{n,k}(f)$, $f\in\mathcal{F}$, determined by this class of functions $\mathcal{F}$ have a good tail behaviour at level $T^{(2k+1)/2k}$ for some $T\ge A_0$, (this property was defined in Chap. 15 in the definition of good tail behaviour for a class of integrals of decoupled U-statistics before the formulation of Propositions 15.3 and 15.4), then the inequality

$$P\left(\sup_{f\in\mathcal{F}}|H_{n,k}(f)|>A^2n^{2k}\sigma^{2(k+1)}\right)<2P\left(\sup_{f\in\mathcal{F}}\left|\bar{W}(f)\right|>\frac{A^2k!}{2}n^{2k}\sigma^{2(k+1)}\right)$$
$$+2^{2k+1}n^{k-1}e^{-\gamma_k A^{1/2k}n\sigma^2/k} \tag{16.8}$$

holds for all $A\ge T$ with the random variables $H_{n,k}(f)$ introduced in the second identity of relation (16.4) and with $\bar{W}(f)$ defined in formula (16.7).

A corollary of Lemma 16.1B will be formulated which can be better applied than the original lemma. Lemma 16.B is a little bit inconvenient, because the expression at the right-hand side of formula (16.8) contains a probability depending on $\sup\limits_{f\in\mathcal{F}} |\bar{W}(f)|$, and $\bar{W}(f)$ is a too complicated expression. Some new formulas (16.9) and (16.10) will be introduced which enable us to rewrite $\bar{W}(f)$ in a slightly simpler form. These formulas yield such a corollary of Lemma 16.B which is more appropriate for our purposes. To work out the details first some diagrams will be introduced.

Let $\mathcal{G} = \mathcal{G}(k)$ denote the set of all diagrams consisting of two rows, such that both rows of these diagrams are the set $\{1,\dots,k\}$, and these diagrams contain some edges $\{(j_1,j_1')\dots,(j_s,j_s')\}$, $0\le s\le k$, connecting a point (vertex) of the first row with a point (vertex) of the second row. The vertices $j_1,\dots,j_s$ which are end points of some edge in the first row are all different, and the same relation holds also for the vertices $j_1',\dots,j_s'$ in the second row. Given a diagram $G\in\mathcal{G}$ let $e(G) = \{(j_1,j_1')\dots,(j_s,j_s')\}$ denote the set of its edges, and let $v_1(G) = \{j_1,\dots,j_s\}$ be the set of those vertices in the first row and $v_2(G) = \{j_1',\dots,j_s'\}$ the set of those vertices in the second row of the diagram G from which an edge of G starts.

Given a diagram $G\in\mathcal{G}$, two sets $V_1,V_2\subset\{1,\dots,k\}$, a function f defined on the space $(X^k\times, Y, \mathcal{X}^k\times\mathcal{Y})$ and a probability measure ρ on $(Y,\mathcal{Y})$ we define the following random variables $H_{n,k}(f|G,V_1,V_2)$ with the help of the random variables $\xi_1^{(j,1)},\dots,\xi_n^{(j,1)},\xi_1^{(j,-1)},\dots,\xi_n^{(j,-1)}$, $1\le j\le k$, and $\varepsilon=(\varepsilon_1,\dots,\varepsilon_n)$ taking part in the definition of the random variables $\bar{W}(f)$:

$$
\begin{aligned}
&H_{n,k}(f|G,V_1,V_2)\\
&\quad=\sum_{\substack{(l_1,\dots,l_k,l_1',\dots,l_k'):\\ 1\le l_j\le n,\, l_j\ne l_{j'} \text{ if } j\ne j',\,1\le j,j'\le k,\\ 1\le l_j'\le n,\, l_j'\ne l_{j'}' \text{ if } j\ne j',\,1\le j,j'\le k,\\ l_j=l_{j'}' \text{ if } (j,j')\in e(G),\, l_j\ne l_{j'}' \text{ if } (j,j')\notin e(G)}} \prod_{j\in\{1,\dots,k\}\setminus v_1(G)} \varepsilon_{l_j} \prod_{j\in\{1,\dots,k\}\setminus v_2(G)} \varepsilon_{l_j'}\\
&\qquad\int f\left(\xi_{l_1}^{(1,\delta_1(V_1))},\dots,\xi_{l_k}^{(k,\delta_k(V_1))},y\right)\\
&\qquad\qquad f\left(\xi_{l_1'}^{(1,\delta_1(V_2))},\dots,\xi_{l_k'}^{(k,\delta_k(V_2))},y\right)\rho(dy), \qquad (16.9)
\end{aligned}
$$

where $\delta_j(V_1)=1$ if $j\in V_1$, $\delta_j(V_1)=-1$ if $j\notin V_1$, and $\delta_j(V_2)=1$ if $j\in V_2$, $\delta_j(V_2)=-1$ if $j\notin V_2$. (Let us observe that if the graph G contains s edges, then the product of the ε-s in (16.9) contains $2(k-s)$ terms, and the number of terms in the sum (16.9) is less than n^{2k-s}.) As the Corollary of Lemma 16.1B will indicate, in the proof of Proposition 15.4 we shall need a good estimate on the tail distribution of the random variables $H_{n,k}(f|G,V_1,V_2)$ for all $f\in\mathcal{F}$ and $G\in\mathcal{G}$, $V_1,V_2\subset\{1,\dots,k\}$. Such an estimate can be obtained by means of Theorem 13.3, the multivariate version of Hoeffding's inequality. But the estimate we get in such a way will be rewritten in a form more appropriate for our inductive procedure. This will be done in the next chapter.

The identity

$$\bar{W}(f) = \sum_{G\in\mathcal{G},\, V_1,V_2\subset\{1,\dots,k\}} (-1)^{|V_1|+|V_2|} H_{n,k}(f|G,V_1,V_2) \tag{16.10}$$

will be proved.

To prove this identity let us write first

$$\bar{W}(f) = \sum_{V_1,V_2\subset\{1,\dots,k\}} (-1)^{|V_1|+|V_2|} \int k!\bar{I}_{n,k}^{(V_1,\varepsilon)}(f,y)k!\bar{I}_{n,k}^{(V_2,\varepsilon)}(f,y)\rho(dy).$$

Let us express the products $k!\bar{I}_{n,k}^{(V_1,\varepsilon)}(f,y)k!\bar{I}_{n,k}^{(V_2,\varepsilon)}(f,y)$ by means of formula (16.5). Let us rewrite this product as a sum of products of the form

$$\prod_{j=1}^{k} \varepsilon_{l_j} f(\cdots) \prod_{j=1}^{k} \varepsilon_{l'_j} f(\cdots),$$

and let us define the following partition of the terms in this sum. The elements of this partition are indexed by the diagrams $G \in \mathcal{G}$, and if we take a diagram $G \in \mathcal{G}$ with the set of edges $e(G) = \{(j_1,j'_1),\dots,(j_s,j'_s)\}$, then the term of this sum determined by the indices $l_1,\dots,l_k,l'_1,\dots,l'_k$ belongs to the element of the partition indexed by this diagram G if and only if $l_{j_u} = l'_{j'_u}$ for all $1 \le u \le s$, and no more numbers between the indices $l_1,\dots,l_k,l'_1\dots,l'_k$ may agree. Since $\varepsilon_{l_{j_u}}\varepsilon_{l'_{j'_u}} = 1$ for all $1 \le u \le s$ and the set of indices of the remaining random variables ε_{l_j} is $\{l_j\colon j \in \{1,\dots,k\} \setminus v_1(G)\}$, the set of indices of the remaining random variables $\varepsilon_{l'_j}$ is $\{l'_j\colon j \in \{1,\dots,k\} \setminus v_2(G)\}$, we get by integrating the product $k!\bar{I}_{n,k}^{(V_1,\varepsilon)}(f,y)k!\bar{I}_{n,k}^{(V_2,\varepsilon)}(f,y)$ with respect to the measure ρ that

$$\int \bar{I}_{n,k}^{(V_1,\varepsilon)}(f,y)\bar{I}_{n,k}^{(V_2,\varepsilon)}(f,y)\rho(dy) = \sum_{G\in\mathcal{G}} H_{n,k}(f|G,V_1,V_2)$$

for all $V_1, V_2 \in \{1,\dots,k\}$. The last two identities imply formula (16.10).

Since the number of terms in the sum of formula (16.10) is less than $2^{4k}k!$, this relation implies that Lemma 16.1B has the following corollary:

Corollary of Lemma 16.1B (A simplified version of the randomization argument of Lemma 16.1B). *Let a set of functions $\mathcal{F}$ satisfy the conditions of Proposition 15.4. Then there exist some constants $A_0 = A_0(k) > 0$ and $\gamma = \gamma_k > 0$ such that if the integrals $H_{n,k}(f)$, $f \in \mathcal{F}$, determined by this class of functions $\mathcal{F}$ have a good tail behaviour at level $T^{(2k+1)/2k}$ for some $T \ge A_0$, then the inequality*

$$P\left(\sup_{f\in\mathscr{F}}|H_{n,k}(f)|>A^2n^{2k}\sigma^{2(k+1)}\right)$$
$$\le 2\sum_{G\in\mathscr{G},\,V_1,V_2\subset\{1,\dots,k\}} P\left(\sup_{f\in\mathscr{F}}|H_{n,k}(f|G,V_1,V_2)|>\frac{A^2n^{2k}\sigma^{2(k+1)}}{2^{4k+1}}\right)$$
$$+2^{2k+1}n^{k-1}e^{-\gamma_k A^{1/2k}n\sigma^2/k} \tag{16.11}$$

holds for all $A\ge T$ with the random variables $H_{n,k}(f)$ and $H_{n,k}(f|G,V_1,V_2)$ defined in formulas (16.4) and (16.9).

In the proof of Lemmas 16.1A and 16.1B the result of the following Lemmas 16.2A and 16.2B will be applied.

Lemma 16.2A. *Let us take $2k$ independent copies*

$$\xi_1^{(j,1)},\dots,\xi_n^{(j,1)}\quad\text{and}\quad \xi_1^{(j,-1)},\dots,\xi_n^{(j,-1)},\quad 1\le j\le k,$$

of a sequence of independent μ distributed random variables $\xi_1,\dots,\xi_n$ together with a sequence of independent random variables $(\varepsilon_1,\dots,\varepsilon_n)$, $P(\varepsilon_l=1)=P(\varepsilon_l=-1)=\frac12$, $1\le l\le n$, which is also independent of the previous sequences.

Let $\mathscr{F}$ be a class of functions which satisfies the conditions of Proposition 15.3. Introduce with the help of the above random variables for all sets $V\subset\{1,\dots,k\}$ and functions $f\in\mathscr{F}$ the decoupled U-statistic

$$\bar I^V_{n,k}(f)=\frac{1}{k!}\sum_{\substack{(l_1,\dots,l_k):\,1\le l_j\le n,\ j=1,\dots,k,\\ l_j\ne l_{j'}\text{ if } j\ne j'}} f\left(\xi_{l_1}^{(1,\delta_1(V))},\dots,\xi_{l_k}^{(k,\delta_k(V))}\right) \tag{16.12}$$

and its "randomized version"

$$\bar I^{(V,\varepsilon)}_{n,k}(f)=\frac{1}{k!}\sum_{\substack{(l_1,\dots,l_k):\,1\le l_j\le n,\ j=1,\dots,k,\\ l_j\ne l_{j'}\text{ if } j\ne j'}} \varepsilon_{l_1}\cdots\varepsilon_{l_k} f\left(\xi_{l_1}^{(1,\delta_1(V))},\dots,\xi_{l_k}^{(k,\delta_k(V))}\right),$$
$$f\in\mathscr{F}, \tag{16.13}$$

where $\delta_j(V)=\pm1$, and we have $\delta_j(V)=1$ if $j\in V$, and $\delta_j(V)=-1$ if $j\in\{1,\dots,k\}\setminus V$.

Then the sets of random variables

$$S(f)=\sum_{V\subset\{1,\dots,k\}}(-1)^{(k-|V|)}k!\bar I^V_{n,k}(f),\quad f\in\mathscr{F}, \tag{16.14}$$

and

$$\bar{S}(f) = \sum_{V\subset\{1,\dots,k\}} (-1)^{(k-|V|)} k! \bar{I}_{n,k}^{(V,\varepsilon)}(f), \quad f \in \mathcal{F}, \tag{16.15}$$

have the same joint distribution.

Lemma 16.2B. *Let us take 2k independent copies*

$$\xi_1^{(j,1)},\dots,\xi_n^{(j,1)} \quad \text{and} \quad \xi_1^{(j,-1)},\dots,\xi_n^{(j,-1)}, \quad 1 \le j \le k,$$

of a sequence of independent, μ distributed random variables $\xi_1,\dots,\xi_n$ together with a sequence of independent random variables $(\varepsilon_1,\dots,\varepsilon_n)$, $P(\varepsilon_l = 1) = P(\varepsilon_l = -1) = \frac{1}{2}$, $1 \le l \le n$, which is independent also of the previous sequences.

Let us consider a class $\mathcal{F}$ of functions $f(x_1,\dots,x_k,y) \in \mathcal{F}$ on a space $(X^k \times Y, \mathcal{X}^k \times \mathcal{Y}, \mu^k \times \rho)$ which satisfies the conditions of Proposition 15.4. For all functions $f \in \mathcal{F}$ and $V \in \{1,\dots,k\}$ consider the decoupled U-statistics $\bar{I}_{n,k}^{V}(f,y)$ defined by formula (16.2) with the help of the random variables $\xi_1^{(j,1)},\dots,\xi_n^{(j,1)}$ and $\xi_1^{(j,-1)},\dots,\xi_n^{(j,-1)}$, and define with their help the random variables

$$W(f) = \int \left[\sum_{V\subset\{1,\dots,k\}} (-1)^{(k-|V|)} k! \bar{I}_{n,k}^{V}(f,y) \right]^2 \rho(dy), \quad f \in \mathcal{F}. \tag{16.16}$$

Then the random vectors $\{W(f)\colon f \in \mathcal{F}\}$ defined in (16.16) and $\{\bar{W}(f)\colon f \in \mathcal{F}\}$ defined in (16.7) have the same distribution.

Proof of Lemmas 16.2A and 16.2B. Lemma 16.2A actually agrees with the already proved Lemma 15.1, only the notation is different. The proof of Lemma 16.2B is also very similar to that of Lemma 15.1. It can be shown that even the following stronger statement holds. For any ± 1 sequence $u = (u_1,\dots,u_n)$ of length n the conditional distribution of the random field $\bar{W}(f)$, $f \in \mathcal{F}$, under the condition $(\varepsilon_1,\dots,\varepsilon_n) = u = (u_1,\dots,u_n)$ agrees with the distribution of the random field $W(f)$, $f \in \mathcal{F}$.

To see this relation let us first observe that the conditional distribution of the field $\bar{W}(f)$ under this condition agrees with the distribution of the random field we get by replacing the random variables ε_l by u_l for all $1 \le l \le n$ in formulas (16.5), (16.6) and (16.7). Beside this, define the vector $(\xi(u)_l^{(j,1)}, \xi(u)_l^{(j,-1)})$, $1 \le j \le k$, $1 \le l \le n$, by the formula $(\xi(u)_l^{(j,1)}, \xi(u)_l^{(j,-1)}) = (\xi_l^{(j,-1)}, \xi_l^{(j,1)})$ for those indices (j,l) for which $u_l = -1$, and $(\xi(u)_l^{(j,1)}, \xi(u)_l^{(j,-1)}) = (\xi_l^{(j,1)}, \xi_l^{(j,-1)})$ for which $u_l = 1$ (independently of the value of the parameter j). Then the joint distribution of the vectors $(\xi(u)_l^{(j,1)}, \xi(u)_l^{(j,-1)})$, $1 \le j \le k$, $1 \le l \le n$, and $(\xi_l^{(j,1)}, \xi_l^{(j,-1)})$, $1 \le j \le k$, $1 \le l \le n$, agree. Hence the joint distribution of the random vectors $\bar{I}_{n,k}^{V}(f,y)$, $f \in \mathcal{F}$, $V \subset \{1,\dots,k\}$ defined in (16.2) and of the random vectors $W(f)$, $f \in \mathcal{F}$, defined in (16.16) do not change if we replace in their definition the random variables $\xi_l^{(j,1)}$ and $\xi_l^{(j,-1)}$ by $\xi(u)_l^{(j,1)}$ and $\xi(u)_l^{(j,-1)}$. But the set of random variables $W(f)$, $f \in \mathcal{F}$, obtained in this way agrees with the set of random variables we introduced to get a set of random variables with the same distribution as the conditional distribution of $\bar W(f)$, $f \in \mathcal{F}$ under the condition $(\varepsilon_1,\dots,\varepsilon_n)=u$. (These random variables are defined as the square integral of the same sum, only the terms of this sum are listed in a different order in the two cases.) These facts imply Lemma 16.2B. □

In the next step we prove the following Lemma 16.3A.

Lemma 16.3A. *Let us consider a class of functions $\mathcal{F}$ satisfying the conditions of Proposition 15.3 with parameter k together with $2k$ independent copies $\xi_1^{(j,1)},\dots,\xi_n^{(j,1)}$ and $\xi_1^{(j,-1)},\dots,\xi_n^{(j,-1)}$, $1\le j\le k$, of a sequence of independent, μ-distributed random variables $\xi_1,\dots,\xi_n$. Take the random variables $\bar I_{n,k}^V(f)$, defined for $f\in\mathcal{F}$ and $V\subset\{1,\dots,k\}$ in formula (16.12). Let*

$$\mathcal{B}=\mathcal{B}(\xi_1^{(j,1)},\dots,\xi_n^{(j,1)},\ 1\le j\le k)$$

denote the σ-algebra generated by the random variables $\xi_1^{(j,1)},\dots,\xi_n^{(j,1)}$, $1\le j\le k$, i.e. by the random variables with upper indices of the form $(j,1)$, $1\le j\le k$. There exists a number $A_0=A_0(k)>0$ such that for all $V\subset\{1,\dots,k\}$, $V\neq\{1,\dots,k\}$, the inequality

$$P\left(\sup_{f\in\mathcal{F}} E\left([k!\bar I_{n,k}^V(f)]^2\,\middle|\,\mathcal{B}\right)>2^{-(3k+3)}A^2n^{2k}\sigma^{2k+2}\right)<n^{k-1}e^{-\gamma_kA^{1/(2k-1)}n\sigma^2/k} \tag{16.17}$$

holds with a sufficiently small $\gamma_k>0$ if $A\ge A_0$.

Proof of Lemma 16.3A. Let us first consider the case $V=\emptyset$. In this case the estimate $E\left((k!\bar I_{n,k}^{\emptyset}(f))^2\,\middle|\,\mathcal{B}\right)=E\left((k!\bar I_{n,k}^{\emptyset}(f))^2\right)\le k!n^k\sigma^2\le 2^kk!n^{2k}\sigma^{2k+2}$ holds for all $f\in\mathcal{F}$. In the above calculation it was exploited that the functions $f\in\mathcal{F}$ are canonical, which implies certain orthogonalities, and beside this the inequality $n\sigma^2\ge\frac12$ holds, because of the relation $n\sigma^2\ge L\log n+\log D$. The above relations imply that for $V=\emptyset$ the probability at the left-hand side of (16.17) equals zero if the number A_0 is chosen sufficiently large. Hence inequality (16.17) holds in this case.

To avoid some complications in the notation let us first restrict our attention to sets of the form $V=\{1,\dots,u\}$ with some $1\le u<k$, and prove relation (16.17) for such sets. For this goal let us introduce the random variables

$$\bar I_{n,k}^V(f,l_{u+1},\dots,l_k)=\frac{1}{k!}\sum_{\substack{(l_1,\dots,l_u):\\ 1\le l_j\le n,\ j=1,\dots,u,\\ l_j\neq l_{j'}\text{ if } j\neq j'\text{ for all }1\le j,j'\le k}} f\left(\xi_{l_1}^{(1,1)},\dots,\xi_{l_u}^{(u,1)},\xi_{l_{u+1}}^{(u+1,-1)},\dots,\xi_{l_k}^{(k,-1)}\right)$$

for all $f \in \mathcal{F}$ and sequences $l(u) = (l_{u+1}, \dots, l_k)$ with the properties $1 \le l_j \le n$ for all $u+1 \le j \le k$ and $l_j \ne l_{j'}$ if $j \ne j'$, i.e. let us fix the last $k-u$ coordinates $\xi_{l_{u+1}}^{(u+1,-1)}, \dots, \xi_{l_k}^{(k,-1)}$ of the random variable $\bar I_{n,k}^V(f)$ and sum up with respect the first u coordinates. Then we can write

$$
\begin{aligned}
&E\left(\bar I_{n,k}^V(f)^2 \middle| \mathcal{B}\right) \\
&= E\left(\left(\sum_{\substack{(l_{u+1},\dots,l_k)\colon 1\le l_j\le n\ j=u+1,\dots,k,\\ l_j\ne l_{j'} \text{ if } j\ne j'}} \bar I_{n,k}^V(f, l_{u+1}, \dots, l_k)\right)^2 \middle| \mathcal{B}\right) \\
&= \sum_{\substack{(l_{u+1},\dots,l_k)\colon 1\le l_j\le n,\ j=u+1,\dots,k,\\ l_j\ne l_{j'} \text{ if } j\ne j'}} E\left(\bar I_{n,k}^V(f, l_{u+1}, \dots, l_k)^2 \middle| \mathcal{B}\right). \qquad (16.18)
\end{aligned}
$$

The last relation follows from the identity

$$
E\left(\bar I_{n,k}^V(f, l_{u+1}, \dots, l_k)\bar I_{n,k}^V(f, l'_{u+1}, \dots, l'_k) \middle| \mathcal{B}\right) = 0
$$

if $(l_{u+1}, \dots, l_k) \ne (l'_{u+1}, \dots, l'_k)$, which holds, since f is a canonical function. We still exploit that the random variables $\xi_l^{(j,1)}$, $1 \le j \le u$ are $\mathcal{B}$ measurable, while the random variables $\xi_{l_j}^{(j,-1)}$, $u+1 \le j \le k$, are independent of the σ-algebra $\mathcal{B}$. These facts enable us to calculate the above conditional expectation in a simple way.

It follows from relation (16.18) that

$$
\begin{aligned}
&\left\{\omega\colon \sup_{f\in\mathcal{F}} E\left([k!\bar I_{n,k}^V(f)]^2 \middle| \mathcal{B}\right)(\omega) > 2^{-(3k+3)}A^2 n^{2k}\sigma^{2k+2}\right\} \\
&\quad\subset \bigcup_{\substack{(l_{u+1},\dots,l_k)\colon\\ 1\le l_j\le n,\ j=u+1,\dots,k.\\ l_j\ne l_{j'} \text{ if } j\ne j'}} \\
&\qquad \left\{\omega\colon \sup_{f\in\mathcal{F}} E\left([k!\bar I_{n,k}^V(f, l_{u+1}, \dots, l_k)]^2 \middle| \mathcal{B}\right)(\omega) > \frac{A^2 n^{2k}\sigma^{2k+2}}{2^{(3k+3)} n^{k-u}}\right\}. \qquad (16.19)
\end{aligned}
$$

The probability of the events in the union at the right-hand side of (16.19) can be estimated with the help of the Corollary of Proposition 15.4 with parameter $u < k$ instead of k. (We may assume that Proposition 15.4 holds for $u < k$.) I claim that this corollary yields that

$$P\left(\sup_{f\in\mathcal{F}} E\left([k!\bar I^V_{n,k}(f,l_{u+1},\dots,l_k)]^2\,|\,\mathcal{B}\right) > \frac{A^2 n^{k+u}\sigma^{2k+2}}{2^{(3k+3)}}\right)$$
$$\le e^{-\gamma_k A^{1/(2u+1)}(n+u-k)\sigma^2} \tag{16.20}$$

with an appropriate $\gamma_k > 0$ for all sequences $(l_{u+1},\dots,l_k)$, $1 \le l_j \le n$, $u+1 \le j \le k$, such that $l_j \ne l_{j'}$ if $j \ne j'$.

Let us show that if a class of functions $f \in \mathcal{F}$ satisfies the conditions of Proposition 15.3, then it also satisfies relation (16.20). For this goal introduce the space $(Y,\mathcal{Y},\rho) = (X^{k-u},\mathcal{X}^{k-u},\mu^{k-u})$, the $k-u$-fold power of the measure space $(X,\mathcal{X},\mu)$, and for the sake of simpler notations write $y = (x_{u+1},\dots,x_k)$ for a point $y \in Y$. Let us also introduce the class of those function $\bar{\mathcal{F}}$ in the space $(X^u\times Y,\mathcal{X}^u\times\mathcal{Y},\mu^u\times\rho)$ consisting of functions $\bar f$ of the form $\bar f(x_1,\dots,x_u,y) = f(x_1,\dots,x_k)$ with $y = (x_{u+1},\dots,x_k)$ and some function $f(x_1,\dots,x_k) \in \mathcal{F}$. If the class of function $\mathcal{F}$ satisfies the conditions of Proposition 15.3 (with parameter k), then the class of functions $\bar{\mathcal{F}}$ satisfies the conditions of Proposition 15.4 with parameter $u < k$. Hence the Corollary of Proposition 15.4 can be applied for the class of functions $\bar{\mathcal{F}}$ by our inductive hypothesis. We shall apply it for decoupled U-statistics with the class of kernel functions $\bar{\mathcal{F}}$ and parameters $n+u-k$ and u (instead of n and k), with the help of the independent random sequences $\xi_l^{(j,1)}$, $1 \le j \le u$, $l \in \{1,\dots,n\}\setminus\{l_{u+1},\dots,l_k\}$ of independent, μ-distributed random variables of length $n+u-k$, where the set of numbers $\{l_{u+1},\dots,l_k\}$ is the set of indices appearing in formula (16.20). This means that we work with random variables $\xi_l^{(j,1)}$ with index l from the set $\{1,\dots,n\}\setminus\{l_{u+1},\dots,l_k\}$ instead of $1 \le u \le n+u-k$. As a consequence, we shall work in the application of Proposition 15.4 with the random variables $\bar I^{l(u)}_{n+u-k,u}(\bar f,y)$ and $H^{l(u)}_{n+u-k,u}(\bar f)$ to be defined below which we get by slightly modifying the definition of $\bar I_{n+u-k,u}(\bar f,y)$ and $H_{n+u-k,u}(\bar f)$ by taking into account the indexation of the random variables $\xi_l^{(j,1)}$.

It can be seen by means of some calculation that the conditional expectation $E\left([k!\bar I^V_{n,k}(f,l_{u+1},\dots,l_k)]^2|\mathcal{B}\right)$ we are working with can be calculated as

$$E\left([k!\bar I^V_{n,k}(f,l_{u+1},\dots,l_k)]^2|\mathcal{B}\right)$$
$$= \int [u!\bar I^{l(u)}_{n+u-k,u}(\bar f,y)]^2\rho(dy) = H^{l(u)}_{n+u-k,u}(\bar f), \tag{16.21}$$

where the function $\bar f \in \bar{\mathcal{F}}$ is defined as $\bar f(x_1,\dots,x_u,y) = f(x_1,\dots,x_k)$ with $y = (x_{u+1},\dots,x_k)$, and the random variables $\bar I^{l(u)}_{n+u-k,u}(\bar f,y)$ and $H^{l(u)}_{n+u-k,u}(\bar f)$ are defined, similarly to (16.2)–(16.4), by the formulas

$$\bar{I}^{l(u)}_{n+u-k,u}(\bar{f},y)$$
$$= \frac{1}{u!} \sum_{\substack{(l_1,\dots,l_u):\, l_j\in\{1,\dots,n\}\setminus\{l_{u+1},\dots,l_k\},\ j=1,\dots,u \\ l_j\neq l_{j'} \text{ if } j\neq j'}} \bar{f}\left(\xi^{(1,1)}_{l_1},\dots,\xi^{(u,1)}_{l_u},y\right)$$

and

$$H^{l(u)}_{n+u-k,u}(\bar{f}) = \int [u!\bar{I}^{l(u)}_{n+u-k,u}(\bar{f},y)]^2\rho(dy), \quad \bar{f}\in\bar{\mathcal{F}}.$$

The value of $H^{l(u)}_{n+u-k,u}(\bar{f})$ depends on the choice of the sequence $l(u)$, but its distribution does not depend on it. Hence we can make the following estimate with the help of the corollary of Proposition (15.4) for $u<k$ and relation (16.21). Choose a sufficiently small $\gamma=\gamma_k>0$. Then we have

$$P\left(\sup_{\bar{f}\in\bar{\mathcal{F}}} E([k!\bar{I}^V_{n,k}(f,l_{u+1},\dots,l_k)]^2|\mathcal{B})\ge\gamma_k^{(4u+2)}A^2(n+u-k)^{2u}\sigma^{2u+2}\right)$$
$$= P\left(\sup_{\bar{f}\in\bar{\mathcal{F}}}(n+u-k)^{-u}H^{l(u)}_{n+u-k,u}(\bar{f})\ge\gamma_k^{(4u+2)}A^2(n+u-k)^u\sigma^{2u+2}\right)$$
$$\le e^{-\gamma_k A^{1/(2u+1)}(n+u-k)\sigma^2} \quad \text{for } A>A_0(u)\gamma_k^{-(4u+2)}. \tag{16.22}$$

It is not difficult to derive formula (16.20) from relation (16.22). It is enough to check that the level $\frac{A^2n^{k+u}\sigma^{2k+2}}{2^{(3k+3)}}$ in the probability at the left-hand side of (16.20) can be replaced by $\gamma_k^{(4u+2)}A^2(n+u-k)^{2u}\sigma^{2u+2}$ if $\gamma_k>0$ is chosen sufficiently small. This statement holds, since $\gamma_k^{(4u+2)}A^2(n+u-k)^{2u}\sigma^{2u+2}\le\gamma_k^{(4u+2)}A^2n^{2u}\sigma^{2u+2}\le\frac{A^2n^{k+u}\sigma^{2k+2}}{2^{(3k+3)}}$ if the constant $\gamma_k>0$ is chosen sufficiently small, since $n\sigma^2>L\log n\le\frac{1}{2}$ by the conditions of Proposition 15.3.

Relations (16.19) and (16.20) imply that

$$P\left(\sup_{f\in\mathcal{F}} E\left([k!\bar{I}^V_{n,k}(f)]^2\middle|\mathcal{B}\right)(\omega)>2^{-(3k+3)}A^2n^{2k}\sigma^{2k+2}\right)$$
$$\le n^{k-u}e^{-\gamma_k A^{1/(2u+1)}(n+u-k)\sigma^2}.$$

Since $e^{-\gamma_k A^{1/(2u+1)}(n+u-k)\sigma^2}\le e^{-\gamma_k A^{1/(2k-1)}n\sigma^2/k}$ if $u\le k-1$, $n\ge k$ and $A>A_0$ with a sufficiently large number A_0, inequality (16.17) holds for all sets V of the form $V=\{1,\dots,u\}$, $1\le u<k$.

The case of a general set $V\subset\{1,\dots,k\}$, $1\le|V|<k$, can be handled similarly, only the notation becomes more complicated. Moreover, the case of general sets V can be reduced to the case of sets of form we have already considered. Indeed, given

some set $V \subset \{1,\dots,k\}$, $1 \le |V| < k$, let us define a new class of function $\mathcal{F}_V$ we get by applying a rearrangement of the indices of the arguments $x_1,\dots,x_k$ of the functions $f \in \mathcal{F}$ in such a way that the arguments indexed by the set V are the first $|V|$ arguments of the functions $f_V \in \mathcal{F}_V$, and put $\bar V = \{1,\dots,|V|\}$. Then the class of functions $\mathcal{F}_V$ also satisfies the condition of Proposition 15.3, and we can get relation (16.17) with the set V by applying it for the set of function $\mathcal{F}_V$ and set $\bar V$. □

Now we prove Lemma 16.1A with the help of Lemma 16.2A, the generalized symmetrization Lemma 15.2 and Lemma 16.3A.

Proof of Lemma 16.1A. First we show with the help of the generalized symmetrization lemma, i.e. of Lemma 15.2 and Lemma 16.3A that

$$P\left(\sup_{f\in\mathcal{F}} n^{-k/2}\left|k!\bar I_{n,k}(f)\right| > An^{k/2}\sigma^{k+1}\right)$$
$$< 2P\left(\sup_{f\in\mathcal{F}} |S(f)| > \frac{A}{2}n^k\sigma^{k+1}\right) + 2^k n^{k-1} e^{-\gamma_k A^{1/(2k-1)} n\sigma^2/k} \qquad (16.23)$$

with the function $S(f)$ defined in (16.14). To prove relation (16.23) introduce the random variables $Z(f) = k!\bar I_{n,k}^{\{1,\dots,k\}}(f)$ and

$$\bar Z(f) = -\sum_{V\subset\{1,\dots,k\},\, V\ne\{1,\dots,k\}} (-1)^{k-|V|} k!\bar I_{n,k}^V(f)$$

for all $f \in \mathcal{F}$, the σ-algebra $\mathcal{B}$ considered in Lemma 16.3A and the set

$$B = \bigcap_{\substack{V\subset\{1,\dots,k\}\\ V\ne\{1,\dots,k\}}} \left\{\omega\colon \sup_{f\in\mathcal{F}} E\left([k!\bar I_{n,k}^V(f)]^2\middle|\mathcal{B}\right)(\omega) \le 2^{-(3k+3)}A^2 n^{2k}\sigma^{2k+2}\right\}.$$

Observe that $S(f) = Z(f) - \bar Z(f)$, $f \in \mathcal{F}$, $B \in \mathcal{B}$, and by Lemma 16.3A the inequality $1 - P(B) \le 2^k n^{k-1} e^{-\gamma_k A^{1/(2k-1)} n\sigma^2/k}$ holds. To prove relation (16.23) apply Lemma 15.2 with the above introduced random variables $Z(f)$ and $\bar Z(f)$, $f \in \mathcal{F}$, (both here and in the subsequent proof of Lemma 16.1B we work with random variables $Z(\cdot)$ and $\bar Z(\cdot)$ indexed by the countable set of functions $f \in \mathcal{F}$, hence the functions $f \in \mathcal{F}$ play the role of the parameters p when Lemma 15.2 is applied) random set B and $\alpha = \frac{A}{2}n^k\sigma^{k+1}$, $u = \frac{A}{2}n^k\sigma^{k+1}$. (At the left-hand side of (16.23) we can replace $k!\bar I_{n,k}(f)$ with $Z(f)$, $f \in \mathcal{F}$, because they have the same joint distribution.) It is enough to show that

$$P\left(|\bar Z(f)| > \frac{A}{2}n^k\sigma^{k+1}\middle|\mathcal{B}\right)(\omega) \le \frac{1}{2} \quad \text{for all } f \in \mathcal{F} \quad \text{if } \omega \in B. \qquad (16.24)$$

But

$$P\left(k!|\bar{I}_{n,k}^{|V|}(f)| > 2^{-(k+1)}An^k\sigma^{k+1}|\mathscr{B}\right)(\omega)$$

$$\le \frac{2^{2(k+1)}E(\bar{I}_{n,k}^{|V|}(f)^2|\mathscr{B})(\omega)}{A^2n^{2k}\sigma^{2(k+1)}} \le 2^{-(k+1)}$$

for all functions $f \in \mathscr{F}$ and sets $V \subset \{1,\dots,k\}$, $V \neq \{1,\dots,k\}$, if $\omega \in B$ by the "conditional Chebishev inequality", hence relations (16.24) and (16.23) hold.

Lemma 16.1A follows from relation (16.23), Lemma 16.2A and the observation that the random variables $\bar{I}_{n,k}^{(V,\varepsilon)}(f)$, $f \in \mathscr{F}$, defined in (16.13) have the same distribution for all $V \subset \{1,\dots,k\}$ as the random variables $\bar{I}_{n,k}^{\varepsilon}(f)$, defined in formula (14.12). Hence Lemma 16.2A and the definition (16.15) of the random variables $\bar{S}(f)$, $f \in \mathscr{F}$, imply the inequality

$$P\left(\sup_{f\in\mathscr{F}} |S(f)| > \frac{A}{2}n^k\sigma^{k+1}\right) = P\left(\sup_{f\in\mathscr{F}} |\bar{S}(f)| > \frac{A}{2}n^k\sigma^{k+1}\right)$$

$$\le 2^k P\left(\sup_{f\in\mathscr{F}} \left|k!\bar{I}_{n,k}^{\varepsilon}(f)\right| > 2^{-(k+1)}An^k\sigma^{k+1}\right).$$

Lemma 16.1A is proved. □

Lemma 16.1B will be proved with the help of the following Lemma 16.3B, which is a version of Lemma 16.3A.

Lemma 16.3B. *Let us consider a class of functions $\mathscr{F}$ satisfying the conditions of Proposition 15.4 together with 2k independent copies*

$$\xi_1^{(j,1)},\dots,\xi_n^{(j,1)}, \textit{ and } \xi_1^{(j,-1)},\dots,\xi_n^{(j,-1)}, \quad 1 \le j \le k,$$

of a sequence of independent, μ-distributed random variables $\xi_1,\dots,\xi_n$. Take the random variables $\bar{I}_{n,k}^V(f,y)$ and $H_{n,k}^V(f)$, $f \in \mathscr{F}$, $V \subset \{1,\dots,k\}$, defined in formulas (16.2) and (16.3) with the help of these quantities. Let

$$\mathscr{B} = \mathscr{B}(\xi_1^{(j,1)},\dots,\xi_n^{(j,1)}, \ 1 \le j \le k)$$

denote the σ-algebra generated by the random variables $\xi_1^{(j,1)},\dots,\xi_n^{(j,1)}$, $1 \le j \le k$, i.e. by those random variables which appear in the definition of the random variables $\bar{I}_{n,k}^V(f,y)$ and $H_{n,k}^V(f)$ introduced in formulas (16.2) and (16.3), and have second argument 1 in their upper index.

(a) There exist some numbers $A_0 = A_0(k) > 0$ and $\gamma = \gamma_k > 0$ such that for all $V \subset \{1,\dots,k\}$, $V \neq \{1,\dots,k\}$, the inequality

$$P\left(\sup_{f\in\mathscr{F}} E(H_{n,k}^V(f)|\mathscr{B}) > \frac{2^{-(4k+4)}}{(k!)^2} A^{(2k-1)/k} n^{2k}\sigma^{2k+2}\right) < n^{k-1} e^{-\gamma_k A^{1/2k} n\sigma^2/k} \tag{16.25}$$

holds if $A \ge A_0$.

(b) Given two subsets $V_1, V_2 \subset \{1,\dots,k\}$ *of the set* $\{1,\dots,k\}$ *define the integrals (of random kernel functions)*

$$H_{n,k}^{(V_1,V_2)}(f) = \int |k!\bar{I}_{n,k}^{V_1}(f,y)k!\bar{I}_{n,k}^{V_2}(f,y)|\rho(dy), \quad f\in\mathscr{F}, \tag{16.26}$$

with the help of the functions $\bar{I}_{n,k}^V(f,y)$ *defined in (16.2). There exist some numbers* $A_0 = A_0(k) > 0$ *and* $\gamma = \gamma_k > 0$ *such that if the integrals* $H_{n,k}(f)$, $f\in\mathscr{F}$, *determined by this class of functions* $\mathscr{F}$ *have a good tail behaviour at level* $T^{(2k+1)/2k}$ *for some* $T \ge A_0$, *then the inequality*

$$P\left(\sup_{f\in\mathscr{F}} E(H_{n,k}^{(V_1,V_2)}(f)|\mathscr{B}) > \frac{2^{-(2k+2)}}{k!} A^2 n^{2k}\sigma^{2k+2}\right) < 2n^{k-1} e^{-\gamma_k A^{1/2k} n\sigma^2/k} \tag{16.27}$$

holds for any pairs of subsets $V_1, V_2 \subset \{1,\dots,k\}$ *with the property that at least one of them does not equal the set* $\{1,\dots,k\}$ *if the number A satisfies the condition* $A > T$.

Proof of Lemma 16.3B. Part (a) of Lemma 16.3B can be proved in almost the same way as Lemma 16.3A. Hence I only briefly explain the main step of the proof. In the case $V = \emptyset$ the identity $E(H_{n,k}^V(f)|\mathscr{B}) = E(H_{n,k}^V(f))$ holds, hence it is enough to show that $E(H_{n,k}^V(f)) \le k!n^k\sigma^2 \le 2^k k! n^{2k}\sigma^{2k+2}$ for all $f\in\mathscr{F}$ under the conditions of Proposition 15.4. (This relation holds, because the functions of the class $\mathscr{F}$ are canonical.) The case of a general set V, $V \ne \emptyset$ and $V \ne \{1,\dots,k\}$, can be reduced to the case $V = \{1,\dots,u\}$ with some $1 \le u < k$.

Given a set $V = \{1,\dots,u\}$, $1 \le u < k$, let us define for all $f\in\mathscr{F}$ and sequences $l(u) = (l_{u+1},\dots,l_k)$ with the properties $1 \le l_j \le n$ for all $u+1 \le j \le k$ and $l_j \ne l_{j'}$ if $j \ne j'$ the random variable

$$\bar{I}_{n,k}^V(f,l_{u+1},\dots,l_k,y) = \frac{1}{k!}\sum_{\substack{(l_1,\dots,l_u):\\ 1\le l_j\le n,\ j=1,\dots,u,\\ l_j\ne l_{j'} \text{ if } j\ne j' \text{ for all } 1\le j,j'\le k}} f\left(\xi_{l_1}^{(1,1)},\dots,\xi_{l_u}^{(u,1)},\xi_{l_{u+1}}^{(u+1,-1)},\dots,\xi_{l_k}^{(k,-1)},y\right).$$

It can be shown, similarly to the proof of relation (16.18) in the proof of Proposition 16.3A that since the functions $f\in\mathscr{F}$ have the canonical property the identity

$$E\left(\bar{H}_{n,k}^{V}(f)\middle|\mathscr{B}\right)=\sum_{\substack{(l_{u+1},\dots,l_k):\\ 1\le l_j\le n,\ j=u+1,\dots,k,\\ l_j\ne l_{j'}\text{ if } j\ne j'}}\int E\left(\left[k!\bar{I}_{n,k}^{V}(f,l_{u+1},\dots,l_k,y)\right]^2\middle|\mathscr{B}\right)\rho(dy)$$

holds, and the proof of part (a) of Lemma 16.3B can be reduced to the inequality

$$P\left(\sup_{f\in\mathscr{F}}E\left(\int[k!\bar{I}_{n,k}^{V}(f,l_{u+1},\dots,l_k,y)]^2\rho(dy)\middle|\mathscr{B}\right)>\frac{A^{(2k-1)/k}n^{k+u}\sigma^{2k+2}}{2^{(4k+4)}(k!)^2}\right)$$
$$\le e^{-\gamma_k A^{(2k-1)/2k(2u+1)}(n+u-k)\sigma^2}$$

with a sufficiently small $\gamma_k > 0$. This inequality can be proved, similarly to relation (16.20) in the proof of Lemma 16.3A with the help of the Corollary of Proposition 15.4. Only here we have to work in the space $(X^u\times\bar{Y},\mathscr{X}^u\times\bar{\mathscr{Y}},\mu^u\times\bar{\rho})$ where $\bar{Y}=X^{k-u}\times Y$, $\bar{\mathscr{Y}}=\mathscr{X}^{k-u}\times\mathscr{Y}$, $\bar{\rho}=\mu^{k-u}\times\rho$ with the class of function $\bar{f}\in\bar{\mathscr{F}}$ consisting of the functions $\bar{f}$ defined by the formula $\bar{f}(x_1,\dots,x_u,\bar{y})=f(x_1,\dots,x_k,y)$ with some $f(x_1,\dots,x_k,y)\in\mathscr{F}$, where $\bar{y}=(x_{u+1},\dots,x_k,y)$. Here we apply the following version of formula (16.21).

$$E\left([k!\bar{I}_{n,k}^{V}(f,l_{u+1},\dots,l_k,y)]^2|\mathscr{B}\right)=\int[u!\bar{I}_{n+u-k,u}^{l(u)}(\bar{f},\bar{y})]^2\bar{\rho}(d\bar{y})=H_{n+u-k,u}^{l(u)}(\bar{f})$$

with the function $\bar{f}\in\bar{\mathscr{F}}$ for which the identity

$$\bar{f}(x_1,\dots,x_u,\bar{y})=f(x_1,\dots,x_k,y)$$

holds with $\bar{y}=(x_{u+1},\dots,x_k,y)$, and we define the random variables $\bar{I}_{n+u-k,u}^{l(u)}(\bar{f},\bar{y})$ and $H_{n+u-k,u}^{l(u)}(\bar{f})$ similarly to the corresponding terms after formula (16.21), only y is replaced by $\bar{y}$, the measure ρ by $\bar{\rho}$, and the presently defined functions $\bar{f}\in\bar{\mathscr{F}}$ are considered. I omit the details.

Part (b) of Lemma 16.3B will be proved with the help of Part (a) and the inequality

$$\sup_{f\in\mathscr{F}}E(H_{n,k}^{(V_1,V_2)}(f)|\mathscr{B})\le\left(\sup_{f\in\mathscr{F}}E(H_{n,k}^{V_1}(f)|\mathscr{B})\right)^{1/2}\left(\sup_{f\in\mathscr{F}}E(H_{n,k}^{V_2}(f)|\mathscr{B})\right)^{1/2}. \tag{16.28}$$

To prove inequality (16.28) observe that the random variables $H_{n,k}^{(V_1,V_2)}(f)$, $H_{n,k}^{V_1}(f)$ and $H_{n,k}^{V_2}(f)$ can be expressed as functions of the random variables $\xi_l^{(j,1)}$, $\xi_l^{(j,-1)}$, $1\le j\le k$, $1\le l\le n$ which are independent of each other, and the random variables $\xi_l^{(j,1)}$ are $\mathscr{B}$ measurable, while the random variables $\xi_l^{(j,-1)}$ are independent of this σ-algebra. Hence we can calculate the conditional expectations $E(H_{n,k}^{(V_1,V_2)}(f)|\mathscr{B})$,

$E(H_{n,k}^{V_1}(f)|\mathscr{B})$ and $E(H_{n,k}^{V_2}(f)|\mathscr{B})$ by putting the value of the random variables $\xi^{(j,1)}(\omega)$ in the appropriate coordinate of the functions expressing these random variables and integrating by the remaining coordinates with respect the distribution of the random variables $\xi_l^{(j,-1)}$. By writing up the above conditional expectations in such a way and applying the Schwarz inequality for them we get the inequality

$$E(H_{n,k}^{(V_1,V_2)}(f)|\mathscr{B}) \le \left(E(H_{n,k}^{V_1}(f)|\mathscr{B})\right)^{1/2} \left(E(H_{n,k}^{V_2}(f)|\mathscr{B})\right)^{1/2} \quad \text{for all } f \in \mathscr{F}.$$

It is not difficult to deduce relation (16.28) from this inequality by showing that it remains valid if we put the $\sup\limits_{f\in\mathscr{F}}$ expressions in it in that way as it is done in (16.28).

In the proof of Part (b) of Lemma 16.3B we may assume that $V_1 \ne \{1,\dots,k\}$. Inequality (16.28) implies that

$$\begin{aligned} &P\left(\sup_{f\in\mathscr{F}} E(H_{n,k}^{(V_1,V_2)}(f)|\mathscr{B}) > \frac{2^{-(2k+2)}}{k!} A^2 n^{2k}\sigma^{2k+2}\right) \\ &\quad\le P\left(\sup_{f\in\mathscr{F}} E(H_{n,k}^{V_1}(f)|\mathscr{B}) > \frac{2^{-(4k+4)}}{(k!)^2} A^{(2k-1)/k} n^{2k}\sigma^{2k+2}\right) \\ &\qquad+ P\left(\sup_{f\in\mathscr{F}} E(H_{n,k}^{V_2}(f)|\mathscr{B}) > A^{(2k+1)/k} n^{2k}\sigma^{2k+2}\right). \end{aligned}$$

Hence if we know that also the inequality

$$P\left(\sup_{f\in\mathscr{F}} E(H_{n,k}^{V_2}(f)|\mathscr{B}) > A^{(2k+1)/k} n^{2k}\sigma^{2k+2}\right) \le n^{k-1} e^{-\gamma_k A^{1/2k} n\sigma^2/k} \tag{16.29}$$

holds, then we can deduce relation (16.27) from the estimate (16.25) and (16.29). Relation (16.29) follows from Part (a) of Lemma 16.3B if $V_2 \ne \{1,\dots,k\}$ and $A \ge 1$, since in this case the level $A^{(2k+1)/k} n^{2k}\sigma^{2k+2}$ can be replaced by the smaller number $2^{-(4k+2)} A^{(2k-1)/k} n^{2k}\sigma^{2k+2}$ in the probability of formula (16.29). In the case $V_2 = \{1,\dots,k\}$ it follows from the conditions of Part (b) of Lemma 16.3B if the number γ_k is chosen so that $\gamma_k \le 1$. Indeed, since $A^{(2k+1)/2k} > T^{(2k+1)/2k}$, and by the conditions of Proposition 15.4 (and as a consequence of Lemma 16.3B) inequality (15.7) holds for all $\bar{A} \ge T^{(2k+1)/2k}$, we can apply this relation for the parameter $A^{(2k+1)/2k}$. In such a way we get inequality (16.29) also for $V_2 = \{1,\dots,k\}$. □

Now we turn to the proof of Lemma 16.1B.

Proof of Lemma 16.1B. By Lemma 16.2B it is enough to prove that relation (16.8) holds if the random variables $\bar{W}(f)$ are replaced in it by the random variables $W(f)$ defined in formula (16.16). We shall prove this by applying the generalized form of the symmetrization lemma, Lemma 15.2, with the choice of $Z(f) = H_{n,k}^{(\bar V,\bar V)}(f)$, $\bar V = \{1,\dots,k\}$, $\bar Z(f) = Z(f) - W(f)$, $f \in \mathscr{F}$, $\mathscr{B} = \mathscr{B}(\xi_1^{(j,1)},\dots,\xi_n^{(j,1)};\ 1 \le j \le k)$, $\alpha = \frac{A^2}{2}n^{2k}\sigma^{2k+2}$, $u = \frac{A^2}{2}n^{2k}\sigma^{2k+2}$ and the set

$$B = \bigcap_{\substack{(V_1,V_2)\colon V_j\in\{1,\dots,k\},\ j=1,2,\\ V_1\ne\{1,\dots,k\} \text{ or } V_2\ne\{1,\dots,k\}}} \left\{\omega\colon \sup_{f\in\mathscr{F}} E(H_{n,k}^{(V_1,V_2)}(f)|\mathscr{B})(\omega) \le \frac{2^{-(2k+2)}}{k!}A^2n^{2k}\sigma^{2k+2}\right\}.$$

By part (b) of Lemma 16.3B the inequality

$$1 - P(B) \le 2^{2k+1}n^{k-1}e^{-\gamma_k A^{1/2k}n\sigma^2/k}$$

holds. Observe that $Z(f) = H_{n,k}^{(\bar V,\bar V)}(f) = H_{n,k}(f)$ for all $f \in \mathscr{F}$. Hence to prove Lemma 16.1B with the help of Lemma 15.2 it is enough to show that

$$P\left(|\bar Z(f)| > \frac{A^2}{2k!}n^{2k}\sigma^{2k+2}\,\middle|\, \mathscr{B}\right)(\omega) \le \frac{1}{2} \quad \text{for all } f \in \mathscr{F} \text{ if } \omega \in B. \tag{16.30}$$

To prove this relation observe that because of the definition of the set B

$$\begin{aligned} &E(|\bar Z(f)||\mathscr{B})(\omega)\\ &\le \sum_{\substack{(V_1,V_2)\colon V_j\in\{1,\dots,k\},\ j=1,2,\\ V_1\ne\{1,\dots,k\} \text{ or } V_2\ne\{1,\dots,k\}}} E(H_{n,k}^{(V_1,V_2)}(f)|\mathscr{B})(\omega) \le \frac{A^2}{4k!}n^{2k}\sigma^{2k+2}\end{aligned}$$

if $\omega \in B$ for all $f \in \mathscr{F}$. Hence the "conditional Markov inequality" implies that $P\left(|\bar Z(f)| > \frac{A^2}{2k!}n^{2k}\sigma^{2(k+1)}\,\middle|\, \mathscr{B}\right)(\omega) \le \frac{2k!E(|\bar Z(f)||\mathscr{B})(\omega)}{A^2n^{2k}\sigma^{2k+2}} \le \frac{1}{2}$ if $\omega \in B$, and inequality (16.30) holds. Lemma 16.1B is proved. □

Chapter 17
The Proof of the Main Result

In this chapter Propositions 15.3 and 15.4 are proved with the help of Lemmas 16.1A and 16.1B. They complete the proof of Theorem 8.4, of the main result in this work.

A.) The proof of Proposition 15.3

The proof of Proposition 15.3 is similar to that of Proposition 7.3. It applies an induction procedure with respect to the order k of the U-statistics. In the proof of Proposition 15.3 for parameter k we may assume that Propositions 15.3 and 15.4 hold for $u < k$. We want to give a good estimate on the expression

$$P\left(\sup_{f\in\mathcal{F}} \left|k!\bar{I}^{\varepsilon}_{n,k}(f)\right| > 2^{-(k+1)}An^k\sigma^{k+1}\right)$$

appearing at the right-hand side of the estimate (16.1) in Lemma 16.1A. To estimate this probability we introduce (using the notation of Proposition 15.3) the functions

$$\begin{aligned} &S^2_{n,k}(f)(x_l^{(j)},\ 1\le l\le n,\ 1\le j\le k)\\ &\quad=\sum_{\substack{(l_1,\dots,l_k):\\ 1\le l_j\le n,\ j=1,\dots,k,\\ l_j\ne l_{j'} \text{ if } j\ne j'}} f^2\left(x_{l_1}^{(1)},\dots,x_{l_k}^{(k)}\right),\quad f\in\mathcal{F}, \qquad (17.1)\end{aligned}$$

with $x_l^{(j)}\in X$, $1\le l\le n$, $1\le j\le k$. We define with the help of this function the following set $H=H(A)\subset X^{kn}$ for all $A>T$ similarly to formula (7.8) in the proof of Proposition 7.3:

$$\begin{aligned} H=H(A)=\Bigg\{&\left(x_l^{(j)},\ 1\le l\le n,\ 1\le j\le k\right):\\ &\sup_{f\in\mathcal{F}} S^2_{n,k}(f)(x_l^{(j)}, 1\le l\le n, 1\le j\le k)>2^kA^{4/3}n^k\sigma^2\Bigg\}. \qquad (17.2)\end{aligned}$$

P. Major, *On the Estimation of Multiple Random Integrals and U-Statistics*, Lecture Notes in Mathematics 2079, DOI 10.1007/978-3-642-37617-7_17,

First we want to show that

$$P(\{\omega\colon (\xi_l^{(j)}(\omega),\ 1 \le j \le n,\ 1 \le j \le k) \in H\}) \le 2^k e^{-A^{2/3k} n\sigma^2} \quad \text{if } A \ge T. \tag{17.3}$$

To prove relation (17.3) we take the Hoeffding decomposition of the U-statistics with kernel functions $f^2(x_1, \dots, x_k)$, $f \in \mathcal{F}$, given in Theorem 9.1, i.e. we write

$$f^2(x_1, \dots, x_k) = \sum_{V \subset \{1,\dots,k\}} f_V(x_j, j \in V), \quad f \in \mathcal{F}, \tag{17.4}$$

with $f_V(x_j, j \in V) = \prod_{j \notin V} P_j \prod_{j \in V} Q_j f^2(x_1, \dots, x_k)$, where P_j and Q_j are the operators defined in formulas (9.1) and (9.2).

The functions f_V appearing in formula (17.4) are canonical (with respect to the measure μ), and the identity $S_{n,k}^2(f)(\xi_l^{(j)}\ 1 \le l \le n, 1 \le j \le k) = k!\bar{I}_{n,k}(f^2)$ holds for all $f \in \mathcal{F}$ with the expression $\bar{I}_{n,k}(\cdot)$ defined in (14.11). By applying the Hoeffding decomposition (17.4) for each term $f^2(\xi_{l_1}^{(1)} \dots, \xi_{l_k}^{(k)})$ in the expression $S_{n,k}^2(f)$ we get that

$$\begin{aligned} &P\left(\sup_{f \in \mathcal{F}} S_{n,k}^2(f)(\xi_l^{(j)},\ 1 \le l \le n,\ 1 \le j \le k) > 2^k A^{4/3} n^k \sigma^2\right) \\ &\qquad \le \sum_{V \subset \{1,\dots,k\}} P\left(\sup_{f \in \mathcal{F}} n^{k-|V|} ||V|!\,\bar{I}_{n,|V|}(f_V)| > A^{4/3} n^k \sigma^2\right) \end{aligned} \tag{17.5}$$

with the functions f_V appearing in formula (17.4). We want to give a good estimate for each term in the sum at the right-hand side in (17.5). For this goal first we show that the classes of functions $\{f_V\colon f \in \mathcal{F}\}$ in the expansion (17.4) satisfy the conditions of Proposition 15.3 for all $V \subset \{1, \dots, k\}$.

The functions f_V are canonical for all $V \subset \{1, \dots, k\}$. It follows from the conditions of Proposition 15.3 that $|f^2(x_1, \dots, x_k)| \le 2^{-2(k+1)}$ and

$$\int f^4(x_1, \dots, x_k)\mu(dx_1) \dots \mu(dx_k) \le 2^{-(k+1)}\sigma^2.$$

Hence relations (9.6) and (9.7) of Theorem 9.2 imply that

$$\left| \sup_{x_j \in X, j \in V} f_V(x_j, j \in V) \right| \le 2^{-(k+2)} \le 2^{-(k+1)}$$

and $\int f_V^2(x_j, j \in V) \prod_{j \in V} \mu(dx_j) \le 2^{-(k+1)}\sigma^2 \le \sigma^2$ for all $V \subset \{1, \dots, k\}$. Finally, to check that the class of functions $\mathcal{F}_V = \{f_V\colon f \in \mathcal{F}\}$ is L_2-dense with exponent L and parameter D observe that for all probability measures ρ on

$(X^k, \mathscr{X}^k)$ and pairs of functions $f, g \in \mathscr{F}$ the inequality $\int (f^2 - g^2)^2 \, d\rho \le 2^{-2k} \int (f - g)^2 \, d\rho$ holds. This implies that if $\{f_1, \dots, f_m\}$, $m \le D\varepsilon^{-L}$, is an ε-dense subset of $\mathscr{F}$ in the space $L_2(X^k, \mathscr{X}^k, \rho)$, then the set of functions $\{2^k f_1^2, \dots, 2^k f_m^2\}$ is an ε-dense subset of the class of functions $\mathscr{F}' = \{2^k f^2 \colon f \in \mathscr{F}\}$, hence $\mathscr{F}'$ is also an L_2-dense class of functions with exponent L and parameter D. Then by Theorem 9.2 the class of functions $\mathscr{F}_V$ is also L_2-dense with exponent L and parameter D for all sets $V \subset \{1, \dots, k\}$.

For $V = \emptyset$, the function f_V is constant, the relation

$$f_V = \int f^2(x_1, \dots, x_k)\mu(\,dx_1) \dots \mu(\,dx_k) \le \sigma^2$$

holds, and $\bar{I}_{|V|}(f_{|V|})| = f_V \le \sigma^2$. Therefore the term corresponding to $V = \emptyset$ in the sum of probabilities at the right-hand side of (17.5) equals zero under the conditions of Proposition 15.3 with the choice of some $A_0 \ge 1$. I claim that the remaining terms in the sum at the right-hand side of (17.5) satisfy the inequality

$$\begin{aligned} &P\left(n^{k-|V|} \sup_{f \in \mathscr{F}} ||V|!\,\bar{I}_{n,|V|}(f_V)| > A^{4/3} n^k \sigma^2\right) \\ &\quad \le P\left(\sup_{f \in \mathscr{F}} ||V|!\,\bar{I}_{n,|V|}(f_V)| > A^{4/3} n^{|V|} \sigma^{|V|+1}\right) \le e^{-A^{2/3k} n\sigma^2} \\ &\quad \text{if } 1 \le |V| \le k. \end{aligned} \tag{17.6}$$

The first inequality in (17.6) holds, since $\sigma^{|V|+1} \le \sigma^2$ for $|V| \ge 1$, and $n \ge k \ge |V|$. The second inequality follows from the inductive hypothesis if $|V| < k$, since in this case the middle expression in (17.6) can be bounded with the help of Proposition 15.3 by $e^{-(A^{4/3})^{1/2|V|} n\sigma^2} \le e^{-A^{2/3k} n\sigma^2}$ if $A_0 = A_0(k)$ in Proposition 15.3 is chosen sufficiently large. In the case $V = \{1, \dots, k\}$ it follows from the inequality $A \ge T$ and the inductive assumption of Proposition 15.3 by which the supremum of decoupled U-statistics determined by such a class of kernel-functions which satisfies the conditions of Proposition 15.3 has a good tail behaviour at level $T^{4/3}$. Relations (17.5) and (17.6) together with the estimate in the case $V = \emptyset$ imply formula (17.3).

By conditioning the probability $P\left(\left|k!\bar{I}^{\varepsilon}_{n,k}(f)\right| > 2^{-(k+2)} A n^{k/2} \sigma^{k+1}\right)$ with respect to the random variables $\xi_l^{(j)}$, $1 \le l \le n$, $1 \le j \le k$ we get with the help of the multivariate version of Hoeffding's inequality (Theorem 13.3) that

$$\begin{aligned} &P\left(\left|k!\bar{I}^{\varepsilon}_{n,k}(f)\right| > 2^{-(k+2)} A n^k \sigma^{k+1} \,\middle|\, \xi_l^{(j)}(\omega) = x_l^{(j)}, 1 \le l \le n, 1 \le j \le k\right) \\ &\quad \le C \exp\left\{-\frac{1}{2}\left(\frac{A^2 n^{2k} \sigma^{2(k+1)}}{2^{2k+4} S^2_{n,k}(f)(x_l^{(j)}, 1 \le l \le n,\, 1 \le j \le k)}\right)^{1/k}\right\} \end{aligned}$$

$$\le Ce^{-2^{-4-4/k}A^{2/3k}n\sigma^2} \quad \text{for all } f\in\mathcal{F}$$
$$\text{if } (x_l^{(j)},\ 1\le l\le n,\ 1\le j\le k)\notin H \tag{17.7}$$

with some appropriate constant $C=C(k)>0$.

Define for all $1\le j\le k$ and sets of points $x_l^{(j)}\in X$, $1\le l\le n$, the probability measures $\rho_j=\rho_{j,(x_l^{(j)},1\le l\le n)}$, $1\le j\le k$ on X, uniformly distributed on the set of points $\{x_l^{(j)},\ 1\le l\le n\}$, i.e. let $\rho_j(x_l^{(j)})=\frac{1}{n}$ for all $1\le l\le n$. Let us also define the product $\rho=\rho(x_l^{(j)},\ 1\le l\le n,\ 1\le j\le k)=\rho_1\times\cdots\times\rho_k$ of these measures on the space $(X^k,\mathcal{X}^k)$. If f is a function on $(X^k,\mathcal{X}^k)$ such that $\int f^2\,d\rho\le\delta^2$ with some $\delta>0$, then

$$\sup_{\varepsilon_1,\dots,\varepsilon_n}|k!\bar I^{\varepsilon}_{n,k}(f)(x_l^{(j)},\ 1\le l\le n,\ 1\le j\le k)|$$
$$\le n^k\int|f(u_1,\dots,u_k)|\rho(du_1,\dots,du_k)\le n^k\left(\int f^2\,d\rho\right)^{1/2}\le n^k\delta,$$

$u_j\in R^k$, $1\le j\le k$, and as a consequence

$$\sup_{\varepsilon_1,\dots,\varepsilon_n}|k!\bar I^{\varepsilon}_{n,k}(f)(x_l^{(j)},\ 1\le l\le n,\ 1\le j\le k) \tag{17.8}$$
$$-k!\bar I^{\varepsilon}_{n,k}(g)(x_l^{(j)},\ 1\le l\le n,\ 1\le j\le k)|$$
$$\le 2^{-(k+2)}An^k\sigma^{k+1}\quad\text{if } \int(f-g)^2\,d\rho\le(2^{-(k+2)}A\sigma^{k+1})^2,$$

where $\bar I^{\varepsilon}_{n,k}(f)(x_l^{(j)},\ 1\le l\le n,\ 1\le j\le k)$ equals the expression $\bar I^{\varepsilon}_{n,k}(f)$ defined in (14.12) if we replace $\xi^{(j)}_{l_j}$ by $x^{(j)}_{l_j}$ for all $1\le j\le k$, and $1\le l_j\le n$ in it, and ρ is the measure $\rho=\rho(x_l^{(j)},\ 1\le l\le n,\ 1\le j\le k)$ defined above.

Remark. Similarly to the remark made in the proof of Proposition 7.3 we may restrict our attention to the case when the random variables $\xi_l^{(j)}$ are non-atomic. A similar statement holds also in the proof of Proposition 15.4,

Let us fix the number $\delta=2^{-(k+2)}A\sigma^{k+1}$, and let us list the elements of the set $\mathcal{F}$ as $\mathcal{F}=\{f_1,f_2,\dots\}$. Put

$$m=m(\delta)=\max(1,D\delta^{-L})=\max(1,D(2^{(k+2)}A^{-(1)}\sigma^{-(k+1)})^L),$$

and choose for all vectors $x^{(n)}=(x_l^{(j)},\ 1\le l\le n,\ 1\le j\le k)\in X^{kn}$ such a sequence of positive integers $p_1(x^{(n)}),\dots,p_m(x^{(n)}))$ for which

$$\inf_{1\le l\le m}\int(f(u)-f_{p_l(x^{(n)})}(u))^2\rho(x^{(n)})(du)\le\delta^2\quad\text{for all } f\in\mathcal{F}\text{ and } x^{(n)}\in X^{kn}.$$

(Here we apply the notation $\rho(x^{(n)}) = \rho(x_l^{(j)}, 1 \le l \le n, 1 \le j \le k)$, which is a probability measure on X^k depending on $x^{(n)}$.) This is possible, since $\mathcal{F}$ is an L_2-dense class with exponent L and parameter D, and we can choose $m = D\delta^{-L}$, if $\delta < 1$, Beside this, we can choose $m = 1$ if $\delta \ge 1$, since $\int |f-g|^2\,d\rho \le \sup|f(x)-g(x)|^2 \le 2^{-2k} \le 1$ for all $f, g \in \mathcal{F}$. Moreover, we have shown in Lemma 7.4A that the functions $p_l(x^{(n)})$, $1 \le l \le m$, can be chosen as measurable functions of the argument $x^{(n)} \in X^{kn}$.

Let us consider the random vector $\xi^{(n)}(\omega) = (\xi_l^{(j)}(\omega), 1 \le l \le n, 1 \le j \le k)$. By arguing similarly as we did in the proof of Proposition 7.3 we get with the help of relation (17.8) and the property of the functions $f_{p_l(x^{(n)})}(\cdot)$ constructed above that

$$\left\{\omega\colon \sup_{f\in\mathcal{F}} |k!\bar{I}^{\varepsilon}_{n,k}(f)(\omega)| \ge 2^{-(k+1)}An^k\sigma^{k+1}\right\}$$
$$\subset \bigcup_{l=1}^{m} \{\omega\colon |k!\bar{I}^{\varepsilon}_{n,k}(f_{p_l(\xi^{(n)}(\omega))})(\omega)| \ge 2^{-(k+2)}An^k\sigma^{(k+1)}\}.$$

The above relation and formula (17.7) imply that

$$P\left(\sup_{f\in\mathcal{F}} \left|k!\bar{I}^{\varepsilon}_{n,k}(f)(\omega)\right| > \frac{An^k\sigma^{k+1}}{2^{(k+1)}} \,\middle|\, \xi_l^{(j)}(\omega) = x_l^{(j)}, 1 \le l \le n, 1 \le j \le k\right)$$
$$\le \sum_{l=1}^{m} P\left(|k!\bar{I}^{\varepsilon}_{n,k}(f_{p_l(\xi^{(n)}(\omega))})(\omega)| > \frac{An^k\sigma^{k+1}}{2^{k+2}} \,\middle|\, \xi_l^{(j)}(\omega) = x_l^{(j)}, 1 \le l \le n, 1 \le j \le k\right)$$
$$\le Cm(\delta)e^{-2^{-4-4/k}A^{2/3k}n\sigma^2} \le C(1 + D(2^{k+2}A^{-1}\sigma^{-(k+1)})^L)e^{-2^{-4-4/k}A^{2/3k}n\sigma^2}$$
$$\text{if } \{x_l^{(j)}, 1 \le l \le n, 1 \le j \le k\} \notin H. \tag{17.9}$$

Relations (17.3) and (17.9) imply that

$$P\left(\sup_{f\in\mathcal{F}} \left|k!\bar{I}^{\varepsilon}_{n,k}(f)\right| > 2^{-(k+1)}An^k\sigma^{k+1}\right) \tag{17.10}$$
$$\le C(1 + D(2^{k+2}A^{-1}\sigma^{-(k+1)})^L)e^{-2^{-4-4/k}A^{2/3k}n\sigma^2} + 2^k e^{-A^{2/3k}n\sigma^2} \quad \text{if } A > T.$$

Proposition 15.3 follows from the estimates (16.1), (17.10) and the condition $n\sigma^2 \ge L\log n + \log D$, $L, D \ge 1$, if $A \ge A_0$ with a sufficiently large number A_0. Indeed, in this case $n\sigma^2 \ge \frac{1}{2}$, $(2^{k+2}A^{-1}\sigma^{-(k+1)})^L \le (\frac{n^{(k+1)/2}}{(2n\sigma^2)^{(k+1)/2}})^L \le n^{L(k+1)/2} = e^{L\log n\cdot(k+1)/2} \le e^{(k+1)n\sigma^2/2}$, $D = e^{\log D} \le e^{n\sigma^2}$, and

$$C(1+D(2^{k+2}A^{-1}\sigma^{-(k+1)})^L)e^{-2^{-4-4/k}A^{2/3k}n\sigma^2} \le \frac{1}{3}e^{-A^{1/2k}n\sigma^2}.$$

The estimation of the remaining terms in the upper bound of the estimates (16.1) and (17.10) leading to the proof of relation (15.5) is simpler. We can exploit that $e^{-A^{2/3k}n\sigma^2} \ll e^{-A^{1/2k}n\sigma^2}$ and as $n^{k-1} \le e^{(k-1)n\sigma^2}$, hence $2^k e^{-A^{2/3k}n\sigma^2} \le \frac{1}{3}e^{-A^{1/2k}n\sigma^2}$, and $2^k n^{k-1} e^{-\gamma_k A^{1/(2k-1)}n\sigma^2/k} \le 2^k e^{(k-1)n\sigma^2} e^{-\gamma_k A^{1/(2k-1)}n\sigma^2/k} \ll e^{-A^{1/2k}n\sigma^2}$ for a large number A. □

Now we turn to the proof of Proposition 15.4.

B.) The proof of Proposition 15.4.

Because of formula (16.11) in the Corollary of Lemma 16.1B to prove Proposition 15.4 i.e. inequality (15.7) it is enough to choose a sufficiently large parameter A_0 and to show that with such a choice the random variables $H_{n,k}(f|G,V_1,V_2)$ defined in formula (16.9) satisfy the inequality

$$P\left(\sup_{f\in\mathcal{F}}|H_{n,k}(f|G,V_1,V_2)| > \frac{A^2n^{2k}\sigma^{2(k+1)}}{2^{4k+1}}\right) \le 2^{k+1}e^{-A^{1/2k}n\sigma^2}$$
$$\text{for all } G\in\mathcal{G} \quad \text{and } V_1,V_2\in\{1,\dots,k\} \quad \text{if } A>T\ge A_0 \qquad (17.11)$$

under the conditions of Proposition 15.4.

Let us first prove formula (17.11) in the case $|e(G)|=k$, i.e. when all vertices of the diagram G are end-points of some edge, and the expression $H_{n,k}(f|G,V_1,V_2)$ contains no "symmetrizing term" ε_j. In this case we apply a special argument to prove relation (17.11).

We will show with the help of the Schwarz inequality that for a diagram G such that $|e(G)|=k$

$$|H_{n,k}(f|G,V_1,V_2)| \qquad (17.12)$$
$$\le \left(\sum_{\substack{(l_1,\dots,l_k):\\ 1\le l_j\le n,\ 1\le j\le k,\\ l_j\ne l_{j'} \text{ if } j\ne j'}} \int f^2(\xi_{l_1}^{(1),\delta_1(V_1))},\dots,\xi_{l_k}^{(k,\delta_k(V_1))},y)\rho(dy)\right)^{1/2}$$
$$\left(\sum_{\substack{(l_1,\dots,l_k):\\ 1\le l_j\le n,\ 1\le j\le k,\\ l_j\ne l_{j'} \text{ if } j\ne j'}} \int f^2(\xi_{l_1}^{(1,\delta_1(V_2))},\dots,\xi_{l_k}^{(k,\delta_k(V_2))},y)\rho(dy)\right)^{1/2}$$

with $\delta_j(V_1) = 1$ if $j \in V_1$, $\delta_j(V_1) = -1$ if $j \notin V_1$, and $\delta_j(V_2) = 1$ if $j \in V_2$, $\delta_j(V_2) = -1$ if $j \notin V_2$.

Relation (17.12) can be proved for instance by bounding first the absolute value of each integral in formula (16.9) by means of the Schwarz inequality, and then by bounding the sum appearing in such a way by means of the inequality $\sum |a_j b_j| \le \left(\sum a_j^2\right)^{1/2} \left(\sum b_j^2\right)^{1/2}$. Observe that in the case $|(e(G)| = k$ the summation in (16.9) is taken for such vectors $(l_1, \dots, l_k, l'_1, \dots, l'_k)$ for which $(l'_1, \dots, l'_k)$ is a permutation of the sequence $(l_1, \dots, l_k)$ determined by the diagram G. Hence the sum we get after applying the Schwarz inequality for each integral in (16.9) has the form $\sum a_j b_j$ where the set of indices j in this sum agrees with the set of vectors $(l_1, \dots, l_k)$ such that $1 \le l_p \le n$ for all $1 \le p \le k$, and $l_p \ne l_{p'}$ if $p \ne p'$.

By formula (17.12)

$$
\begin{aligned}
&\left\{\omega\colon \sup_{f\in\mathcal{F}} |H_{n,k}(f|G,V_1,V_2)(\omega)| > \frac{A^2 n^{2k}\sigma^{(2(k+1)}}{2^{4k+1}}\right\}\\
&\subset \Bigg\{\omega\colon \sup_{f\in\mathcal{F}} \sum_{\substack{(l_1,\dots,l_k):\\ 1\le l_j\le n,\, 1\le j\le k,\\ l_j\ne l_{j'} \text{ if } j\ne j'}} \int f^2(\xi_{l_1}^{(1,\delta_1(V_1))}(\omega),\dots,\xi_{l_k}^{(k,\delta_k(V_1))}(\omega),y)\rho(dy)\\
&\qquad > \frac{A^2 n^{2k}\sigma^{2(k+1)}k!}{2^{4k+1}}\Bigg\}\\
&\quad\cup \Bigg\{\omega\colon \sup_{f\in\mathcal{F}} \sum_{\substack{(l_1,\dots,l_k):\\ 1\le l_j\le n,\, 1\le j\le k,\\ l_j\ne l_{j'} \text{ if } j\ne j'}} \int f^2(\xi_{l_1}^{(1,\delta_1(V_2))}(\omega),\dots,\xi_{l_k}^{(k,\delta_k(V_2))}(\omega),y)\rho(dy)\\
&\qquad > \frac{A^2 n^{2k}\sigma^{2(k+1)}k!}{2^{4k+1}}\Bigg\},
\end{aligned}
$$

hence

$$
\begin{aligned}
&P\left(\sup_{f\in\mathcal{F}} |H_{n,k}(f|G,V_1,V_2)| > \frac{A^2 n^{2k}\sigma^{2(k+1)}}{2^{4k+1}}\right)\\
&\le 2P\left(\sup_{f\in\mathcal{F}} \left| \sum_{\substack{(l_1,\dots,l_k):\\ 1\le l_j\le n,\, 1\le j\le k,\\ l_j\ne l_{j'} \text{ if } j\ne j'}} h_f(\xi_{l_1}^{(1,1)},\dots,\xi_{l_k}^{(k,1)})\right| > \frac{A^2 n^{2k}\sigma^{2(k+1)}}{2^{4k+1}}\right)\\
&= 2P\left(\sup_{f\in\mathcal{F}} |k!\bar I_{n,k}(h_f)| > \frac{A^2 n^{2k}\sigma^{2(k+1)}}{2^{4k+1}}\right), \qquad (17.13)
\end{aligned}
$$

where $\bar{I}_{n,k}(h_f)$, $f \in \mathscr{F}$, are the decoupled U-statistics defined in (14.11) with the kernel functions $h_f(x_1,\dots,x_k) = \int f^2(x_1,\dots,x_k,y)\rho(dy)$ and the random variables $\xi_l^{(j,1)}$, $1 \le j \le k$, $1 \le l \le n$. (In this upper bound we could get rid of the terms $\delta_j(V_1)$ and $\delta_j(V_2)$, i.e. of the dependence of the expression $H_{n,k}(f|G,V_1,V_2)$ on the sets V_1 and V_2, since the probability of the events in the previous formula do not depend on them.)

I claim that

$$P\left(\sup_{f\in\mathscr{F}} |k!\bar{I}_{n,k}(h_f)| \ge 2^k A n^k \sigma^2\right) \le 2^k e^{-A^{1/2k} n\sigma^2} \quad \text{for } A \ge A_0 \tag{17.14}$$

if the constant $A_0 = A_0(k)$ is chosen sufficiently large in Proposition 15.4. Relation (17.14) together with the relation $A^2\frac{n^{2k}\sigma^{2(k+1)}}{2^{4k+1}} \ge 2^k A n^k\sigma^2$ (if $A > A_0$ with a sufficiently large A_0) imply that the probability at the right-hand side of (17.13) can be bounded by $2^{k+1}e^{-A^{1/2k}n\sigma^2}$, and the estimate (17.11) holds in the case $|e(G)| = k$.

Relation (17.14) is similar to relation (17.3) (together with the definition of the random set H in formula (17.2)), and a modification of the proof of the latter estimate yields the proof also in this case. Indeed, it follows from the conditions of Proposition 15.4 that $0 \le \int h_f(x_1,\dots,x_k)\mu(dx_1)\dots\mu(dx_k) \le \sigma^2$ for all $f \in \mathscr{F}$, and it is not difficult to check that $\sup|h_f(x_1,\dots,x_k)| \le 2^{-2(k+1)}$, and the class of functions $\mathscr{H} = \{2^k h_f,\ f \in \mathscr{F}\}$ is an L_2-dense class with exponent L and parameter D. Hence by applying the Hoeffding decomposition of the functions h_f, $f \in \mathscr{F}$, similarly to formula (17.4) we get for all $V \subset \{1,\dots,k\}$ such a set of functions $\{(h_f)_V,\ f \in \mathscr{F}\}$, which satisfies the conditions of Proposition 15.3. Hence a natural adaptation of the estimate given for the expression at the right-hand side of (17.5) (with the help of (17.6) and the investigation of $|V|!\,\bar{I}_{|V|}(f_V)$ for $V = \emptyset$) yields the proof of formula (17.14). We only have to replace $S_{n,k}(f)$ by $k!\bar{I}_{n,k}(h_f)$, then $|V|!\,\bar{I}_{n,|V|}(f_V)$ by $|V|!\,\bar{I}_{n,|V|}((h_f)_V)$ and the levels $2^k A^{4/3} n^k\sigma^2$ in (17.3) and $A^{4/3}n^k\sigma^2$ in (17.5) by $2^k A n^k \sigma^2$ and $A n^k\sigma^2$ respectively. Let us observe that each term of the upper bound we get in such a way can be directly bounded, since during the proof of Proposition 15.4 for parameter k we may assume that the result of Proposition 15.3 holds also for this parameter k.

In the case of a diagram $G \in \mathscr{G}$ such that $e(G) < k$ formula (17.11) will be proved with the help of the multivariate version of Hoeffding's inequality, Theorem 13.3. In the proof of this case an expression, analogous to $S^2_{n,k}(f)$ defined in formula (17.1) will be introduced and estimated for all sets $V_1, V_2 \subset \{1,\dots,k\}$ and diagrams $G \in \mathscr{G}$ such that $|e(G)| < k$. To define it first some notations will be introduced.

Let us consider the set $J_0(G) = J_0(G,k,n)$,

$$\begin{aligned} J_0(G) = \{(l_1,\dots,l_k,l'_1,\dots,l'_k)\colon\ & 1 \le l_j, l'_j \le n,\ 1 \le j \le k,\ l_j \ne l_{j'} \text{ if } j \ne j', \\ & l'_j \ne l'_{j'} \text{ if } j \ne j', l_j = l'_{j'} \text{ if } (j,j') \in e(G),\ l_j \ne l'_{j'} \text{ if } (j,j') \notin e(G)\}. \end{aligned}$$

The set $J_0(G)$ contains those sequences $(l_1,\dots,l_k,l'_1,\dots,l'_k)$ which appear as indices in the summation in formula (16.9) for a fixed diagram G. We also introduce an appropriate partition of it.

For this aim let us first define the sets

$$M_1(G)=\{j(1),\dots,j(k-|e(G)|)\}=\{1,\dots,k\}\setminus v_1(G),$$
$$j(1)<\dots<j(k-|e(G)|),$$

and

$$M_2(G)=\{\bar j(1),\dots,\bar j(k-|e(G|)\}=\{1,\dots,k\}\setminus v_2(G),$$
$$\bar j(1)<\dots<\bar j(k-|e(G|),$$

the sets of those vertices of the first and second row of the diagram G, indexed in increasing order, from which no edge starts. Let us also introduce the set $V(G)=V(G,n,k)$, which consists of the restriction of the vectors $(l_1,\dots,l_k,l'_1,\dots,l'_k)\in J_0(G)$ to the coordinates indexed by the elements of the set $M_1(G)\cup M_2(G)$. Formally,

$$\begin{aligned}V(G)=\{&(l_{j(1)},\dots,l_{j(k-|e(G)|)},l'_{\bar j(1)},\dots,l'_{\bar j(k-|e(G)|)})\colon 1\le l_{j(p)},l'_{\bar j(p)}\le n,\\ &1\le p\le k-|e(G)|,\ l_{j(p)}\ne l_{j(p')},\ l'_{\bar j(p)}\ne l'_{\bar j(p')}\\ &\text{if } p\ne p',\ 1\le p,p'\le k-|e(G)|,\\ &l_{j(p)}\ne l'_{\bar j(p')},\ 1\le p,p'\le k-|e(G)|\}.\end{aligned}$$

The elements of $V(G)$ are vectors with elements indexed by the set $M_1(G)\cup M_2(G)$, which take different integer values between 1 and n.

We write all vectors $v=(l_{j(1)},\dots,l_{j(k-|e(G)|)},l'_{\bar j(1)},\dots,l'_{\bar j(k-|e(G)|)})\in V(G)$ in the form $v=(v^{(1)},v^{(2)})$ with $v^{(1)}=(l_{j(1)},\dots,l_{j(k-|e(G)|)})$ and $v^{(2)}=(l'_{\bar j(1)},\dots,l'_{\bar j(k-|e(G)|)})$, i.e. $v^{(1)}$ contains the first $k-|e(G)|$ coordinates of v with indices of the set $M_1(G)$, and $v^{(1)}$ contains the last $k-|e(G)|$ coordinates of v with indices of the set $M_2(G)$. We define with their help the set $E_G(v)$ which consists of those vectors $\ell=(l_1,\dots,l_k,l'_1,\dots,l'_k)\in J_0(G)$ whose restrictions to the coordinates with indices in $M_1(G)$ and $M_2(G)$ equal $v^{(1)}$ and $v^{(2)}$ respectively. More explicitly, we put

$$\begin{aligned}E_G(v)=\{&(l_1,\dots,l_k,l'_1,\dots,l'_k)\colon 1\le l_j\le n,\ 1\le l'_{\bar j}\le n,\ \text{for } 1\le j,\bar j\le k,\\ &l_j\ne l_{j'} \text{ if } j\ne j',\ l'_{\bar j}\ne l'_{\bar j'} \text{ if } \bar j\ne\bar j',\\ &l_j=l'_{\bar j} \text{ if } (j,\bar j)\in e(G) \text{ and } l_j\ne l'_{\bar j} \text{ if } (j,\bar j)\notin e(G), \text{ and}\\ &l_{j(r)}=v(r),\ l'_{\bar j(r)}=\bar v(r),\ 1\le r\le k-|e(G)|\},\quad \text{for all } v\in V(G),\end{aligned}$$

where $\{j(1),\dots,j(k-|e(G)|)\}=M_1(G)$, $\{\bar j(1),\dots,\bar j(k-|e(G)|)\}=M_2(G)$, $v=(v^{(1)},v^{(2)})$ with $v^{(1)}=(v(1),\dots,v(k-|e(G)|))$ and $v^{(2)}=(\bar v(1),\dots,\bar v(k-|e(G)|))$ in the last line of this definition. Beside this, let us define

$$E_G^1(v)=\{(l_1,\dots,l_k)\colon (l_1,\dots,l_k,l'_1,\dots,l'_k)\in E_G(v)\}$$

and

$$E_G^2(v)=\{(l'_1,\dots,l'_k)\colon (l_1\dots,l_k,l'_1,\dots,l'_k)\in E_G(v)\}.$$

The vectors $\ell=(l_1,\dots,l_k,l'_1,\dots,l'_k)\in E_G(v)$ can be characterized in the following way. For $j\in M_1(G)$ their coordinates l_j agree with the corresponding elements of $v^{(1)}$, for $\bar j\in M_2(G)$ their coordinates $l'_{\bar j}$ agree with the corresponding elements of $v^{(2)}$. The indices of the remaining coordinates of ℓ can be partitioned into pairs $(j_s,\bar j_{s'})$, $1\le s,s'\le |e(G)|$ in such a way that $(j_s,\bar j_{s'})\in e(G)$. The identity $l_{j_s}=l'_{\bar j_{s'}}$ holds if $(j_s,\bar j_{s'})\in e(G)$, and if $(j_s,\bar j_{s'})\notin e(G)$, then the coordinates l_{j_s} and $l'_{\bar j_{s'}}$ are different. Otherwise, the coordinates l_{j_s} and $l'_{\bar j_{s'}}$ can be freely chosen from the set $\{1,\dots,n\}\setminus\{v^{(1)},v^{(2)}\}$. The sets $E_G^1(v)$ and $E_G^2(v)$ consist of the vectors containing the first k and the second k coordinates of the vectors $\ell\in E_G(v)$.

The sets $E_G(v)$, $v\in V(G)$, constitute a partition of the set $J_0(G)$, and the random variables $H_{n,k}(f|G,V_1,V_2)$ defined in (16.9) can be rewritten with their help as

$$\begin{aligned}H_{n,k}(f|G,V_1,V_2)(\omega)=&\sum_{v=(v^{(1)},v^{(2)})\in V(G)}\prod_{s=1}^{k-|e(G)|}\varepsilon_{l_{j(s)}}(\omega)\prod_{s=1}^{k-|e(G)|}\varepsilon_{l'_{\bar j(s)}}(\omega)\\&\sum_{(l_1,\dots,l_k,l'_1\dots,l'_k)\in E_G(v)}\int f(\xi_{l_1}^{(1,\delta_1(V_1))}(\omega),\dots,\xi_{l_k}^{(k,\delta_k(V_1))}(\omega),y)\\&f(\xi_{l'_1}^{(1,\delta_1(V_2))}(\omega),\dots,\xi_{l'_k}^{(k,\delta_k(V_2))}(\omega),y)\rho(dy),\end{aligned}\tag{17.15}$$

where $\delta_j(V_1)=1$ if $j\in V_1$, $\delta_j(V_1)=-1$ if $j\notin V_1$, and $\delta_j(V_2)=1$ if $j\in V_2$, $\delta_j(V_2)=-1$ if $j\notin V_2$.

Let us fix some diagram $G\in\mathcal{G}$ and sets $V_1,V_2\subset\{1,\dots,k\}$. We will prove the inequality

$$P\left(S^2(\mathcal{F}|G,V_1,V_2)>2^{2k}A^{8/3}n^{2k}\sigma^4\right)\le 2^{k+1}e^{-A^{2/3k}n\sigma^2}\quad\text{if } A\ge A_0 \text{ and } e(G)<k\tag{17.16}$$

for the random variable

$$\begin{aligned}S^2(\mathcal{F}|G,V_1,V_2)=\sup_{f\in\mathcal{F}}\sum_{v\in V(G)}\Bigg(&\sum_{(l_1,\dots,l_k,l'_1,\dots,l'_k)\in E_G(v)}\int f(\xi_{l_1}^{(1,\delta_1(V_1))},\dots,\xi_{l_k}^{(k,\delta_k(V_1))},y)\\&f(\xi_{l'_1}^{(1,\delta_1(V_2))},\dots,\xi_{l'_k}^{(k,\delta_k(V_2))},y)\rho(dy)\Bigg)^2,\end{aligned}\tag{17.17}$$

where $\delta_j(V_1) = 1$ if $j \in V_1$, $\delta_j(V_1) = -1$ if $j \notin V_1$, and $\delta_j(V_2) = 1$ if $j \in V_2$, $\delta_j(V_2) = -1$ if $j \notin V_2$. The random variable $S^2(\mathcal{F}|G, V_1, V_2)$ defined in (17.17) plays a similar role in the proof of Proposition 15.4 as the random variable $\sup\limits_{f \in \mathcal{F}} S^2_{n,k}(f)$ with $S^2_{n,k}(f)$ defined in formula (17.1) played in the proof of Proposition 15.3.

To prove formula (17.16) let us first fix some $v \in V(G)$, and let us show that the following inequality, similar to relation (17.12) holds.

$$\left(\sum_{(l_1,\dots,l_k,l'_1,\dots,l'_k) \in E_G(v)} \int f(\xi_{l_1}^{(1,\delta_1(V_1))}, \dots, \xi_{l_k}^{(k,\delta_k(V_1))}, y) \right.$$
$$\left. f(\xi_{l'_1}^{(1,\delta_1(V_2))}, \dots, \xi_{l'_k}^{(k,\delta_k(V_2))}, y)\rho(dy) \right)^2$$
$$\le \left(\sum_{(l_1,\dots,l_k) \in E^1_G(v)} \int f^2(\xi_{l_1}^{(1,\delta_1(V_1))}, \dots, \xi_{l_k}^{(k,\delta_k(V_1))}, y)\rho(dy) \right)$$
$$\left(\sum_{(l'_1,\dots,l'_k) \in E^2_G(v)} \int f^2(\xi_{l'_1}^{(1,\delta_1(V_2))}, \dots, \xi_{l'_k}^{(k,\delta_k(V_2))}, y)\rho(dy) \right) \tag{17.18}$$

for all $f \in \mathcal{F}$ and $v \in V(G)$. Indeed, observe that for a vector $\bar{v} = (\bar{v}_1, \bar{v}_2) \in E_G(v)$ with $\bar{v}_1 \in E^1_G(v)$ and $\bar{v}_2 \in E^2_G(v)$, the coordinates of the vector $\bar{v}_1$ in the set $M_1(G)$ and the coordinates of the vector $\bar{v}_2$ in the set $M_2(G)$ are prescribed, while the coordinates of $\bar{v}_1$ in the set $v_1(G)$ are given by a permutation of the coordinates $\bar{v}_2$ in the set $v_2(G)$. (The sets $v_1(G)$ and $v_2(G)$ were defined before the introduction of formula (16.9) as the sets of those vertices in the first and second row of the diagram G respectively from which an edge of G starts.) This permutation is determined by the diagram G. Inequality (17.18) can be proved on the basis of the above observation similarly to formula (17.12).

We shall prove with the help of formula (17.18) the following inequality.

$$S^2(\mathcal{F}|G, V_1, V_2) \tag{17.19}$$
$$\le \sup_{f \in \mathcal{F}} \sum_{v \in V(G)} \left(\sum_{(l_1,\dots,l_k) \in E^1_G(v)} \int f^2(\xi_{l_1}^{(1,\delta_1(V_1))}, \dots, \xi_{l_k}^{(k,\delta_k(V_1))}, y)\rho(dy) \right)$$
$$\left(\sum_{(l'_1,\dots,l'_k) \in E^2_G(v)} \int f^2(\xi_{l'_1}^{(1,\delta_1(V_2))}, \dots, \xi_{l'_k}^{(k,\delta_k(V_2))}, y)\rho(dy) \right)$$

$$\le \sup_{f\in\mathcal{F}} \left(\sum_{\substack{(l_1,\dots,l_k):\\ 1\le l_j\le n,\, 1\le j\le k,\\ l_j\ne l_{j'} \text{ if } j\ne j'}} \int f^2(\xi_{l_1}^{(1,\delta_1(V_1))},\dots,\xi_{l_k}^{(k,\delta_k(V_1))},y)\rho(dy) \right)$$

$$\sup_{f\in\mathcal{F}} \left(\sum_{\substack{(l'_1,\dots,l'_k):\\ 1\le l'_j\le n,\, 1\le j\le k,\\ l'_j\ne l'_{j'} \text{ if } j\ne j'}} \int f^2(\xi_{l'_1}^{(1,\delta_1(V_2))},\dots,\xi_{l'_k}^{(k,\delta_k(V_2))},y)\rho(dy) \right).$$

The first inequality of (17.19) is a simple consequence of formula (17.18) and the definition of the random variable $S^2(\mathcal{F}|G,V_1,V_2)$. To check its second inequality let us observe that it can be reduced to the simpler relation where the expression $\sup\limits_{f\in\mathcal{F}}$ is omitted at each place. The simplified inequality obtained after the omission of the expressions sup can be checked by carrying out a term by term multiplication between the products of sums appearing in (17.19). At both sides of the inequality a sum consisting of terms of the form

$$\int f^2(\xi_{l_1}^{(1,\delta_1(V_1))},\dots,\xi_{l_k}^{(k,\delta_k(V_1))},y)\rho(dy)$$
$$\int f^2(\xi_{l'_1}^{(1,\delta_1(V_2))},\dots,\xi_{l'_k}^{(k,\delta_k(V_2))},y)\rho(dy), \tag{17.20}$$

appears. It is enough to check that if a term of this form appears in the middle term of the simplified version formula of (17.19), then it appears with multiplicity 1, and it also appears at the right-hand side of this formula. To see this, observe that each term of the form (17.20) which appears in the sum we get by carrying out the multiplications in middle term of (17.19) determines uniquely the index $v = (v^{(1)},v^{(2)}) \in V(G)$ in the outer sum of the middle term in the inequality (17.19). Indeed, if the random variables defining this expression of the form (17.20) have indices $\ell = (l_1,\dots,l_k,l'_1,\dots,l'_k)$, then this vector ℓ uniquely determines the vector $v = (v^{(1)},v^{(2)}) \in V(G)$, since $v^{(1)}$ must agree with the restriction of the vector $l = (l_1,\dots,l_k)$ to the coordinates with indices in $M_1(G)$ and $v^{(2)}$ must agree with the restriction of the vector $l' = (l'_1,\dots,l'_k)$ to the coordinates with indices in $M_2(G)$. Beside this, by carrying out the multiplication at the right-hand side of (17.19) we get such a sum which contains all such terms of the form (17.20) which appeared in the sum expressing the middle term in inequality (17.19). The above arguments imply inequality (17.19).

Relation (17.19) implies that

$$P(S^2(\mathcal{F}|G,V_1,V_2)) > 2^{2k}A^{8/3}n^{2k}\sigma^4) \le 2P\left(\sup_{f\in\mathcal{F}} k!\bar{I}_{n,k}(h_f) > 2^k A^{4/3}n^k\sigma^2\right),$$

where $\bar{I}_{n,k}(h_f)$, $f\in\mathcal{F}$, are the decoupled U-statistics defined in (14.11) with the kernel functions $h_f(x_1,\dots,x_k)=\int f^2(x_1,\dots,x_k,y)\rho(dy)$ and the random variables $\xi_l^{(j,1)}$, $1\le j\le k$, $1\le l\le n$. (Here we exploited that in the last formula $S^2(\mathcal{F}|G,V_1,V_2)$ is bounded by the product of two random variables whose distributions do not depend on the sets V_1 and V_2.) Thus to prove inequality (17.16) it is enough to show that

$$2P\left(\sup_{f\in\mathcal{F}} k!\bar{I}_{n,k}(h_f) > 2^k A^{4/3}n^k\sigma^2\right) \le 2^{k+1}e^{-A^{2/3k}n\sigma^2} \quad \text{if } A\ge A_0. \tag{17.21}$$

Actually formula (17.21) follows from the already proven formula (17.14), only the parameter A has to be replaced by $A^{4/3}$ in it.

With the help of relation (17.16) the proof of Proposition 15.4 can be completed similarly to Proposition 15.3. The following version of inequality (17.7) can be proved with the help of the multivariate version of Hoeffding's inequality (Theorem 13.3) and the representation of the random variable $H_{n,k}(f|G,V_1,V_2)$ in the form (17.15).

$$\begin{aligned}
&P\left(|H_{n,k}(f|G,V_1,V_2)| > \frac{A^2}{2^{4k+2}}n^{2k}\sigma^{2(k+1)}\,\middle|\, \xi_l^{j,\pm1},\ 1\le l\le n,\ 1\le j\le k\right)(\omega)\\
&\quad\le Ce^{-2^{-(6+2/k)}A^{2/3k}n\sigma^2} \qquad\qquad (17.22)\\
&\quad\text{if}\quad S^2(\mathcal{F}|G,V_1,V_2)(\omega)\le 2^{2k}A^{8/3}n^{2k}\sigma^4 \text{ and } A\ge A_0
\end{aligned}$$

with an appropriate constant $C=C(k)>0$ for all $f\in\mathcal{F}$ and $G\in\mathcal{G}$ such that $|e(G)|<k$ and $V_1,V_2\subset\{1,\dots,k\}$. [Observe that the conditional probability estimated in (17.22) can be represented in the following way. In a point $\omega\in\Omega$ fix the values of $\xi_l^{(j,\pm1)}(\omega)$ for all indices $1\le l\le n$ and $1\le j\le k$ in the random variable $H_{n,k}(f|G,V_1,V_k)$, and the conditional probability in this point ω equals the probability that the random variable (depending on the random variables ε_l, $1\le l\le n$), obtained in such a way is greater than $\frac{A^2}{2^{4k+2}k!}n^{2k}\sigma^{2(k+1)}$.]

Indeed, in this case the conditional probability considered in (17.22) can be bounded because of the multivariate version of Hoeffding's inequality by

$$C\exp\left\{-\frac12\left(\frac{A^4n^{4k}\sigma^{4(k+1)}}{2^{8k+4}S^2(\mathcal{F}|G,V_1,V_2)}\right)^{1/2j}\right\}\le C\exp\left\{-\frac12\left(\frac{A^{4/3}n^{2k}\sigma^{4k}}{2^{10k+4}}\right)^{1/2j}\right\}$$

with an appropriate $C = C(k) > 0$, where $2j = 2k - 2|e(G)|$, and $0 \le |e(G)| \le k-1$. Since $j \le k$, $n\sigma^2 \ge \frac{1}{2}$, and also $\frac{A^{4/3}}{2^{10k+4}} \ge 2$ if A_0 is chosen sufficiently large we can write in the above upper bound for the left-hand side of (17.22) $j = k$, and in such a way we get inequality (17.22).

The next inequality, in which we estimate $\sup\limits_{f \in \mathscr{F}} H_{n,k}(f|G, V_1, V_2)$, is a natural version of formula (17.9) in the proof of Proposition 15.3. We shall show that

$$P\left(\sup_{f\in\mathscr{F}} |H_{n,k}(f|G,V_1,V_2)| > \frac{A^2}{2^{4k+1}} n^{2k}\sigma^{2(k+1)} \middle| \xi_l^{(j,\pm1)}, 1 \le l \le n, 1 \le j \le k \right)(\omega)$$
$$\le C\left(1 + D\left(\frac{2^{4k+3}}{A^2\sigma^{2(k+1)}}\right)^L\right) e^{-2^{-(6+2/k)}A^{2/3k}n\sigma^2}$$
$$\text{if } S^2(\mathscr{F}|G,V_1,V_2))(\omega) \le 2^{2k}A^{8/3}n^{2k}\sigma^4 \text{ and } A \ge A_0 \tag{17.23}$$

for all $G \in \mathscr{G}$ such that $|e(G)| < k$ and $V_1, V_2 \subset \{1,\dots,k\}$.

To prove formula (17.23) let us fix two sets $V_1, V_2 \subset \{1,\dots,k\}$ and a diagram G such that $|e(G)| < k$. We shall define for all vectors $x^{(n)} = (x_l^{(j,1)}, x_l^{(j,-1)}, 1 \le l \le n, 1 \le j \le k) \in X^{2kn}$ some probability measure $\alpha(x^{(n)})$ on the space $X^k \times Y$ (with the space Y which appears in the formulation of Proposition 15.4) with which we can work so as we did with the probability measures $\nu(x^{(n)})$ and $\rho(x^{(n)})$ in the proof of Propositions 7.3 and 15.3.

To do this we define first for a vector $x^{(n)} = (x_l^{(j,1)}, x_l^{(j,-1)}, 1 \le l \le n, 1 \le j \le k) \in X^{2kn}$ and for all $1 \le j \le k$ two probability measures $\nu_j^{(1)} = \nu_j^{(1)}(x^{(n)}, V_1)$ and $\nu_j^{(2)} = \nu_j^{(2)}(x^{(n)}, V_2)$ in the space $(X, \mathscr{X})$ in the following way. The measures $\nu_j^{(1)}(x^{(n)}, V_1)$ and $\nu_j^{(2)}(x^{(n)}, V_2)$ are uniformly distributed in the set of points $x_l^{(j,\delta_j(V_1))}$, $1 \le l \le n$ and $x_l^{(j,\delta_j(V_2))}$, $1 \le l \le n$, respectively. More explicitly, we define for all $1 \le j \le k$ (and sets V_1 and V_2) the probability measures $\nu_j^{(1)}\left(\{x_l^{(j,\delta_j(V_1))}\}\right) = \frac{1}{n}$ and $\nu_j^{(2)}\left(\{x_l^{(j,\delta_j(V_2))}\}\right) = \frac{1}{n}$ for all $1 \le l \le n$, where $\delta_j(V_1) = 1$ if $j \in V_1$, $\delta_j(V_1) = -1$ if $j \notin V_1$, and similarly $\delta_j(V_2) = 1$ if $j \in V_2$ and $\delta_j(V_2) = -1$ if $j \notin V_2$. Let us consider the product measures $\alpha_1 = \alpha_1(x^{(n)}, V_1) = \nu_1^{(1)} \times \dots \times \nu_k^{(1)} \times \rho$ and $\alpha_2 = \alpha_2(x^{(n)}, V_2) = \nu_1^{(2)} \times \dots \times \nu_k^{(2)} \times \rho$ on the product space $(X^k \times Y, \mathscr{X}^k \times \mathscr{Y})$, where ρ is that probability measure on $(Y, \mathscr{Y})$ which appears in Proposition 15.4. With the help of the measures α_1 and α_2 we define the measure $\alpha = \alpha(x^{(n)}) = \alpha(x^{(n)}, V_1, V_2) = \frac{\alpha_1+\alpha_2}{2}$ in the space $(X^k \times Y, \mathscr{X}^k \times \mathscr{Y})$. Let us also define the measure $\tilde\alpha = \tilde\alpha(x^{(n)}) = \tilde\alpha(x^{(n)}, V_1, V_2) = \nu_1^{(1)} \times \cdots \nu_k^{(1)} \times \nu_1^{(2)} \times \cdots \nu_k^{(2)} \times \rho$ in the space $(X^{2k} \times Y, \mathscr{X}^{2k} \times \mathscr{Y})$.

Define $H_{n,k}(f|G, V_1, V_2)$ as a function in the product space $(X^{2kn}, \mathscr{X}^{2kn})$ (with arguments $x_l^{(j,1)}$ and $x_l^{(j,-1)}$, $1 \le j \le k$, $1 \le l \le n$) by means of formula (17.15) by replacing the random variables $\xi_{l_j}^{(j,\delta_j(V_1))}(\omega)$ by $x_{l_j}^{(j,\delta_j(V_1))}$ and the random variables $\xi_{l'_j}^{(j,\delta_j(V_2))}(\omega)$ by $x_{l'_j}^{(j,\delta_j(V_2))}$ in it for all $1 \le j \le k$ and $1 \le l_j, l'_j \le n$. (We consider the value of the coefficients $\varepsilon_{l_{j(s)}}$ and $\varepsilon_{l'_{\bar{j}(s)}}$ in (17.5) fixed.) With such a notation we can write for any pairs $f, g \in \mathcal{F}$ and $x^{(n)} = (x_l^{(j,1)}, x_l^{(j,-1)}, 1 \le j \le k, 1 \le l \le n) \in X^{2kn}$, by exploiting the properties of the above defined measure $\tilde{\alpha}$ the inequality

$$
\begin{aligned}
&\sup_{\varepsilon_1,\dots,\varepsilon_n} |H_{n,k}(f|G,V_1,V_2)(x^{(n)}) - H_{n,k}(g|G,V_1,V_2)(x^{(n)})| \\
&\quad \le \sum_{v=(v^{(1)},v^{(2)})\in V(G)} \quad \sum_{(l_1,\dots,l_k,l'_1,\dots,l'_k)\in E_G(v)} \\
&\qquad \int |f(x_{l_1}^{(1,\delta_1(V_1))},\dots,x_{l_k}^{(k,\delta_k(V_1))},y) f(x_{l'_1}^{(1,\delta_1(V_2))},\dots,x_{l'_k}^{(k,\delta_k(V_2))},y) \\
&\qquad -g(x_{l_1}^{(1,\delta_1(V_1))},\dots,x_{l_k}^{(k,\delta_k(V_1))},y) g(x_{l'_1}^{(1,\delta_1(V_2))},\dots,x_{l'_k}^{(k,\delta_k(V_2))},y)|\rho(dy) \\
&\quad \le n^{2k} \int |f(x_1,\dots,x_k,y) f(x_{k+1},\dots,x_{2k},y) \\
&\qquad -g(x_1,\dots,x_k,y) g(x_{k+1},\dots,x_{2k},y)|\tilde{\alpha}(\,dx_1,\dots,\,dx_{2k},\,dy). \qquad (17.24)
\end{aligned}
$$

Beside this, since both $\sup|f(x_1,\dots,x_k,y)| \le 1$ and $\sup|g(x_1,\dots,x_k,y)| \le 1$, we have

$$
\begin{aligned}
&|f(x_1,\dots,x_k,y) f(x_{k+1},\dots,x_{2k},y) - g(x_1,\dots,x_k,y) g(x_{k+1},\dots,x_{2k},y)| \\
&\quad \le |f(x_1,\dots,x_k,y)||f(x_{k+1},\dots,x_{2k},y) - g(x_{k+1},\dots,x_{2k},y)| \\
&\qquad +|g(x_{k+1},\dots,x_{2k})||f(x_1,\dots,x_k,y) - g(x_1,\dots,x_k,y)| \\
&\quad \le |f(x_{k+1},\dots,x_{2k},y) - g(x_{k+1},\dots,x_{2k},y)| \\
&\qquad +|f(x_1,\dots,x_k,y) - g(x_1,\dots,x_k,y)|.
\end{aligned}
$$

It follows from this inequality, formula (17.24) and the definition of the measures $\tilde{\alpha}$, α_1, α_2 and α that

$$
\begin{aligned}
&\sup_{\varepsilon_1,\dots,\varepsilon_n} |H_{n,k}(f|G,V_1,V_2)(x^{(n)}) - H_{n,k}(g|G,V_1,V_2)(x^{(n)})| \\
&\quad \le n^{2k} \int (|f(x_{k+1},\dots,x_{2k},y) - g(x_{k+1},\dots,x_{2k},y)| \\
&\qquad +|f(x_1,\dots,x_k,y) - g(x_1,\dots,x_k,y)|)\tilde{\alpha}(\,dx_1,\dots,\,dx_{2k},\,dy) \\
&\quad = n^{2k} \int |f(x_1,\dots,x_k,y) - g(x_1,\dots,x_k,y)| \\
&\qquad (\alpha_1(\,dx_1,\dots,\,dx_k,\,dy) + \alpha_2(\,dx_1,\dots,\,dx_k,\,dy)) \qquad (17.25) \\
&\quad = 2n^{2k} \int |f(x_1,\dots,x_k,y) - g(x_1,\dots,x_k,y)|\alpha(\,dx_1,\dots,\,dx_k,\,dy)
\end{aligned}
$$

$$\leq 2n^{2k}\left(\int |f(x_1,\dots,x_k,y)-g(x_1,\dots,x_k,y)|^2\alpha(\,dx_1,\dots,\,dx_k,\,dy)\right)^{1/2}$$

with the previously defined probability measure $\alpha=\alpha(x^{(n)})$. Put $\delta=\frac{A^2\sigma^{2(k+1)}}{2^{4k+3}}$, list the elements of $\mathcal{F}$ as $\mathcal{F}=\{f_1,f_2,\dots\}$, and choose such a set of indices $p_1(x^{(n)}),\dots,p_m(x^{(n)})$ taking positive integer values with $m=\max(1,D\delta^{-L})$ elements for which

$$\min_{1\le l\le m}\int (f(u)-f_{p_l(x^{(n)})}(u))^2\alpha(x^{(n)})(\,du)\le\delta^2\quad\text{for all } f\in\mathcal{F} \text{ and } x^{(n)}\in X^{2kn}.$$

(Here integration is taken with respect to $u\in X^k\times Y$.)

Such a choice of the indices $p_l(x^{(n)})$, $1\le l\le m$, is possible, since $\mathcal{F}$ is L_2-dense with exponent L and parameter D. Moreover, by Lemma 7.4B we may chose the functions $p_l(x^{(n)})$, $1\le l\le m$, as measurable functions of their argument $x^{(n)}\in X^{2kn}$.

Put $\xi^{(n)}(\omega)=(\xi_l^{(j,\pm1)}(\omega),\ 1\le l\le n,\ 1\le j\le k)$. By arguing similarly as we did in the proof of Propositions 7.3 and (15.3) we get with the help of relation (17.25) and the property of the functions $f_{p_l(x^{(n)})}(\cdot)$ constructed above that

$$\left\{\omega\colon\ \sup_{f\in\mathcal{F}}|H_{n,k}(f|G,V_1,V_2)(\omega)|\ge\frac{A^2n^{2k}\sigma^{2(k+1)}}{2^{(4k+1)}}\right\}$$
$$\subset\bigcup_{l=1}^m\left\{\omega\colon\ |H_{n,k}(f_{p_l(\xi^{(n)}(\omega)}|G,V_1,V_2)(\omega)(\omega)|\ge\frac{A^2n^{2k}\sigma^{2(k+1)}}{2^{(4k+2)}}\right\}.$$

Hence

$$P\left(\sup_{f\in\mathcal{F}}|H_{n,k}(f|G,V_1,V_2)|>\frac{A^2n^{2k}\sigma^{2(k+1)}}{2^{4k+1}}\,\middle|\,\xi_l^{(j,\pm1)},1\le l\le n,\ 1\le j\le k\right)(\omega)$$
$$\le\sum_{l=1}^m P\left(|H_{n,k}(f_{p_l(\xi^{(n)}(\omega))}|G,V_1,V_2)|>\frac{A^2n^{2k}\sigma^{2(k+1)}}{2^{4k+1}}\,\middle|\right.$$
$$\left.\xi_l^{(j,\pm1)},\ 1\le l\le n,\ 1\le j\le k\right)(\omega)$$

for almost all ω. The last inequality together with (17.22) and the inequality $m=\max(1,D\delta^{-L})\le1+D\left(\frac{2^{4k+3}}{A^2\sigma^{2(k+1)}}\right)^L$ imply relation (17.23).

It follows from relations (17.16) and (17.23) that

$$P\left(\sup_{f\in\mathcal{F}}|H_{n,k}(f|G,V_1,V_2)|>\frac{A^2n^{2k}\sigma^{2(k+1)}}{2^{4k+1}}\right)\le2^{k+1}e^{-A^{2/3k}n\sigma^2}$$

$$+C\left(1+D\left(\frac{2^{4k+3}}{A^2\sigma^{2(k+1)}}\right)^L\right)e^{-2^{-(6+2/k)}A^{2/3k}n\sigma^2} \quad \text{if } A \ge A_0$$

for all $V_1, V_2 \subset \{1,\dots,k\}$ and diagram $G \in \mathcal{G}$ such that $|e(G)| \le k-1$. This inequality implies that relation (17.11) holds also in the case $|e(G)| \le k-1$ if the constants A_0 is chosen sufficiently large in Proposition 15.4, and this completes the proof of Proposition 15.4. To prove relation (17.11) in the case $|e(G)| \le k-1$ with the help of the last inequality it is enough to show that $D(\frac{2^{4k+3}}{A^2\sigma^{2(k+1)}})^L \le e^{\text{const.}\, n\sigma^2}$ if $A > A_0$ with a sufficiently large A_0, since this implies that the second term at the right-hand of our last estimation is not too large.

This relation follows from the inequality $n\sigma^2 \ge L\log n + \log D$ which implies that

$$\left(\frac{2^{4k+3}}{A^2\sigma^{2(k+1)}}\right)^L \le \left(\frac{n^{(k+1)}}{(2n\sigma^2)^{(k+1)}}\right)^L \le n^{(k+1)L} = e^{(k+1)L\log n} \le e^{(k+1)n\sigma^2}$$

if A_0 is sufficiently large, and $D = e^{\log D} \le e^{n\sigma^2}$. □

Chapter 18
An Overview of the Results and a Discussion of the Literature

I discuss briefly the problems investigated in this work, recall some basic results related to them, and also give some references. I also write about the background of these problems which may explain the motivation for their study. I list the remarks following the subsequent chapters in this work. Chapter 1 is an introductory text, the real work starts at Chap. 2.

CHAPTER 2

I met the main problem considered in this work when I tried to adapt the method of proof of the central limit theorem for maximum-likelihood estimates to some more difficult questions about so-called non-parametric maximum likelihood estimate problems. The Kaplan–Meyer estimate for the empirical distribution function with the help of censored data investigated in the second chapter is an example for such problems. It is not a maximum-likelihood estimate in the classical sense, but it can be considered as a non-parametric version of it. In the estimation of the distribution function with the help of censored data we cannot apply the classical maximum likelihood method, since in this problem we have to choose our estimate from a too large class of distribution functions. The main problem is that there is no dominating measure with respect to which all candidates which may appear as our estimate have a density function. A natural way to overcome this difficulty is to choose an appropriate smaller class of distribution functions, to compare the probability of the appearance of the sample we observed with respect to all distribution functions of this class and to choose that distribution function as our estimate for which this probability takes its maximum.

The Kaplan–Meyer estimate can be found on the basis of the above principle in the following way: Let us estimate the distribution function $F(x)$ of the censored data simultaneously together with the distribution function $G(x)$ of the censoring data. (We have a sample of size n and know which sample elements are censored and which are censoring data.) Let us consider the class of such pairs of estimates $(F_n(x), G_n(x))$ of the pair $(F(x), G(x))$ for which the distribution function $F_n(x)$ is concentrated in the censored sample points and the distribution function $G_n(x)$ is

P. Major, *On the Estimation of Multiple Random Integrals and U-Statistics*, Lecture Notes in Mathematics 2079, DOI 10.1007/978-3-642-37617-7_18,

concentrated in the censoring sample points; more precisely, let us also assume that if the largest sample point is a censored point, then the distribution function $G_n(x)$ of the censoring data takes still another value which is larger than any sample point, and if it is a censoring point then the distribution function $F_n(x)$ of the censored data takes still another value larger than any sample point. (This modification at the end of the definition is needed, since if the largest sample point is from the class of censored data, then the distribution $G(x)$ of the censoring data in this point must be strictly less than 1, and if it is from the class of censoring data, then the value of the distribution function $F(x)$ of the censored data must be strictly less than 1 in this point.) Let us take this class of pairs of distribution functions $(F_n(x), G_n(x))$, and let us choose that pair of distribution functions of this class as the (non-parametric maximum likelihood) estimate with respect to which our observation has the greatest probability.

The above extremum problem about a pair of distribution functions $(F_n(x), G_n(x))$ can be solved explicitly, (see [28]), and it yields the estimate of $F_n(x)$ written down in formula (2.3). (The function $G_n(x)$ satisfies a similar relation, only the random variables X_j and Y_j and the events $\delta_j = 1$ and $\delta_j = 0$ have to be replaced in it.) If we want to prove that the estimate of the distribution function we found in such a way satisfies the central limit theorem, then we can do this with the help of a good adaptation of the method applied in the study of maximum likelihood estimates. We apply an appropriate linearization procedure, and there is only one really hard part of the proof. We have to show that this linearization procedure gives a small error. This problem led to the study of a good estimate on the tail distribution of the integral of an appropriate function of two variables with respect to the product of a normalized empirical measure with itself. Moreover, as a more detailed investigation showed, we actually need the solution of a more general problem where we have to bound the tail distribution of the supremum of a class of such integrals. The main subject of this work is to solve the above problems in a more general setting, to estimate not only twofold, but also k-fold random integrals and the supremum of such integrals for an appropriate class of kernel functions with respect to a normalized empirical distribution for all $k \ge 1$.

The proof of the limit theorem for the Kaplan–Meyer estimate explained in this work applied the explicit form of this estimate. It would be interesting to find such a modification of this proof which only exploits that the Kaplan–Meyer estimate is the solution of an appropriate extremum problem. We may expect that such a proof can be generalized to a general result about the limit behaviour for a wide class of non-parametric maximum likelihood estimates. Such a consideration was behind the remark of Richard Gill I quoted at the end of Chap. 2.

A detailed proof together with a sharp estimate on the speed of convergence for the limit behaviour of the Kaplan–Meyer estimate based on the ideas presented in Chap. 2 is given in paper [40]. Paper [41] explains more about its background, and it also discusses the solution of some other non-parametric maximum likelihood problems. The results about multiple integrals with respect to a normalized empirical distribution function needed in these works were proved in [33]. These results were satisfactory for the study in [40], but they also have some drawbacks. They do not show that if the random integrals we are considering have small variances, then they satisfy better estimates. Beside this, if we consider the supremum of random integrals of an appropriate class of functions, then these results can be applied only in very special cases. Moreover, the method of proof of [33] did not allow a real generalization of these results. Hence I had to find a different approach when I tried to generalize them.

I do not know of other works where the distribution of multiple random integrals with respect to a normalized empirical distribution is studied. On the other hand, there are some works where a similar problem is investigated about the distribution of (degenerate) U-statistics. The most important results obtained in this field are contained in the book of de la Peña and Giné *Decoupling, From Dependence to Independence* [9]. The problems about the behaviour of degenerate U-statistics and multiple integrals with respect to a normalized empirical distribution function are closely related, but the explanation of their relation is far from trivial. The main difference between them is that integration with respect to $\mu_n - \mu$ instead of the empirical distribution μ_n means of some sort of normalization, while this normalization is missing in the definition of U-statistics. I return to this question later.

Let me finish my discussion about Chap. 2 with some personal remarks. Here I investigated a special problem. But in my opinion the method applied in this chapter works well in several similar problems about the limit behaviour of a non-linear functional of independent identically distributed random variables. In the study of such problems we express the non-linear functional we are investigating as an integral with respect to the normalized empirical distribution determined by the random variables we are working with plus some negligibly small error terms. Then we have to describe the limit behaviour of the random integral we got, and this can be done with the help of some classical results of probability theory. Beside this we have to show that the remaining error terms are really small. This can be done, but at this point the results discussed in this work play a crucial role. I believe that a similar picture arises in many cases. In certain problems it may happen that the main term is not a onefold, but a multiple integral with respect to the normalized empirical distribution. But the limit distribution of such functionals can also be described. This is the content of Theorem 10.4′ proved in Appendix C.

Chapter 3

The main part of this work starts at Chap. 3. A general overview of the results without the hard technical details can be found in [36].

First the estimation of sums of independent random variables or of onefold random integrals with respect to a normalized empirical distribution and the supremum of such expressions is investigated in Chaps. 3 and 4. This question has a fairly big literature. I would mention first of all the books *A course on empirical processes* [13], *Real Analysis and Probability* [14] and *Uniform Central Limit Theorems* [15] of R.M. Dudley. These books contain a much more detailed description of the empirical processes than the present work together with a lot of interesting results.

In Chap. 3 I presented the proof of some classical results about the tail behaviour of sums of independent and bounded random variables with expectation zero. They are Bernstein's and Bennett's inequalities. Their proofs can be found at many places, e.g. in Theorem 1.3.2 of [6, 15]. We are also interested in the question when these results give such an estimate that the central limit theorem suggests. Actually, as it is explained in Chap. 3, Bennett's inequality gives such a bound that the Poissonian approximation of partial sums of independent random variables suggests. Bernstein's inequality provides an estimate suggested by the central limit theorem if the variance of the sum we consider is not too small. The results in Chap. 3 explain these statements more explicitly. If the variance of the sum is too small, then Bennett's inequality provides a slight improvement of Bernstein's inequality. Moreover, as Example 3.3 shows, Bennett's inequality is essentially sharp in this case. But these results are much weaker than the estimates suggested by a normal comparison.

The relative weakness of Bernstein's and Bennett's inequality for random sums with small variance had a deep consequence in our investigation about the supremum (of appropriate classes) of sums of independent random variables. Because of the weakness of these estimates in certain cases we had to find a new method. We could overcome the difficulty we met with the help of a symmetrization argument which is explained in Chap. 7. But to apply this method we needed another result, known under the name Hoeffding's inequality. It yields an estimate about the tail behaviour of linear combinations of independent Rademacher functions. This result always provides such a good bound as the central limit theorem suggests. This is the reason why I discuss this inequality at the end of Chap. 3, in Theorem 3.4. It is also a classical result whose proof can be found for instance in [25].

The content of Chap. 3 can be found in the literature, e.g. in [13]. The main difference between my discussion and that of earlier works is that I put more emphasis on the investigation of the question when the estimates on the tail distribution of partial sums of independent random variables are similar to their Gaussian counterpart. I had a good reason to discuss this question in more detail. I was also interested in the estimation of the tail distribution of the supremum of partial sums of independent random variables, and in the study of this problem we have to understand when the classical methods related to Gaussian random variables can be applied and when we have to look for a new approach.

Chapter 4

Chapter 4 contains the one-variate version of our main result about the supremum of the integrals of a class $\mathcal{F}$ of functions with respect to a normalized empirical measure together with an equivalent statement about the tail distribution of the supremum of a class of random sums defined with the help of a sequence of independent and identically distributed random variables and a class of functions $\mathcal{F}$ with some nice properties. These results are formulated in Theorems 4.1 and 4.1′. They appeared in [33]. Also a Gaussian version of them is presented in Theorem 4.2 about the distribution of the supremum of a Gaussian random field with some appropriate properties. A deeper version of Theorem 4.2 is studied in paper [12].

The content of these results can be so interpreted that if we take the supremum of random integrals or of random sums determined by a nice class of functions $\mathcal{F}$ in the way described in Chap. 4, then the tail distribution of this supremum satisfies an almost as good estimate as the "worst element" of the random variables taking part in this supremum. But such a result holds only if we consider the value of this tail distribution at a sufficiently large level, since—as some concentration inequalities imply—the supremum of these random sums are larger than the expected value of this supremum with probability almost one. I also discussed a result in Example 4.3 which shows that some rather technical conditions of Theorem 4.1 cannot be omitted.

The most important condition in Theorem 4.1 was that the class of functions $\mathcal{F}$ we considered in it is L_2-dense. This property was introduced before the formulation of Theorem 4.1. One may ask whether one can prove a better version of this result, which states a similar bound for a different, possibly larger class of functions $\mathcal{F}$. It is worth mentioning that Talagrand proved results similar to Theorem 4.1 for different classes of functions $\mathcal{F}$ in his book [58]. These classes of functions are very different of ours, and Talagrand's results seem to be incomparable with ours. I return to this question later in the discussion of Chaps. 6 and 7, which deal with the proof of the results of Chap. 4. In the remaining part of the discussion of Chap. 4 I write about the notion of countably approximable classes of random variables and its role in the present work.

In the first formulation of our results we have imposed the condition that the class of functions $\mathcal{F}$ is countable, i.e. we take the supremum of countably many random variables. In the proofs this condition was heavily exploited. On the other hand, in some important applications we also need results about the supremum of a possibly non-countable set of random variables. To handle such cases I introduced the notion of countably approximable classes of random variables and proved that in the results of this work the condition about countability can be replaced by the weaker condition that the supremum of countably approximable classes is taken. R.M. Dudley worked out a different method to handle the supremum of possibly non-countably many random variables, and generally his method is applied in the literature. The relation between these two methods deserves some discussion.

To understand the problem we are discussing let us first recall that if we take a class of random variables S_t, $t \in T$, indexed by some index set T, then for all sets A measurable with respect to the σ-algebra generated by the random variables S_t, $t \in T$, there exists a countable subset $T' = T'(A) \subset T$ such that the set A is measurable also with respect to the smaller σ-algebra generated by the random variable S_t, $t \in T'$. Beside this, if the finite dimensional distributions of the random variables S_t, $t \in T$, are given, then by the results of classical measure theory the probability of all events measurable with respect to the σ-algebra generated by these random variables S_t, $t \in T$, is also determined. But it may happen that we want to deal with such events whose probability cannot be defined in such a way. In particular, if T is a non-countable set, then the events $\left\{\omega\colon \sup\limits_{t\in T} S_t(\omega) > u\right\}$ are non-measurable with respect to the above σ-algebra, and generally we cannot speak of their probabilities. To overcome this difficulty Dudley worked out a theory which enabled him to work also with outer measures. His theory is based on some rather deep results of the analysis. It can be found for instance in his book [15].

I restricted my attention to such cases when after the completion of the probability measure P we can also speak of the real (and not only outer) probabilities $P\left(\sup\limits_{t\in T} S_t > u\right)$. I tried to find appropriate conditions under which these probabilities really exist. More explicitly, I was interested in the case when for all $u > 0$ there exists some set $A = A_u$ measurable with respect to the σ-algebra generated by the random variables S_t, $t \in T$, such that the symmetric difference of the sets A_u and $\left\{\omega\colon \sup\limits_{t\in T} S_t(\omega) > u\right\}$ is contained in a set which is measurable with respect to the σ-algebra generated by the random variables S_t, $t \in T$, and it has probability zero. In such a case the probability $P\left(\sup\limits_{t\in T} S_t > u\right)$ can be defined as $P(A_u)$. This approach led me to the definition of countable approximable classes of random variables. If this property holds, then we can speak about the probability of the event that the supremum of the random variables we are interested in is larger than some fixed value. I proved a simple but useful result in Lemma 4.4 which provides a condition for the validity of this property. In Lemma 4.5 I proved with its help that an important class of functions is countably approximable. It seems that this property can be proved for many other interesting classes of functions with the help of Lemma 4.4, but I did not investigate this question in more detail.

The problem we met here is not an abstract, technical difficulty. Indeed, the distribution of the supremum of uncountably many random variables can become different if we modify each random variable on a set of probability zero, although their finite dimensional distributions remain the same after such an operation. Hence, if we are interested in the probability of the supremum of a non-countable set of random variables with prescribed finite dimensional distributions we have to tell more explicitly which version of this set of random variables we consider. It is natural to look for such an appropriate version of the random field S_t, $t \in T$, whose "trajectories" $S_t(\omega)$, $t \in T$, have nice properties for all elementary events $\omega \in \Omega$. Lemma 4.4 can be interpreted as a result in this spirit. The condition given for the countable approximability of a class of random variables at the end of this lemma can be considered as a smoothness type condition about the "trajectories" of the random field we consider. This approach shows some analogy to some important problems in the theory of stochastic processes when a regular version of a stochastic process is considered, and the smoothness properties are investigated for the trajectories of this version.

In our problems the version of the set of random variables S_t, $t \in T$, we work with appears in a simple and natural way. In these problems we have finitely many random variables $\xi_1, \dots, \xi_n$ at the start, and all random variables $S_t(\omega)$, $t \in T$, we are considering can be defined individually for each ω as a function of these random variables $\xi_1(\omega), \dots, \xi_n(\omega)$. We take the version of the random field $S_t(\omega)$, $t \in T$, we get in such a way and want to show that it is countably approximable. In Chap. 4

this property is proved in an important model, probably in the most important model in possible applications we are interested in. In more complicated situations when our random variables are defined not as a function of finitely many sample points, for instance in the case when we define our set of random variables by means of integrals with respect to a Gaussian random field it is harder to find the right regular version of our sets of random variables. In this case the integrals we consider are defined only with probability 1, and it demands some extra work to find their right version. But in the problems studied in this work the above sketched approach is satisfactory for our purposes, and it is simpler than that of Dudley; we do not have to follow his rather difficult technique. On the other hand, I must admit that I do not know the precise relation between the approach of this work and that of Dudley.

Chapter 5

In Chap. 4 the notion of L_p-dense classes, $1 \le p < \infty$, also has been introduced. The notion of L_2-dense classes appeared in the formulation Theorems 4.1 and 4.1′. It can be considered as a version of the ε-entropy, discussed at many places in the literature. (See e.g. [13] or [14].) On the other hand, there seems to be no standard definition of the ε-entropy. The term of L_2-dense classes seemed to be the appropriate object to work with in this lecture note. To apply the results related to L_2-dense classes we also need some knowledge about how to check this property in concrete models. For this goal I discussed here Vapnik–Červonenkis classes, a popular and important notion of modern probability theory. Several books and papers, (see e.g. the books [15, 49, 59] and the references in them) deal with this subject. An important result in this field is Sauer's lemma, (Theorem 5.1) which together with some other results, like Theorem 5.3 imply that several interesting classes of sets or functions are Vapnik–Červonenkis classes.

I put the proof of these results to the Appendix, partly because they can be found in the literature, partly because in this work Vapnik–Červonenkis classes play a different and less important role than at other places. Here Vapnik–Červonenkis classes are applied to show that certain classes of functions are L_2-dense. At this point a result of Dudley formulated in Theorem 5.2 plays an important role. It implies that a Vapnik–Červonenkis class of functions with absolute value bounded by a fixed constant is an L_1, and as a consequence also an L_2-dense class of functions. The proof of this important result which seems to be less known even among experts of this subject than it would deserve is contained in the main text. Dudley's original result was formulated in the special case when the functions we consider are indicator functions of some sets. But its proof contains all important ideas needed in the proof of Theorem 5.2. A proof of the result in the form formulated in this work can be found in [49]. This book also contains the other results of this chapter about Vapnik–Červonenkis classes.

Chapters 6 and 7

Theorem 4.2, which is the Gaussian counterpart of Theorems 4.1 and 4.1′ is proved in Chap. 6 by means of a natural and important technique, called the chaining argument. This means the application of an inductive procedure, in which an appropriate sequence of finite subsets of the original set of random variables is introduced, and a good estimate is given on the supremum of the random variables in these subsets by means of an inductive procedure. The subsets became denser subsets of the original set of the random variables at each step of this procedure. This chaining argument is a popular method in certain investigations. It is hard to say with whom to attach it. Its introduction may be connected to some works of R.M. Dudley. It is worth mentioning that Talagrand [58] worked out a sharpened version of it which yields in the study of certain problems a sharper and more useful estimate. But it seems to me that in the study of the problems of this work this improvement has a limited importance, it turns out to be useful in the study of different problems.

Theorem 4.2 can be proved by means of the chaining argument, but this method is not strong enough to supply a proof of Theorem 4.1. It provides only a weak estimate in this case, because there is no good estimate on the probability that a sum of independent random variables is greater than a prescribed value if these random variables have too small variances. As a consequence, the chaining argument supplies a much weaker estimate than the result we want to prove under the conditions of Theorem 4.1. Lemma 6.1 contains the result the chaining argument yields under these conditions. In Chap. 6 still another result, Lemma 6.2 is formulated. It can be considered as a special case of Theorem 4.1 where only the supremum of partial sums with small variances is estimated. We also show in this chapter that Propositions 6.1 and 6.2 together imply Theorem 4.1. The proof is not difficult, despite of some non-attractive details. It has to be checked that the parameters in Propositions 6.1 and 6.2 can be fitted to each other.

Proposition 6.2 is proved in Chap. 7. It is based on a symmetrization argument. This proof applies the ideas of a paper of Kenneth Alexander [3], and although its presentation is different from Alexander's approach, it can be considered as a version of his proof. It may be worth mentioning that the symmetrization arguments were first applied in the theory of Vapnik–Červonenkis classes to get some useful estimates (see e.g. [49]). But it turned out that an appropriate refinement of this method supplies sharper results if we are working with L_2-dense classes instead of Vapnik–Červonenkis classes of functions.

A similar problem should also be mentioned at this place. M. Talagrand wrote a series of papers about concentration inequalities, (see e.g. [55] or [56]), and his research was also continued by some other authors. I would mention the works of M. Ledoux [30] and P. Massart [43]. Concentration inequalities give a bound about the difference between the supremum of a set of appropriately defined random variables and the expected value of this supremum. They express how strongly this supremum is concentrated around its expected value. Such results are closely related to Theorem 4.1, and the discussion of their relation deserves some attention. A typical concentration inequality is the following result of Talagrand [56].

Theorem 18.1 (Theorem of Talagrand). *Consider n independent and identically distributed random variables $\xi_1, \dots, \xi_n$ with values in some measurable space $(X, \mathscr{X})$. Let $\mathscr{F}$ be some countable family of real-valued measurable functions of*

$(X,\mathcal{X})$ such that $\|f\|_\infty\le b<\infty$ for every $f\in\mathcal{F}$. Let $Z=\sup\limits_{f\in\mathcal{F}}\sum\limits_{i=1}^n f(\xi_i)$ and $v=E\left(\sup\limits_{f\in\mathcal{F}}\sum\limits_{i=1}^n f^2(\xi_i)\right)$. Then for every positive number x

$$P(Z\ge EZ+x)\le K\exp\left\{-\frac{1}{K'}\frac{x}{b}\log\left(1+\frac{xb}{v}\right)\right\}$$

and

$$P(Z\ge EZ+x)\le K\exp\left\{-\frac{x^2}{2(c_1v+c_2bx)}\right\},$$

where K, K', c_1 and c_2 are universal positive constants. Moreover, the same inequalities hold when replacing Z by $-Z$.

Theorem 18.1 yields, similarly to Theorem 4.1, an estimate about the distribution of the supremum for a class of sums of independent random variables. (The paper of P. Massart [43] contains a similar estimate which is better for our purposes. The main difference between these two estimates is that the bound given by Massart depends on $\sigma^2=\sup\limits_{f\in\mathcal{F}}\sum\limits_{i=1}^n \operatorname{Var} f(\xi_i)$ instead of $v=E\left(\sup\limits_{f\in\mathcal{F}}\sum\limits_{i=1}^n f^2(\xi_i)\right)$.) Theorem 18.1 can be considered as a generalization of Bernstein's and Bennett's inequalities when the distribution of the supremum of partial sums (and not only the distribution of one partial sum) is estimated. A remarkable feature of this result is that it assumes no condition about the structure of the class of functions $\mathcal{F}$ (like the condition of L_2-dense property of the class $\mathcal{F}$ imposed in Theorem 4.1). On the other hand, the estimates in Theorem 18.1 contain the quantity $EZ=E\left(\sup\limits_{f\in\mathcal{F}}\sum\limits_{i=1}^n f(\xi_i)\right)$. Such an expectation of some supremum appears in all concentration inequalities. As a consequence, they are useful only if we can bound the expected value of the supremum we want to estimate. It is difficult to find a good bound on this expected value in the general case. Paper [18] provides a useful estimate on it if the expected value of the supremum of random sums is considered under the conditions of Theorem 4.1. But I preferred a direct proof of this result. Let me remark that because of the above mentioned concentration inequality the condition $u\ge \text{const.}\,\sigma\log^{1/2}\frac{2}{\sigma}$ with some appropriate constant which cannot be dropped from Theorem 4.1 can be interpreted so that under the conditions of Theorem 4.1 $\text{const.}\,\sigma\log^{1/2}\frac{2}{\sigma}$ is an upper bound for the expected value of the supremum we investigated in this result. Example 4.3 implies that if the conditions of Theorem 4.1 are violated, then the expected value of the above supremum may be larger.

It is also worth mentioning Talagrand's work [58] which contains several interesting results similar to Theorem 4.1. But despite their formal similarity, they are essentially different from the results of this work. This difference deserves a special discussion.

Talagrand proved in [58] by working out a more refined, better version of the chaining argument a sharp upper bound for the expected value $E \sup_{t \in T} \xi_t$ of the supremum of countably many (jointly) Gaussian random variable with zero expectation. This result is sharp. Indeed, Talagrand proved also a lower bound for this expected value, and the quotient of his upper and lower bound is bounded by a universal constant. By applying similar arguments he also gave an upper bound for $E \sup_{f \in \mathcal{F}} \sum_{k=1}^{N} f(\xi_k)$ in Proposition 2.7.2 of his book if $\xi_1, \dots, \xi_N$ is a sequence of independent, identically distributed random variables with some known distribution μ, and $\mathcal{F}$ is a class of functions with some nice properties. Then he proved in Chap. 3 of this book some estimates with the help of this result for certain models which solved some problems that could not be solved with the help of the original version of the chaining argument.

Let me make a short comparison between our Theorem 4.1 and Talagrand's result. Talagrand investigated in his book [58] the expected value of the supremum of partial sums, while we gave an estimate on its tail distribution. But this is not an essential difference. Talagrand's results also give an estimate on the tail distribution of the supremum by means of concentration inequalities, and actually his proofs also provide a direct estimate for the tail distribution we are interested in without the application of these results. The main difference between the two works is that Talagrand's method gives a sharp estimate for different classes of functions $\mathcal{F}$.

Talagrand could prove sharp results in such cases when the class of functions $\mathcal{F}$ for which the supremum is taken consists of smooth functions. An example for such classes of functions which he thoroughly investigated is the class of Lipschitz 1 functions. In particular, in Chap. 3 of his book [58] he proved that if $\xi_1, \dots, \xi_n$ is a sequence of independent random variables, uniformly distributed in the unit square $D = [0,1] \times [0,1]$, and $\mathcal{F}$ is the class of Lipschitz 1 functions on the unit square D such that $\int_D f \, d\lambda = 0$ for all $f \in \mathcal{F}$, where λ denotes the Lebesgue measure on D, then $E \sup_{f \in \mathcal{F}} \sum_{l=1}^{n} f(\xi_l) \le L\sqrt{n \log n}$ with a universal constant L. He was interested in this result, because it is equivalent to a theorem of Ajtai–Komlós–Tusnády [2]. (See Chap. 3 of [58] for details.) On the other hand, we can give sharp results in such cases when $\mathcal{F}$ consists of non-smooth functions, (see Example 5.5), and Talagrand's method does not work in the study of such problems.

This difference in the conditions of the results in these two books is not a small technical detail. Talagrand heavily exploited in his proof that he worked with such classes of functions $\mathcal{F}$ from which he could select a subclass of functions of $\mathcal{F}$ of relatively small cardinality which is dense in $\mathcal{F}$ not only in the $L_2(\mu)$-norm with the probability measure μ he was working with, but also in the supremum norm. He needed this property, because this enabled him to get sharp estimates on the tail distribution of the differences of functions he had to work with by means of Bernstein's inequality. The smallness of the supremum norm of these random variables was useful, since it implied that Bernstein's inequality provides a sharp estimate in a large domain. Talagrand needed such sharp estimates to apply (a refined version of) the chaining argument. On the other hand, we considered such classes of functions $\mathcal{F}$ which may have no small subclasses which are dense in $\mathcal{F}$ in the supremum norm.

I would characterize the difference between the results of the two works in the following way. Talagrand proved the sharpest possible estimates which can be obtained by a refinement of the chaining argument, while our main problem was to get sharp estimates also in such cases when the chaining argument does not work. Let me remark that we could prove our results only for such classes of functions $\mathcal{F}$ which are L_2-dense. (See Theorem 4.1.) In the Gaussian counterpart of this result, in Theorem 4.2, it was enough to impose that $\mathcal{F}$ is an L_2-dense class with respect to a fixed probability measure μ. We needed the extra condition about L_2-dense property to prove sharp results about the tail distribution of supremum of partial sums when the chaining argument does not work.

CHAPTER 8

The main results of this work are presented in Chap. 8. One of them is Theorem 8.3 which is a multivariate version of Bernstein's inequality (Theorem 3.1) about degenerate U-statistics. A weaker version of this result was first proved in a paper of Arcones and Giné in [4]. In the present form it was proved in my paper [39]. Its version about multiple integrals with respect to a normalized empirical measure formulated in Theorem 8.1 is proved in [35]. This paper contains a direct proof. On the other hand, Theorem 8.1 can be derived from Theorem 8.3 by means of Theorem 9.4 of this paper. Theorem 8.5 is the natural Gaussian counterpart of Theorem 8.3. The limit theorem about degenerate U-statistics, Theorem 10.4 (and its version about limit theorems for multiple integrals with respect to normalized empirical measures, presented in Theorem 10.4′ of Appendix C was discussed in this work to explain better the relation between degenerate U-statistics (or multiple integrals with respect to normalized empirical measures) and multiple Wiener–Itô integrals. A proof of this result based on similar ideas as that discussed here can be found in [16]. Theorem 6.6 of my lecture note [32] contains such a weaker version of Theorem 8.5 which does not take into account the variance of the random integral we are considering.

Example 8.7 is a natural supplement of Theorem 8.5. It shows that the estimate of Theorem 8.5 is sharp if only the variance of a Wiener–Itô integral is known. At the end of Chap. 13 I also mentioned the results of papers [1, 29] without proof which also have some relation to this problem. I discussed mainly the content of [29], and explained its relation to some results discussed in this work. The proof of these papers apply a method different of those in this work. I make some comments about them in the discussion of Chap. 13.

Theorems 8.2 and 8.4 which are the natural multivariate counterparts of Theorems 4.1 and 4.1′ yield an estimate about the supremum of (degenerate) U-statistics or of multiple random integrals with respect to a normalized empirical measure when the class of kernel functions in these U-statistics or random integrals satisfy some conditions. They were proved in my paper [37]. Actually I consider these theorems the hardest and most important results of this lecture note. Earlier Arcones and Giné proved a weaker version of this result in paper [5], but their work did not help in the proof of the results of this note. The proofs of the present note were based on an adaptation of Alexander's method [3] to the multivariate case. Theorem 8.6 is the natural Gaussian counterpart of Theorems 8.2 and 8.4.

Example 8.8 in Chap. 8 shows that the condition $u \leq \text{const.}\, n\sigma^3$ imposed in Theorem 8.3 in the case $k = 2$ cannot be dropped. The paper of Arcones and Giné [4] contains another example explained by Talagrand to the authors of that paper which also has a similar consequence. But that example does not provide such an explicit comparison of the upper and lower bound on the probability investigated in Theorem 8.3 as Example 8.8. Similar examples could be constructed for all $k \geq 1$.

Example 8.8 shows that at high levels only a very weak (and from practical point of view not really important) improvement of the estimation on the tail distribution of degenerate U-statistics is possible. But probably there exists a multivariate version of Bennett's inequality, i.e. of Theorem 3.2 which provides such an estimate. Moreover, there is some hope to get a similar strengthened form of Theorems 8.2 and 8.4 (or of Theorem 4.2 in the one-dimensional case). This question is not investigated in the present work.

Chapter 9

Chapter 9 deals with the properties of U-statistics. Its first result, Theorem 9.1, is a classical result. It is the so-called Hoeffding decomposition of U-statistics to the sum of degenerate statistics. Its proof first appeared in the paper [24], but it can be found at many places. The explanation of this work contains some ideas similar to [54]. I tried to explain that Hoeffding's decomposition is the natural multivariate version of the (trivial) decomposition of sums of independent random variables to sums of independent random variables *with expectation zero* plus the sum of the expectations of the original random variables. Moreover, even the proof of Hoeffding's decomposition shows some similarity to this simple decomposition.

Theorem 9.2 and Proposition 9.3 can be considered as a continuation of the investigation about the Hoeffding decomposition. They tell us how some properties of the kernel function of the original U-statistic are inherited in the properties of the kernel functions of the degenerate U-statistics taking part in its Hoeffding decomposition. In several applications of Hoeffding's decomposition we need such results.

The last result of Chap. 9, Theorem 9.4, enables us to reduce the estimation of multiple random integrals with respect to normalized empirical measures to the estimation of degenerate U-statistics. This result is a version of Hoeffding's decomposition, where instead of U-statistics multiple integrals with respect to a normalized empirical distribution are decomposed to the sum of *degenerate U-statistics*. In these two decompositions the same degenerate U-statistics appear. The main difference between them is that in the decomposition of the random integrals in Theorem 9.4 the coefficients of the degenerate U-statistics are relatively small. The appearance of small coefficients in this decomposition is due to the cancellation effect caused by integration with respect to a *normalized* empirical measure $\sqrt{n}(\mu_n - \mu)$. Theorem 9.4 was proved in [37]. The proof in this note is essentially different of the original proof in [37], and it is simpler.

Some remarks related to Chaps. 10–12

Theorem 8.1 can be derived from Theorems 8.3 and 8.2 from Theorem 8.4 by means of Theorem 9.4. The proof of the latter results is simpler. Chapters 10–12 contain the results needed in the proof of Theorem 8.3 and of its Gaussian counterpart Theorems 8.5 and 8.6. They are proved by means of good estimates on the high moments of degenerate U-statistics and multiple Wiener–Itô integrals. The classical proof of the one-variate counterparts of these results is based on a good estimate of the moment generating function. This method had to be replaced by the estimation of high moments, because the moment generating function of a k-fold Wiener–Itô integral is divergent for all non-zero parameters if $k \ge 3$, (this is a consequence of Theorem 13.6), and this property of Wiener–Itô integrals is also reflected in the behaviour of degenerate U-statistics. On the other hand, we can give good estimates on the tail distribution of a random variable if we have good estimates on its high moments. The results of Chaps. 10–12 enable us to prove good moment estimates.

I know of two deep and interesting methods to study high moments of multiple Wiener–Itô integrals. The first of them is called Nelson's inequality named after Edward Nelson who published it in his paper [45]. This inequality simply implies Theorem 8.5 about multiple Wiener–Itô integrals, although with worse constants. Later Leonhard Gross discovered a deep and useful generalization of this result which he published in the work *Logarithmic Sobolev inequalities* [21]. Gross considered in his paper a *stationary* Markov process $X(t)$, $t \ge 0$, and gave a good bound on the L_p-norm of functions of the form $U_t(f)(x) = E(f(X(t)|X(0) = x)$, where the L_p-norm is taken with respect to the distribution of the random variable $X(0)$. The proof of this L_p-norm estimate is based on the study of the infinitesimal operator of the Markov process. Gross' results provide Nelson's inequality, if they are applied for the Ornstein–Uhlenbeck process.

Gross' investigation in [21] revealed very much about the behaviour of Markov processes. The book [46] is partly based on this method. Gross' approach turned out to be very fruitful in the study of several hard problems of the probability theory and statistical physics. (See e.g [22] or [30]). It also provides a good estimate for the high moments of Wiener–Itô integrals.

There is another useful method to study Wiener–Itô integrals due to Kyoshi Itô and Roland L'vovich Dobrushin. This seemed to me more useful if we want estimate the high moments not only of Wiener–Itô integrals but also of degenerate U-statistics. I applied this method in Chaps. 10–12. I showed how we can get with its help results that enable us to prove good moment estimates both for Wiener–Itô integrals and degenerate U-statistics. The main step in this approach is the proof of a so-called diagram formula which makes possible to rewrite a product of Wiener–Itô integrals as a sum of Wiener–Itô integrals. Moreover, this result also has a natural counterpart for the products of degenerate U-statistics.

Chapter 10

In Chap. 10 I discuss a method related to Kyoshi Itô and Roland L'vovich Dobrushin. This is the theory of multiple Wiener–Itô integrals with respect to a white noise. This integral was introduced in paper [26]. It is useful, because every random variable which is measurable with respect to the σ-algebra generated by the Gaussian random variables of the underlying white noise and has finite second moment can be written as the sum of Wiener–Itô integrals of different order. Moreover, if only Wiener–Itô integrals of symmetric kernel functions are taken, then this representation is unique. Actually this result was originally proved by Norbert Wiener [60]. This representation also appeared in physics under the name Fock space. It plays an important role in quantum physics. Let me briefly explain the reason for the name white noise for the appropriate notion introduced in Chap. 10.

The notion of white noise was originally introduced at a heuristic level as the derivative of the trajectories of a Wiener process. But as these trajectories are non-differentiable the introduction of this notion demands a better explanation. A natural way to overcome the difficulties is to consider the derivative of a trajectory of a Wiener process as a generalized random function, and to take its integral on all measurable sets. In such a way we get a collection of Gaussian random variables $\xi(A)$ with expectation zero, indexed by the measurable sets A. These random variables have correlation function $E\xi(A)\xi(B) = \lambda(A \cap B)$, where $\lambda(\cdot)$ denotes the Lebesgue measure. In such a way we get a correct definition of the white noise which preserves the heuristic content of the original approach. In the definition of general white noise we allow to work with an arbitrary measure μ and not only with the Lebesgue measure λ. If we have a white noise we would like to have a tool that enables us to study not only the Gaussian random variables measurable with respect to the σ-algebra generated by the random variables of the white noise but all random variables measurable with respect to this σ-algebra. The Wiener–Itô integrals were defined with such a goal.

An important result of the theory of Wiener–Itô integrals, the so-called diagram formula, formulated in Theorem 10.2, expresses products of Wiener–Itô integrals as a sum of such integrals. This result which shows some similarity to the Feynman diagrams applied in the statistical physics was proved in [11]. Actually this paper discussed a modified version of Wiener–Itô integrals which is more appropriate to study the action of shift operators for non-linear functionals of a stationary Gaussian field. But these modified Wiener–Itô integrals can be investigated in almost the same way as the original ones. The diagram formula has a simple consequence formulated in Corollary of Theorem 10.2 of this note. It enables us to calculate the expectation of products of Wiener–Itô integrals. It yields an explicit formula for them. This result was applied in the proof of Theorem 8.5, i.e. in the estimation of the tail-distribution of Wiener–Itô integrals. Itô's formula for multiple Wiener–Itô integrals (Theorem 10.3) was proved in [26].

Actually the above results about Wiener–Itô integrals would have been sufficient for our purposes. But I also presented some other results for the sake of completeness. In particular, I discussed some results about Hermite polynomials. Wiener–Itô integrals are closely related to Hermite polynomials or to their multivariate version, to the so-called Wick polynomials. (See e.g. [32] or [42] for the definition of Wick polynomials.) Appendix C contains the most important properties of Hermite polynomials needed in the study of Wiener–Itô integrals. In particular, it contains the proof of Proposition C2 about the completeness of the Hermite polynomials in the Hilbert space of the functions square integrable with respect to the standard Gaussian distribution. This result can be found for instance in Theorem 5.2.7 of [53]. In the present proof I wanted to show that this result is closely related to the so-called moment problem, i.e. to the question when a distribution is determined by its moments uniquely. The method of proof described in this note can be applied with some refinement to prove some generalizations of Proposition C2 about the completeness of orthogonal polynomials with respect to more general weight functions.

On the other hand, I did not try to give a complete picture about Wiener–Itô integrals. The reader interested in it may consult with the book of S. Janson [27]. There are also other interesting and important topics related to Wiener–Itô integrals not discussed in this work. In some investigations of probability theory and statistical physics it is useful to study not only moments but also cumulants (called also semiinvariants in the literature) of Wiener–Itô integrals. It is also useful to study the moments and cumulants of polynomials and Wick polynomials of Gaussian random vectors. The book of Malyshev and Minlos [42] contains many interesting results about this subject.

Another interesting and popular subject not discussed in this work is the problem of limit theorems for Wiener–Itô integrals. In particular, one is interested in the question when a sequence of such random integrals satisfies the central limit theorem. The study of such problems heavily exploits the diagram formula, or more precisely its consequence about the calculation of moments and cumulants. In some works, see e.g. [46] or [48] this subject is worked out in detail. Moreover, a popular subject of recent research is the study of the speed of convergence in the central limit theorem. In such investigations the so-called Stein method turned out to be very useful. In its application the integral of sufficiently smooth test functions with respect to the distribution we are investigating are estimated together with the integral of their derivative (with respect to the same distribution). In a somewhat surprising way it turned out that if we are studying the central limit theorem for Wiener–Itô integrals with the help of the Stein method, then the role of the derivative of a function is taken by the so-called Malliavin derivative. (See [46].) So the theory of Malliavin calculus, see [47], became very important in such research. But this problem is a bit far from the main subject of this work, hence I do not go into the details.

Chapters 11 and 12

The diagram formula has a natural and useful analogue both for degenerate U-statistics and multiple integrals with respect to a normalized empirical measure. They enable us to rewrite the product of degenerate U-statistics and multiple integrals as the sum of such expressions. Actually the proof of these results is simpler than the proof of the original diagram formula for Wiener–Itô integrals. They make possible to adapt several useful methods of the study of non-linear functionals of Gaussian random fields to the study of non-linear functionals of normalized empirical measures. But to apply them we also need some good estimate on the L_2-norm of the kernel functions of the random integrals or U-statistics appearing in the diagram formula. Hence we also proved such results.

A version of the diagram formula was proved for degenerate U-statistics in [39] and for multiple random integrals with respect to a normalized empirical measures in [35]. Let me remark that in the formulation of the result in the work [39] a different notation was applied than in the present note. In that paper I wanted to formulate such a version of the diagram formula for U-statistics where we work with diagrams similar to those introduced in the study of Wiener–Itô integrals. I could do this only in a somewhat artificial way. In this work I formulated the diagram formula for U-statistics with the help of diagrams of a more general form. I introduced the notion of chains and coloured chains, and defined (coloured) diagrams with their help. The formulation of the results with the help of such more general diagrams seems to be more natural. I met some works where similar diagrams were introduced, see e.g. [48], but I did not meet works where also the coloured diagrams introduced in this work were applied. It is possible that this happened so, because I do not know the literature well enough, but this also may have a different cause.

In the work [48] the diagram formula was applied for the calculation of moments and cumulants, and if we are working only with them, then the results of this work can also be formulated with the help of so-called closed diagrams, and no coloured diagrams are needed. They are needed if we want to express the product of U-statistics as a sum of U-statistics. It may also be interesting that the results considered in [48] are based on some combinatorial arguments worked out in [50].

There are some works like [48], where diagram formulas are considered for other models too, e.g. in models where we integrate with respect to a normalized Poisson process. Nevertheless, in my opinion the results about the diagram formula for the products of Wiener–Itô integrals and in particular their modified versions for the products of integrals with respect to normalized Poisson processes, normalized empirical distribution or for the product of U-statistics did not get such an attention in the literature as they would deserve. An interesting paper in this direction is that of Surgailis [51], where a version of the diagram formula is proved for Poissonian integrals. It may be worth mentioning that the diagram formula for Poisson integrals shows a very strong similarity to the diagram formula for the product of integrals with respect to normalized empirical distributions. (Integrals with respect to normalized empirical distribution were discussed only at an informal level in this work.)

The Hermite polynomials and their multivariate versions, the Wick polynomials have their counterparts when instead of Wiener–Itô integrals we consider more general classes of random integrals. Itô's formula creates a relation between Wiener–Itô integrals and Hermite polynomials or their multivariate versions, the Wick polynomials. The relation between Wiener–Itô integrals and Hermite polynomials has a natural counterpart in the study of other multiple random integrals. In such a way a new notion, the Appell polynomials appeared in the literature. (See e.g. [52].)

Chapter 13

Theorems 8.3, 8.5 and 8.7 were proved on the basis of the results of Chaps. 10–12 in Chap. 13. These proofs are slight modifications of those given in [39]. An earlier proof of a result similar to Theorem 8.3 based on a different method was given by Arcones and Giné in [4]. Theorem 8.3 is a slightly stronger estimate than that of Arcones and Giné. It provides at not too high levels an estimate with almost as good constants in the exponent as the corresponding estimate about Wiener–Itô integrals in Theorem 8.5. Chapter 13 also contains the proof of a multivariate version of Hoeffding's inequality formulated in Theorem 13.3. This result is needed in the symmetrization argument applied in the proof of Theorem 8.4. A weaker version of it (an estimate with a worse constant in the exponent) which would be satisfactory for our purposes simply follows from a classical result, called Borell's inequality, which was proved in [8]. But since the methods needed to prove this result are not discussed in this note, and I was interested in a proof which yields an estimate with the best possible constant in the exponent I chose another proof, given in [38]. It is based on the results of Chap. 10–12. Later I have learned that this estimate is contained in an implicit form also in the paper [7] of Aline Bonami.

In Part B of Chap. 13 I discussed some results related to the problems considered in this work. I would like to make some comments about the result of R. Latała presented in Theorem 13.7. The estimates of this result depend on such quantities which are hard to calculate. Hence they have a limited importance in the problems I had in mind when working on this lecture note. On the other hand, such results and the methods behind them may be interesting in the study of some problems of statistical physics, e.g. in the problems discussed in [57]. I would like to remark that Latała's proof works only for decoupled and not for usual U-statistics. Formally, this is not a restriction, because the results of de la Peña and Montgomery–Smith (see [10]) enable us to extend their validity also for usual U-statistics. Nevertheless, the lack of a direct proof of this estimate for U-statistics disturbs me a bit, because this means for me that we do not really understand this result. I have some ideas how to get the desired proof, but it demands some time and energy to work out the details.

Chapter 14

Chapters 14–17 are devoted to the proof of Theorems 8.4 and 8.6. They are based on a similar argument as their one-variate counterparts, Theorems 4.1 and 4.2. The proof of Theorem 8.6 about the supremum of Wiener–Itô integrals is based, similarly to the proof of Theorem 4.2, on the chaining argument. In the proof of Theorem 8.4 the chaining argument yields only a weaker result formulated in Proposition 14.1 which helps to reduce Theorem 8.4 to the proof of Proposition 14.2. In the one-variate case a similar approach was applied. In that case the proof of Theorem 4.1 was reduced to that of Proposition 6.2 by means of Proposition 6.1. The next step in the proof of Theorem 8.4 has no one-variate counterpart. The notion of so-called decoupled U-statistics was introduced, and Proposition 14.2 was reduced to a similar result about decoupled U-statistics formulated in Proposition 14.2′.

The adjective "decoupled" in the expression decoupled U-statistic refers to the fact that it is such a version of a U-statistic where independent copies of a sequence of independent and identically distributed random variables are put into different coordinates of the kernel function. Their study is a popular subject of some mathematical schools. In particular, the main topic of the book [9] is a comparison of the properties of U-statistics and decoupled U-statistics. A result of de la Peña and Montgomery–Smith [10] formulated in Theorem 14.3 helps in reducing some problems about U-statistics to a similar problem about decoupled U-statistics. In this lecture note the proof of Theorem 14.3 is given in Appendix D. It follows the argument of the original proof, but several steps are worked out in detail where the authors gave only a very short explanation. Paper [10] also contains some kind of converse results to Theorem 14.3, but as they are not needed in the present work, I omitted their discussion.

Decoupled U-statistics behave similarly to the original U-statistics. Beside this, some symmetrization arguments become considerably simpler if we are working with decoupled U-statistics instead of the original ones, because decoupled U-statistics have more independence property. This can be exploited in some investigations. For example the proof of Proposition 14.2′ is simpler than a direct proof of Proposition 14.2. On the other hand, Theorem 14.3 enables us to reduce the proof of Proposition 14.2 to that of Proposition 14.2′, and we have exploited this possibility. Let me finally remark that although our proofs could be simplified with the help of decoupled U-statistics, they could have been done also without it. But this would demand a much more complicated notation that would have made the proof much less transparent. Hence I have decided to introduce decoupled U-statistics and to work with them.

Chapters 15–17

The proof of Theorem 8.4 was reduced to that of Proposition 14.2′ in Chap. 14. Chapters 15–17 deal with the proof of this result. The original proof was given in my paper [37]. It is similar to that of its one-variate version, Proposition 6.2, but some additional difficulties have to be overcome. The main difficulty appears when we want to find the multivariate analogue of the symmetrization argument which could be carried out in the one-variate case by means of Lemmas 7.1 and 7.2.

In the multivariate case Lemma 7.1 is not sufficient for our purposes. So we work instead with a generalized version of this result, formulated in Lemma 15.2. The proof of Lemma 15.2 is not hard. It is a simple and natural modification of the proof of Lemma 7.1. The real difficulty arises when we want to apply it in the proof of Proposition 14.2′. When we applied the symmetrization argument Proposition 6.2 in the proof of Lemma 7.1 we worked with two independent sequences of random variables Z_n and $\bar{Z}_n$. In the analogous symmetrization argument Lemma 15.2, applied in the proof of Proposition 14.2′, we had to work with two not necessarily independent sequences of random variables Z_p and $\bar{Z}_p$. This has the consequence that it is much harder to check condition (15.3) needed in the application of Lemma 15.2 than the analogous condition (7.1) in Lemma 7.1. The hardest problems in the proof of Proposition 14.2′ appear at this point.

Proposition 14.2′ was proved by means of an inductive procedure formulated in Proposition 15.3, which is the multivariate analogue of Proposition 7.3. A basic ingredient of both proofs was a symmetrization argument. But while this symmetrization argument could be simply carried out in the one-variate case, its adaptation to the multivariate case was a most serious problem. To overcome this difficulty another inductive statement was formulated in Proposition 15.4. Propositions 15.3 and 15.4 could be proved simultaneously by means of an appropriate inductive procedure. Their proofs were based on a refinement of the arguments in the proof of Proposition 7.3. But some new difficulties arose. In the proof of Proposition 7.3 we could simply apply Lemma 7.2, and it provided the necessary symmetrization argument. On the other hand, the verification of the corresponding symmetrization argument in the proof of Propositions 15.3 and 15.4 was much harder. Actually this was the subject of Chap. 16. After this we could prove Propositions 15.3 and 15.4 in Chap. 17 similarly to Proposition 7.3, although some additional technical difficulties arose also at this point. Here we needed the multivariate version of Hoeffding's inequality, formulated in Theorem 13.3 and some properties of the Hoeffding decomposition of U-statistics proved in Chap. 9.

Appendix A
The Proof of Some Results About Vapnik–Červonenkis Classes

Proof of Theorem 5.1. (Sauer's lemma). This result has several different proofs. Here I write down a relatively simple proof of P. Frankl and J. Pach which appeared in [17]. It is based on some linear algebraic arguments.

The following equivalent reformulation of Sauer's lemma will be proved. Let us take a set $S = S(n)$ consisting of n elements and a class $\mathscr{E}$ of subsets of S consisting of m subsets $E_1, \dots, E_m \subset S$. Assume that $m \ge m_0 + 1$ with $m_0 = m_0(n,k) = \binom{n}{0} + \binom{n}{1} + \cdots + \binom{n}{k-1}$. Then there exists a set $F \subset S$ of cardinality k which is shattered by the class of sets $\mathscr{E}$. Actually, it is enough to show that there exists a set F of cardinality greater than or equal to k which is shattered by the class of sets $\mathscr{E}$, because if a set has this property, then all of its subsets have it. This latter statement will be proved.

To prove this statement let us first list the subsets $X_0, \dots, X_{m_0}$ of the set S of cardinality less than or equal to $k-1$, and correspond to all sets $E_i \in \mathscr{E}$ the vector $e_i = (e_{i,1}, \dots, e_{i,m_0})$, $1 \le i \le m$, with elements

$$e_{i,j} = \begin{cases} 1 & \text{if } X_j \subseteq E_l \\ 0 & \text{if } X_j \not\subseteq E_i \end{cases} \qquad 1 \le i \le m, \text{ and } 1 \le j \le m_0.$$

Since $m > m_0$, the vectors $e_1, \dots, e_m$ are linearly dependent. Because of the definition of the vectors e_i, $1 \le i \le m$, this can be expressed in the following way: There is a non-zero vector $(f(E_1), \dots, f(E_m))$ such that

$$\sum_{E_i \colon E_i \supseteq X_j} f(E_i) = 0 \quad \text{for all } 1 \le j \le m_0. \tag{A.1}$$

Let F, $F \subset S$, be a *minimal* set with the property

$$\sum_{E_i \colon E_i \supseteq F} f(E_i) = \alpha \ne 0. \tag{A.2}$$

P. Major, *On the Estimation of Multiple Random Integrals and U-Statistics*, Lecture Notes in Mathematics 2079, DOI 10.1007/978-3-642-37617-7,

Such a set F really exists, since every maximal element of the family $\{E_i\colon 1 \le i \le m,\ f(E_i) \ne 0\}$ satisfies relation (A.2). The requirement that F should be a minimal set means that if F is replaced by some $H \subset F$, $H \ne F$, at the left-hand side of (A.2), then this expression equals zero. The inequality $|F| \ge k$ holds because of relation (A.1) and the definition of the sets X_j.

Introduce the quantities

$$Z_F(H) = \sum_{E_i\colon E_i \cap F = H} f(E_i)$$

for all $H \subseteq F$.

Then $Z_F(F) = \alpha$, and for any set of the form $H = F \setminus \{x\}$, $x \in F$,

$$Z_F(H) = \sum_{E_i\colon E_i \cap F = H} f(E_i) = \sum_{E_i\colon E_i \supseteq H} f(E_i) - \sum_{E_i\colon E_i \supseteq F} f(E_i) = 0 - \alpha = -\alpha$$

because of the minimality property of the set F.

Moreover, the identity

$$Z_F(H) = (-1)^p \alpha \quad \text{for all } H \subseteq F \text{ such that } |H| = |F| - p,\ 0 \le p \le |F|. \tag{A.3}$$

holds. To show relation (A.3) observe that

$$Z_F(H) = \sum_{E_i\colon E_i \cap F = H} f(E_i) = \sum_{j=0}^{p} (-1)^j \sum_{G\colon H \subset G \subset F,\ |G| = |H| + j} \ \sum_{E_i\colon E_i \supseteq G} f(E_i) \tag{A.4}$$

for all sets $H \subset F$ with cardinality $|H| = |F| - p$. Identity (A.4) holds, since the term $f(E_i)$ is counted at the right-hand side of (A.4) $\sum\limits_{j=0}^{l} (-1)^j \binom{l}{j} = (1-1)^l = 0$ times if $E_i \cap F = G$ with some $H \subset G \subseteq F$ with $|G| = |H| + l$ elements, $1 \le l \le p$, while in the case $E_i \cap F = H$ it is counted once. Relation (A.4) together with (A.2) and the minimality property of the set F imply relation (A.3).

It follows from relation (A.3) and the definition of the function $Z_F(H)$ that for all sets $H \subseteq F$ there exists some set E_i such that $H = E_i \cap F$, i.e. F is shattered by $\mathscr{E}$. Since $|F| \ge k$, this implies Theorem 5.1. □

Proof of Theorem 5.3. Let us fix an arbitrary set $F = \{x_1, \dots, x_{k+1}\}$ of the set X, and consider the set of vectors $\mathscr{G}_k(F) = \{(g(x_1), \dots, g(x_{k+1}))\colon g \in \mathscr{G}_k\}$ of the $k+1$-dimensional space R^{k+1}. By the conditions of Theorem 5.3 $\mathscr{G}_k(F)$ is an at most k-dimensional subspace of R^{k+1}. Hence there exists a non-zero vector $a = (a_1, \dots, a_{k+1})$ such that $\sum\limits_{j=1}^{k+1} a_j g(x_j) = 0$ for all $g \in \mathscr{G}_k$. We may assume that the set $A = A(a) = \{j\colon a_j < 0,\ 1 \le j \le k+1\}$ is non-empty, by multiplying the vector a by -1 if it is necessary.

Thus the identity

$$\sum_{j\in A} a_j g(x_j) = \sum_{j\in\{1,\dots,k+1\}\setminus A} (-a_j) g(x_j), \qquad \text{for all } g \in \mathcal{G}_k \tag{A.5}$$

holds. Put $B = \{x_j\colon j \in A\}$. Then $B \subset F$, and $F \setminus B \ne \{x\colon g(x) \ge 0\} \cap F$ for all $g \in \mathcal{G}_k$. Indeed, if there were some $g \in \mathcal{G}_k$ such that $F \setminus B = \{x\colon g(x) \ge 0\} \cap F$, then the left-hand side of (A.5) would be strictly positive (as $a_j < 0$, $g(x_j) < 0$ if $j \in A$, and $A \ne \emptyset$) its right-hand side would be non-positive for this $g \in \mathcal{G}_k$, and this is a contradiction.

The above proved property means that $\mathcal{D}$ shatters no set $F \subset X$ of cardinality $k+1$. Hence Theorem 5.1 implies that $\mathcal{D}$ is a Vapnik–Červonenkis class. □

Appendix B
The Proof of the Diagram Formula for Wiener–Itô Integrals

We start the proof of Theorem 10.2A (the diagram formula for the product of two Wiener–Itô integrals) with the proof of inequality (10.13). To show that this relation holds let us observe that the Cauchy inequality yields the following bound on the function $F_\gamma(f,g)$ defined in (10.11) (with the notation introduced there):

$$\begin{aligned} &F_\gamma^2(f,g)(x_{(1,j)},x_{(2,j')},\ (1,j)\in V_1(\gamma),\ (2,j')\in V_2(\gamma)) \\ &\le \int f^2(x_{\alpha_\gamma(1,1)},\dots,x_{\alpha_\gamma(1,k)}) \prod_{(2,j)\in\{(2,1),\dots,(2,l)\}\setminus V_2(\gamma)} \mu(\,dx_{(2,j)}) \\ &\qquad \int g^2(x_{(2,1)},\dots,x_{(2,l)}) \prod_{(2,j)\in\{(2,1),\dots,(2,l)\}\setminus V_2(\gamma)} \mu(\,dx_{(2,j)}). \qquad \text{(B.1)} \end{aligned}$$

The expression at the right-hand side of inequality (B.1) is the product of two functions with different arguments. The first function has arguments $x_{(1,j)}$ with $(1,j)\in V_1(\gamma)$ and the second one $x_{(2,j')}$ with $(2,j')\in V_2(\gamma)$. By integrating both sides of inequality (B.1) with respect to these arguments we get inequality (10.13).

Relation (10.14) will be proved first for the product of the Wiener–Itô integrals of two elementary functions. Let us consider two (elementary) functions $f(x_1,\dots,x_k)$ and $g(x_1,\dots,x_l)$ given in the following form: Let some disjoint sets $A_1,\dots,A_M$, $\mu(A_s)<\infty$, $1\le s\le M$, be given together with some real numbers $c(s_1,\dots,s_k)$ indexed with such k-tuples $(s_1,\dots,s_k)$, $1\le s_j\le M$, $1\le j\le k$, for which the numbers $s_1,\dots,s_k$ in a k-tuple are all different. Put $f(x_1,\dots,x_k)=c(s_1,\dots,s_k)$ if $(x_1,\dots,x_k)\in A_{s_1}\times\cdots\times A_{s_k}$ with some vector $(s_1,\dots,s_k)$ with different coordinates, and let $f(x_1,\dots,x_k)=0$ if $(x_1,\dots,x_k)$ is outside of these rectangles. Take similarly some disjoint sets $B_1,\dots,B_{M'}$, $\mu(B_t)<\infty$, $1\le t\le M'$, and some real numbers $d(t_1,\dots,t_l)$, indexed with such l-tuples $(t_1,\dots,t_l)$, $1\le t_{j'}\le M'$, $1\le j'\le l$, for which the numbers $t_1,\dots,t_l$ in an l-tuple are different. Put $g(x_1,\dots,x_l)=d(t_1,\dots,t_l)$ if $(x_1,\dots,x_l)\in B_{t_1}\times\cdots\times B_{t_l}$ with edges indexed with some of the above introduced l-tuples, and let $g(x_1,\dots,x_l)=0$ otherwise.

P. Major, *On the Estimation of Multiple Random Integrals and U-Statistics*, Lecture Notes in Mathematics 2079, DOI 10.1007/978-3-642-37617-7,

Let us take some small number $\varepsilon > 0$ and rewrite the above introduced functions $f(x_1,\dots,x_k)$ and $g(x_1,\dots,x_l)$ with the help of this number $\varepsilon > 0$ in the following way. Divide the sets $A_1,\dots,A_M$ to smaller sets $A_1^\varepsilon,\dots,A_{M(\varepsilon)}^\varepsilon$, $\bigcup\limits_{s=1}^{M(\varepsilon)} A_s^\varepsilon = \bigcup\limits_{s=1}^{M} A_s$, in such a way that all sets $A_1^\varepsilon,\dots,A_{M(\varepsilon)}^\varepsilon$ are disjoint, and $\mu(A_s^\varepsilon) \le \varepsilon$, $1 \le s \le M(\varepsilon)$. Similarly, take sets $B_1^\varepsilon,\dots,B_{M'(\varepsilon)}^\varepsilon$, $\bigcup\limits_{t=1}^{M'(\varepsilon)} B_t^\varepsilon = \bigcup\limits_{t=1}^{M'} B_t$, in such a way that all sets $B_1^\varepsilon,\dots,B_{M'(\varepsilon)}^\varepsilon$ are disjoint, and $\mu(B_t^\varepsilon) \le \varepsilon$, $1 \le t \le M'(\varepsilon)$. Beside this, let us also demand that two sets A_s^ε and B_t^ε, $1 \le s \le M(\varepsilon)$, $1 \le t \le M'(\varepsilon)$, are either disjoint or they agree. Such a partition exists because of the non-atomic property of measure μ. The above defined functions $f(x_1,\dots,x_k)$ and $g(x_1,\dots,x_l)$ can be rewritten by means of these new sets A_s^ε and B_t^ε. Namely, let $f(x_1,\dots,x_k) = c^\varepsilon(s_1,\dots,s_k)$ on the rectangles $A_{s_1}^\varepsilon \times\cdots\times A_{s_k}^\varepsilon$ with $1 \le s_j \le M(\varepsilon)$, $1 \le j \le k$, with different indices $s_1,\dots,s_k$, where $c^\varepsilon(s_1,\dots,s_k) = c(p_1,\dots,p_k)$ with those indices $(p_1,\dots,p_k)$ for which $A_{s_1}^\varepsilon \times\cdots\times A_{s_k}^\varepsilon \subset A_{p_1}\times\cdots\times A_{p_k}$. The function f disappears outside of these rectangles. The function $g(x_1,\dots,x_l)$ can be written similarly in the form $g(x_1,\dots,x_l) = d^\varepsilon(t_1,\dots,t_l)$ on the rectangles $B_{t_1}^\varepsilon\times\cdots\times B_{t_l}^\varepsilon$ with $1 \le t_{j'} \le M'(\varepsilon)$, $1 \le j' \le l$, and different indices, $t_1,\dots,t_l$. Beside this, the function g disappears outside of these rectangles.

The above representation of the functions f and g through a parameter ε is useful, since it enables us to give a good asymptotic formula for the product $k!Z_{\mu,k}(f)l!Z_{\mu,l}(g)$ which yields the diagram formula for the product of Wiener–Itô integrals of elementary functions with the help of a limiting procedure $\varepsilon \to 0$.

Fix a small number $\varepsilon > 0$, take the representation of the functions f and g with its help, and write

$$k!Z_{\mu,k}(f)l!Z_{\mu,l}(g) = \sum_{\gamma\in\Gamma(k,l)} Z_\gamma(f,g,\varepsilon) \tag{B.2}$$

with

$$\begin{aligned} Z_\gamma(f,g,\varepsilon) = {\sum}^\gamma\, & c^\varepsilon(s_1,\dots,s_k)d^\varepsilon(t_1,\dots,t_l) \\ & \mu_W(A_{s_1}^\varepsilon)\dots\mu_W(A_{s_k}^\varepsilon)\mu_W(B_{t_1}^\varepsilon)\dots\mu_W(B_{t_l}^\varepsilon), \end{aligned} \tag{B.3}$$

where $\Gamma(k,l)$ denotes the class of diagrams introduced before the formulation of Theorem 10.2A, and $\sum^\gamma$ denotes summation for $k+l$-tuples $(s_1,\dots,s_k,t_1,\dots,t_l)$ such that $1 \le s_j \le M(\varepsilon)$, $1 \le j \le k$, $1 \le t_{j'} \le M'(\varepsilon)$, $1 \le j' \le l$, and $A_{s_j}^\varepsilon = B_{t_{j'}}^\varepsilon$ if $((1,j),(2,j')) \in E(\gamma)$, i.e. if it is an edge of γ, and otherwise all sets $A_{s_j}^\varepsilon$ and $B_{t_{j'}}^\varepsilon$ are disjoint. (This sum also depends on ε.) In the case of an empty sum $Z_\gamma(f,g,\varepsilon)$ equals zero.

We write the expression $Z_\gamma(f,g,\varepsilon)$ for all $\gamma\in\Gamma(k,l)$ in the form

$$Z_\gamma(f,g,\varepsilon) = Z_\gamma^{(1)}(f,g,\varepsilon) + Z_\gamma^{(2)}(f,g,\varepsilon), \quad \gamma\in\Gamma(k,l), \tag{B.4}$$

with

$$Z_\gamma^{(1)}(f,g,\varepsilon) = \sum\nolimits^\gamma c^\varepsilon(s_1,\dots,s_k)d^\varepsilon(t_1,\dots,t_l) \prod_{j:(1,j)\in V_1(\gamma)} \mu_W(A^\varepsilon_{s_j}) \prod_{j':(2,j')\in V_2(\gamma)} \mu_W(B^\varepsilon_{t_{j'}}) \prod_{j':(2,j')\in\{(2,1),\dots,(2,l)\}\setminus V_2(\gamma)} \mu(B^\varepsilon_{t_{j'}}) \tag{B.5}$$

and

$$Z_\gamma^{(2)}(f,g,\varepsilon) = \sum\nolimits^\gamma c^\varepsilon(s_1,\dots,s_k)d^\varepsilon(t_1,\dots,t_l) \prod_{j:(1,j)\in V_1(\gamma)} \mu_W(A^\varepsilon_{s_j}) \prod_{j':(2,j')\in V_2(\gamma)} \mu_W(B^\varepsilon_{t_{j'}}) \left[\prod_{j:(1,j)\in\{(1,1),\dots,(1,k)\}\setminus V_1(\gamma)} \mu_W(A^\varepsilon_{s_j}) \prod_{j':(2,j')\in\{(2,1),\dots,(2,l)\}\setminus\in V_2(\gamma)} \mu_W(B^\varepsilon_{t_{j'}}) - \prod_{j':(2,j')\in\{(2,1),\dots,(2,l)\}\setminus V_2(\gamma)} \mu(B^\varepsilon_{t_{j'}})\right], \tag{B.6}$$

where $V_1(\gamma)$ and $V_2(\gamma)$ (introduced before formula (10.9) during the preparation to the formulation of Theorem 10.2A) are the sets of vertices in the first and second row of the diagram γ from which no edge starts.

I claim that there is some constant $C>0$ not depending on ε such that

$$E\left(|\gamma|!Z_{\mu,|\gamma|}(F_\gamma(f,g)) - Z_\gamma^{(1)}(f,g,\varepsilon)\right)^2 \le C\varepsilon \quad \text{for all } \gamma\in\Gamma(k,l) \tag{B.7}$$

with the Wiener–Itô integral with the kernel function $F_\gamma(f,g)$ defined in (10.9), (10.10) and (10.11), and

$$E\left(Z_\gamma^{(2)}(f,g,\varepsilon)\right)^2 \le C\varepsilon \quad \text{for all } \gamma\in\Gamma(k,l). \tag{B.8}$$

Relations (B.2), (B.4), (B.7) and (B.8) imply relation (10.14) if f and g are elementary functions. Indeed, (B.4), (B.7) and (B.8) imply that

$$\lim_{\varepsilon\to 0}\left\| |\gamma|!Z_{\mu,|\gamma|}(F_\gamma(f,g)) - Z_\gamma(f,g,\varepsilon)\right\|_2 \to 0 \quad \text{for all } \gamma\in\Gamma(k,l),$$

and this relation together with (B.2) yield relation (10.14) with the help of a limiting procedure $\varepsilon\to 0$.

To prove relation (B.7) let us introduce the function

$$\begin{aligned}
&F_\gamma^\varepsilon(f,g)(x_{(1,j)},x_{(2,j')},\ (1,j)\in V_1(\gamma),\ (2,j')\in V_2(\gamma))\\
&\quad = F_\gamma(f,g)(x_{(1,j)},x_{(2,j')},\ (1,j)\in V_1(\gamma),\ (2,j')\in V_2(\gamma))\\
&\qquad \text{if } x_{(1,j)}\in A^\varepsilon_{s_j},\ \text{ for all } (1,j)\in V_1(\gamma),\\
&\qquad\quad x_{(2,j')}\in B^\varepsilon_{t_{j'}},\ \text{ for all } (2,j')\in V_2(\gamma)),\quad \text{and}\\
&\qquad\qquad \text{all sets } A^\varepsilon_{s_j},\ (1,j)\in V_1(\gamma),\ \text{and } B^\varepsilon_{t_{j'}},\ (2,j')\in V_2(\gamma) \text{ are different.}
\end{aligned}$$

with the function $F_\gamma(f,g)$ defined in (10.10) and (10.11), and put

$$F_\gamma^\varepsilon(f,g)(x_{(1,j)},x_{(2,j')},\ (1,j)\in V_1(\gamma),\ (2,j')\in V_2(\gamma))=0 \quad \text{otherwise.}$$

The function $F_\gamma^\varepsilon(f,g)$ is elementary, and a comparison of its definition with relation (B.5) and the definition of the function $F_\gamma(f,g)$ yields that

$$Z_\gamma^{(1)}(f,g,\varepsilon)=|\gamma|!Z_{\mu,|\gamma|}(F_\gamma^\varepsilon(f,g)). \tag{B.9}$$

The function $F_\gamma^\varepsilon(f,g)$ slightly differs from $F_\gamma(f,g)$, since the function $F_\gamma(f,g)$ may not disappear in such points $(x_{(1,j)},x_{(2,j')},\ (1,j)\in V_1(\gamma),\ (2,j')\in V_2(\gamma))$ for which there is some pair (j,j') with the property $x_{(1,j)}\in A^\varepsilon_{s_j}$ and $x_{(2,j')}\in B^\varepsilon_{t_{j'}}$ with some sets $A^\varepsilon_{s_j}$ and $B^\varepsilon_{t_{j'}}$ such that $A^\varepsilon_{s_j}=B^\varepsilon_{t_{j'}}$, while $F_\gamma^\varepsilon(f,g)$ must be zero in such points. On the other hand, in the case $|\gamma|=\max(k,l)-\min(k,l)$, i.e. if one of the sets $V_1(\gamma)$ or $V_2(\gamma)$ is empty, $F_\gamma(f,g)=F_\gamma^\varepsilon(f,g)$, $Z_\gamma^{(1)}(f,g,\varepsilon)=|\gamma|!Z_{\mu,|\gamma|}(F_\gamma(f,g))$, and relation (B.7) clearly holds for such diagrams γ.

In the case $|\gamma|=\max(k,l)-\min(k,l)>0$ we prove a good estimate on the measure of the set where $F_\gamma\neq F_\gamma^\varepsilon$ with respect to an appropriate power of the measure μ. Relation (B.7) will be proved with the help of this estimate and formula (B.9).

Let us define the sets $A=\bigcup\limits_{s=1}^{M(\varepsilon)} A_s^\varepsilon$ and $B=\bigcup\limits_{t=1}^{M'(\varepsilon)} B_t^\varepsilon$. These sets A and B do not depend on the parameter ε. Beside this $\mu(A)<\infty$, and $\mu(B)<\infty$. Define for all pairs (j_0,j_0') such that $(1,j_0)\in V_1(\gamma)$, $(2,j_0')\in V_2(\gamma)$ the set

$$\begin{aligned}
D(j_0,j_0')=\{&(x_{(1,j)},x_{(2,j')},\ (1,j)\in V_1(\gamma),\ (2,j')\in V_2(\gamma)):\\
&x_{(1,j_0)}\in A_s^\varepsilon, x_{(2,j_0')}\in B_t^\varepsilon \text{ with some } 1\le s\le M(\varepsilon) \text{ and } 1\le t\le M'(\varepsilon)\\
&\text{such that } A_s^\varepsilon=B_t^\varepsilon,\quad \text{and}\quad x_{(1,j)}\in A \text{ for all } (1,j)\in V_1(\gamma),\\
&\text{and } x_{(2,j')}\in B \text{ for all } (2,j')\in V_2(\gamma)\}.
\end{aligned}$$

Introduce the notation $x^\gamma=(x_{(1,j)},x_{(2,j')}),\ (1,j)\in V_1(\gamma),\ (2,j')\in V_2(\gamma))$, and consider only such vectors x^γ whose coordinates satisfy the conditions $x_{(1,j)}\in A$

for all $(1, j) \in V_1(\gamma)$ and $x_{(2,j')} \in B$ for all $(2, j') \in V_2(\gamma)$. Put

$$D_\gamma = \{x^\gamma\colon F_\gamma^\varepsilon(f, g)(x^\gamma) \neq F_\gamma(f, g)(x^\gamma)\}.$$

The relation $D_\gamma \subset \bigcup_{j=1}^{k} \bigcup_{j'=1}^{l} D(j_0, j_0')$ holds, since if $F_\gamma^\varepsilon(f, g)(x^\gamma) \neq F_\gamma(f, g)(x^\gamma)$ for some vector x^γ, then it has some coordinates $(1, j_0) \in V_1(\gamma)$ and $(2, j_0') \in V_2(\gamma)$ such that $x_{(1,j_0)} \in A_s^\varepsilon$ and $x_{(1,j_0')} \in B_t^\varepsilon$ with some sets $A_s^\varepsilon = B_t^\varepsilon$, and the relation in the last line of the definition of $D(j_0, j_0')$ must also hold for such a vector x^γ, since otherwise $F_\gamma(f, g)(x_\gamma) = 0 = F_\gamma^\varepsilon(f, g)(x_\gamma)$.

I claim that there is some constant C_1 such that

$$\mu^{|V_1(\gamma)|+|V_2(\gamma)|}(D(j_0, j_0')) \le C_1\varepsilon \quad \text{for all sets } D(j_0, j_0'),$$

where $\mu^{|V_1(\gamma)|+|V_2(\gamma)|}$ denotes the direct product of the measure μ on some copies of the original space $(X, \mathcal{X})$ indexed by $(1, j) \in V_1(\gamma)$ and $(2, j') \in V_2(\gamma)$. To see this relation one has to observe that $\sum\limits_{A_s^\varepsilon = B_t^\varepsilon} \mu(A_s^\varepsilon)\mu(B_t^\varepsilon) \le \sum \varepsilon\mu(A_s^\varepsilon) = \varepsilon\mu(A)$. Thus the set $D(j_0, j_0')$ can be covered by the direct product of a set whose μ measure is not greater than $\varepsilon\mu(A)$ and of a rectangle whose edges are either the set A or the set B.

The above relations imply that

$$\mu^{|V_1(\gamma)|+|V_2(\gamma)|}(D_\gamma) \le C_2\varepsilon \tag{B.10}$$

with some constant $C_2 > 0$.

Relation (B.9), estimate (B.10), the property (c) formulated in Theorem 10.1 for Wiener–Itô integrals and the observation that the function $F_\gamma(f, g)$ is bounded in supremum norm if f and g are elementary functions imply the inequality

$$\begin{aligned}
&E\left(|\gamma|! Z_{\mu,|\gamma|}(F_\gamma(f, g)) - Z_\gamma^{(1)}(f, g, \varepsilon)\right)^2 \\
&\quad = |\gamma!|^2 E\left(Z_{\mu,|\gamma|}(F_\gamma(f, g) - F_\gamma^\varepsilon(f, g))\right)^2 \le |\gamma|! \|F_\gamma(f, g) - F_\gamma^\varepsilon(f, g)\|_2^2 \\
&\quad \le K\mu^{|V_1(\gamma)|+|V_2(\gamma)|}(D_\gamma) \le C\varepsilon.
\end{aligned}$$

Hence relation (B.7) holds.

To prove relation (B.8) we rewrite $E\left(Z_\gamma^{(2)}(f, g, \varepsilon)\right)^2$ in the following form:

$$\begin{aligned}
E\left(Z_\gamma^{(2)}(f, g, \varepsilon)\right)^2 = {\sum}^\gamma {\sum}^\gamma\, & c^\varepsilon(s_1, \dots, s_k) d^\varepsilon(t_1, \dots, t_l) c^\varepsilon(\bar{s}_1, \dots, \bar{s}_k) \\
& d^\varepsilon(\bar{t}_1, \dots, \bar{t}_l) EU(s_1, \dots, s_k, t_1, \dots, t_l, \bar{s}_1, \dots, \bar{s}_k, \bar{t}_1, \dots, \bar{t}_l)
\end{aligned} \tag{B.11}$$

with

$$
\begin{aligned}
&U(s_1,\dots,s_k,t_1,\dots,t_l,\bar s_1,\dots,\bar s_k,\bar t_1,\dots,\bar t_l)\\
&=\prod_{j:(1,j)\in V_1(\gamma)}\mu_W(A^\varepsilon_{s_j})\prod_{j':(2,j')\in V_2(\gamma)}\mu_W(B^\varepsilon_{t_{j'}})\\
&\qquad\prod_{\bar j:(1,\bar j)\in V_1(\gamma)}\mu_W(A^\varepsilon_{\bar s_{\bar j}})\prod_{\bar j':(2,\bar j')\in V_2(\gamma)}\mu_W(B^\varepsilon_{\bar t_{\bar j'}})\\
&\qquad\Bigg[\prod_{j:(1,j)\in\{(1,1),\dots,(1,k)\}\setminus V_1(\gamma)}\mu_W(A^\varepsilon_{s_j})\prod_{j':(2,j')\in\{(2,1),\dots,(2,l)\}\setminus\in V_2(\gamma)}\mu_W(B^\varepsilon_{t_{j'}})\\
&\qquad-\prod_{j':(2,j')\in\{(2,1),\dots,(2,l)\}\setminus V_2(\gamma)}\mu(B^\varepsilon_{t_{j'}})\Bigg]\\
&\qquad\Bigg[\prod_{\bar j:(1,\bar j)\in\{(1,1),\dots,(1,k)\}\setminus V_1(\gamma)}\mu_W(A^\varepsilon_{\bar s_{\bar j}})\prod_{\bar j':(2,\bar j')\in\{(2,1),\dots,(2,l)\}\setminus\in V_2(\gamma)}\mu_W(B^\varepsilon_{\bar t_{\bar j'}})\\
&\qquad-\prod_{\bar j':(2,\bar j')\in\{(2,1),\dots,(2,l)\}\setminus V_2(\gamma)}\mu(B^\varepsilon_{\bar t_{\bar j'}})\Bigg].
\end{aligned}
\tag{B.12}
$$

The double sum $\sum^\gamma\sum^\gamma$ in (B.11) has to be understood in the following way. The first summation is taken for vectors $(s_1,\dots,s_k,t_1,\dots,t_l)$, and $\sum^\gamma$ is defined in the same way as in formula (B.3). The second summation is taken for vectors $(\bar s_1,\dots,\bar s_k,\bar t_1,\dots,\bar t_l)$, and again the summation $\sum^\gamma$ is taken as in (B.3), only here $\bar s_j$ plays the role of s_j and $\bar t_{j'}$ plays the role of $t_{j'}$.

Relation (B.8) will be proved by means of some estimates about the expectation of the above defined random variable $U(\cdot)$ which will be presented in the following Lemma B. To formulate this result I introduce the following Properties A and B.

Property A. *A sequence $s_1,\dots,s_k,t_1,\dots,t_l,\bar s_1,\dots,\bar s_k,\bar t_1,\dots,\bar t_l$, with elements $1\le s_j,\bar s_{\bar j}\le M(\varepsilon)$, for $1\le j,\bar j\le k$, and $1\le t_j,\bar t_{\bar j'}\le M'(\varepsilon)$ for $1\le j',\bar j'\le l$, satisfies Property A (depending on a fixed diagram γ and number $\varepsilon>0$) if the sequence of sets $A^\varepsilon_{s_j}$, $(1,j)\in V_1(\gamma)$, $B^\varepsilon_{t_{j'}}$, $(2,j')\in V_2(\gamma)$, and the sequence of sets $A^\varepsilon_{\bar s_{\bar j}}$, $(1,\bar j)\in V_1(\gamma)$, $B^\varepsilon_{\bar t_{\bar j'}}$, $(2,\bar j')\in V_2(\gamma)$, agree. (Here we say that two sequences agree if they contain the same elements in a possibly different order.)*

Property B. *A sequence $s_1,\dots,s_k,t_1,\dots,t_l,\bar s_1,\dots,\bar s_k,\bar t_1,\dots,\bar t_l$, with elements $1\le s_j,\bar s_{\bar j}\le M(\varepsilon)$, for $1\le j,\bar j\le k$, and $1\le t_j,\bar t_{\bar j'}\le M'(\varepsilon)$ for $1\le j',\bar j'\le l$, satisfies Property B (depending on a fixed diagram γ and number $\varepsilon>0$) if the sequences of sets*

$$A^\varepsilon_{s_j},\ (1,j)\in\{(1,1),\dots,(1,k)\}\setminus V_1(\gamma),\quad B^\varepsilon_{t_{j'}},(2,j')\in\{(2,1),\dots,(2,l)\}\setminus V_2(\gamma),$$

and

$$A^{\varepsilon}_{\bar{s}_{\bar{j}}},\ (1,\bar{j}) \in \{(1,1),\dots,(1,k)\}\setminus V_1(\gamma), \quad B^{\varepsilon}_{\bar{t}_{\bar{j}'}},\ (2,\bar{j}') \in \{(2,1),\dots,(2,l)\}\setminus V_2(\gamma),$$

have at least one common element.

(In the above definitions two sets A^{ε}_s and B^{ε}_t are identified if $A^{\varepsilon}_s = B^{\varepsilon}_t$.)

Now I formulate the following

Lemma B. *Let us consider the function $U(\cdot)$ introduced in formula (B.12). Assume that its arguments $s_1,\dots,s_k,t_1,\dots,t_l,\bar{s}_1,\dots,\bar{s}_k,\bar{t}_1,\dots,\bar{t}_l$ are chosen in such a way that the function $U(\cdot)$ with these arguments appears in the double sum $\sum^{\gamma}\sum^{\gamma}$ in formula (B.11), i.e. $A^{\varepsilon}_{s_j} = B^{\varepsilon}_{t_{j'}}$ if $((1,j),(2,j')) \in E(\gamma)$, otherwise all sets $A^{\varepsilon}_{s_j}$ and $B^{\varepsilon}_{t_{j'}}$ are disjoint, and an analogous statement holds if the coordinates $s_1,\dots,s_k,t_1,\dots,t_l$ are replaced by $\bar{s}_1,\dots,\bar{s}_k$ and $\bar{t}_1,\dots,\bar{t}_l$.*

If the sequence of the arguments in $U(\cdot)$ does not satisfies either Property A or Property B, then

$$EU(s_1,\dots,s_k,t_1,\dots,t_l,\bar{s}_1,\dots,\bar{s}_k,\bar{t}_1,\dots,\bar{t}_l) = 0. \tag{B.13}$$

If the sequence of the arguments in $U(\cdot)$ satisfies both Property A and Property B, then

$$|EU(s_1,\dots,s_k,t_1,\dots,t_l,\bar{s}_1,\dots,\bar{s}_k,\bar{t}_1,\dots,\bar{t}_l)| \le C\varepsilon \prod{}' \mu(A^{\varepsilon}_{\bar{s}_{\bar{j}}})\mu(B^{\varepsilon}_{\bar{t}_{\bar{j}'}}) \tag{B.14}$$

with some appropriate constant $C = C(k,l) > 0$ depending only on the number of variables k and l of the functions f and g. The prime in the product $\prod'$ at the right-hand side of (B.14) means that in this product the measure μ of those sets $A^{\varepsilon}_{\bar{s}_{\bar{j}}}$ and $B^{\varepsilon}_{\bar{t}_{\bar{j}'}}$ are considered, whose indices are listed among the arguments $\bar{s}_{\bar{j}}$ or $\bar{t}_{\bar{j}'}$ of $U(\cdot)$, and the measure μ of each such set appears exactly once. (This means that if $A^{\varepsilon}_{\bar{s}_{\bar{j}}} = B^{\varepsilon}_{\bar{t}_{\bar{j}'}}$ then one of the terms between $\mu(A^{\varepsilon}_{\bar{s}_{\bar{j}}})$ and $\mu(B^{\varepsilon}_{\bar{t}_{\bar{j}'}})$ is omitted from the product. For the sake of definitiveness let us preserve the set $\mu(A^{\varepsilon}_{\bar{s}_{\bar{j}}})$ in such a case.)

Remark. The content of Lemma B is that most terms in the double sum in formula (B.11) equal zero, and even the non-zero terms are small.

The proof of Lemma B. Let us prove first relation (B.13) in the case when Property A does not hold. It will be exploited that for disjoint sets the random variables $\mu_W(A_s)$ and $\mu_W(B_t)$ are independent, and this provides a good factorization of the expectation of certain products.

Let us carry out the multiplications in the expression $U(\cdot)$ defined (B.12). We get a sum consisting of four terms. We show that each of them has zero expectation. Indeed, if a sequence $s_1,\dots,s_k,t_1,\dots,t_l,\bar{s}_1,\dots,\bar{s}_k,\bar{t}_1,\dots,\bar{t}_l$ does not satisfy Property A, but it satisfies the remaining conditions of Lemma B, then each term in the sum expressing $U(\cdots)$ with these arguments is a product which contains a factor $\mu_W(A^{\varepsilon}_{s_{j_0}})$, $(1,j_0) \in V_1(\gamma)$ with the following property. It is independent of all those terms in this product which are in the following list: $\mu_W(A^{\varepsilon}_{s_j})$ with some

$j \neq j_0$, $1 \le j \le k$, or $\mu_W(B^\varepsilon_{t_{j'}})$, $1 \le j \le l$, or $\mu_W(A^\varepsilon_{s_{\bar j}})$ with $(1,\bar j) \in V_1(\gamma)$, or $\mu_W(B^\varepsilon_{t_{\bar j'}})$ with $(2,\bar j') \in V_2(\gamma)$. We will show with the help of this property that the expectation of the terms we consider can be written in the form of a product either with a factor of the form $E\mu_W(A^\varepsilon_{s_{j0}}) = 0$ or with a factor of the form $E\mu_W(A^\varepsilon_{s_{j0}})^3 = 0$. Hence this expectation equals zero.

Indeed, although the above properties do not exclude the existence of a set $A^\varepsilon_{t_{\bar j'}}$, $(1,\bar j') \in \{(1,1),\dots,(1,k) \setminus V_1(\gamma)$ or $B^\varepsilon_{t_{\bar j'}}$, $(2,\bar j') \in \{(2,1),\dots,(2,l)\} \setminus V_2(\gamma)$ such that $\mu_W(A^\varepsilon_{t_{\bar j'}})$ or $\mu_W(B^\varepsilon_{t_{\bar j'}})$, is not independent of $\mu_W(A^\varepsilon_{s_{j0}})$, but this can only happen if $A^\varepsilon_{t_{\bar j}} = B^\varepsilon_{t_{\bar j'}} = A^\varepsilon_{s_{j0}}$. This implies that in such a case when our term does not contain a factor of the form $E\mu_W(A^\varepsilon_{s_{j0}})$, then it contains a factor of the form $E\mu_W(A^\varepsilon_{s_{j0}})^3 = 0$. Hence $EU(\cdot) = 0$ if the arguments of $U(\cdot)$ do not satisfy Property A.

To finish the proof of relation (B.13) it is enough consider the case when the arguments of $U(\cdot)$ satisfy Property A, but they do not satisfy Property B. The validity of Property A implies that the sets $\{A^\varepsilon_{s_j},\ j \in V_1(\gamma)\} \cup \{B^\varepsilon_{t_{j'}},\ j' \in V_2(\gamma)\}$ and $\{A^\varepsilon_{\bar s_j},\ j \in V_1(\gamma)\} \cup \{B^\varepsilon_{\bar t_{j'}},\ j' \in V_2(\gamma)\}$ agree. The conditions of Lemma B also imply that the elements of these sets are disjoint of the sets $A^\varepsilon_{s_j}$, $B^\varepsilon_{t_{j'}}$, $A^\varepsilon_{s_{\bar j}}$ and $B^\varepsilon_{t_{\bar j'}}$ with indices $(1,j),(1,\bar j) \in \{(1,1),\dots,(1,k)\} \setminus V_1(\gamma)$ and $(2,j'),(2,\bar j') \in \{(2,1),\dots,(2,l)\} \setminus V_2(\gamma)$. If Property B does not hold, then we can divide the latter class of sets into two disjoint subclasses in an appropriate way. The first subclass consists of the sets $A^\varepsilon_{s_j}$ and $B^\varepsilon_{t_{j'}}$, and the second one of the sets $A^\varepsilon_{s_{\bar j}}$ and $B^\varepsilon_{t_{\bar j'}}$ with indices such that $(1,j),(1,\bar j) \in \{(1,1),\dots,(1,k)\} \setminus V_1(\gamma)$ and $(2,j'),(2,\bar j') \in \{(2,1),\dots,(2,l)\} \setminus V_2(\gamma)$. These facts imply that $EU(\cdot)$ has a factorization, which contains the term

$$E\Bigg[\prod_{j:(1,j)\in\{(1,1),\dots,(1,k)\}\setminus V_1(\gamma)} \mu_W(A^\varepsilon_{s_j}) \prod_{j':(2,j')\in\{(2,1),\dots,(2,l)\}\setminus \in V_2(\gamma)} \mu_W(B^\varepsilon_{t_{j'}}) - \prod_{j':(2,j')\in\{(2,1),\dots,(2,l)\}\setminus V_2(\gamma)} \mu(B^\varepsilon_{s_{j'}})\Bigg] = 0,$$

hence relation (B.13) holds also in this case. The last expression has zero expectation, since if we take such pairs $A^\varepsilon_{s_j}$, $B^\varepsilon_{t'_j}$ for the sets appearing in it for which that $((1,j),(2,j')) \in E(\gamma)$, i.e. these vertices are connected with an edge of γ, then $A^\varepsilon_{s_j} = B^\varepsilon_{t'_j}$ in a pair, and elements in different pairs are disjoint. This observation allows a factorization in the product whose expectation is taken, and then the identity $E\mu_W(A^\varepsilon_{s_j})\mu_W(B^\varepsilon_{t_{j'}}) = \mu(A^\varepsilon_{s_j})$ implies the desired identity.

To prove relation (B.14) if the arguments of the function $U(\cdot)$ satisfy both Properties A and B consider the expression (B.12) which defines $U(\cdot)$, carry out the term by term multiplication between the two differences at the end of this formula, take expectation for each term of the sum obtained in such a way and factorize them.

Since $E\mu_W(A)^2 = \mu(A)$, $E\mu_W(A)^4 = 3\mu(A)^2$ for all sets $A \in \mathscr{X}$, $\mu(A) < \infty$, some calculation shows that each term can be expressed as constant times a product whose elements are those probabilities $\mu(A^\varepsilon_{\bar{s}_{\bar{j}}})$ and $\mu(B^\varepsilon_{\bar{t}_{\bar{j}'}})$ or their square which appear at the right-hand side of (B.14). Moreover, since the arguments of $U(\cdot)$ satisfy Property B, there will be at least one term of the form $\mu(A^\varepsilon_s)^2$ in this product. Since $\mu(A^\varepsilon_s)^2 \le \varepsilon\mu(A^\varepsilon_s)$, these calculations provide formula (B.14). Lemma B is proved. □

Relation (B.11) implies that

$$E\left(Z^{(2)}_\gamma(f,g,\varepsilon)\right)^2 \le K\sum\nolimits^\gamma \sum\nolimits^\gamma |EU(s_1,\dots,s_k,t_1,\dots,t_l,\bar{s}_1,\dots,\bar{s}_k,\bar{t}_1,\dots,\bar{t}_l)| \tag{B.15}$$

with some appropriate $K > 0$. By Lemma B it is enough to sum up only for such terms $U(\cdot)$ in (B.15) whose arguments satisfy both Properties A and B. Moreover, each such term can be bounded by means of inequality (B.14). Let us write up the upper bound we get on $E\left(Z^{(2)}_\gamma(f,g,\varepsilon)\right)^2$ in such a way. We get a sum consisting of terms of the form $\mu(A^\varepsilon_{s_1})\cdots\mu(A^\varepsilon_{s_p})\mu(B^\varepsilon_{t_1})\cdots\mu(B^\varepsilon_{t_q})$ multiplied by constant times ε. The sets A^ε_s and B^ε_t whose measure μ appears in such a term are disjoint. Beside this $1 \le p \le k$, and $1 \le q \le l$.

In the above indicated estimation of $E\left(Z^{(2)}_\gamma(f,g,\varepsilon)\right)^2$ with the help of formula (B.15) and Lemma B we have exploited the following fact. A term

$$\mu(A^\varepsilon_{s_1})\cdots\mu(A^\varepsilon_{s_p})\mu(B^\varepsilon_{t_1})\cdots\mu(B^\varepsilon_{t_q})$$

with prescribed indices $s_1,\dots,s_p$ and $t_1,\dots,t_q$ came up in the sum at the right-hand of our bound as a contribution of only finitely many expressions $|EU(\cdots)|$. Hence we get this term in the upper bound with a multiplying coefficient bounded by constant times ε.

We also have $\sum\limits_{s=1}^{M(\varepsilon)}\mu(A^\varepsilon_s) + \sum\limits_{t=1}^{M'(\varepsilon)}\mu(B^\varepsilon_t) = \mu(A)+\mu(B) < \infty$. The above relations imply that

$$E\left(Z^{(2)}_\gamma(f,g,\varepsilon)\right)^2 \le C_1\varepsilon \sum_{\substack{1\le p\le k\\ 1\le q\le l}} \sum_{\substack{1\le s_l\le M\\ 1\le l\le p}} \sum_{\substack{1\le t_l\le M'\\ 1\le l\le q}} \mu(A^\varepsilon_{s_1})\cdots\mu(A^\varepsilon_{s_p})\mu(B^\varepsilon_{t_1})\cdots\mu(B^\varepsilon_{t_q})$$

$$\le C_2\varepsilon\sum_{j=1}^{(k+l)}(\mu(A)+\mu(B))^j \le C\varepsilon.$$

Hence relation (B.8) holds.

To prove Theorem 10.2A in the general case take for all pairs of functions $f \in \mathscr{H}_{\mu,k}$ and $g \in \mathscr{H}_{\mu,l}$ two sequences of elementary functions $f_n \in \bar{\mathscr{H}}_{\mu,k}$ and $g_n \in$

$\bar{\mathcal{H}}_{\mu,l}$, $n = 1, 2, \dots$, such that $\|f_n - f\|_2 \to 0$ and $\|g_n - g\|_2 \to 0$ as $n \to \infty$. It is enough to show that

$$E|k!Z_{\mu,k}(f)l!Z_{\mu,l}(g) - k!Z_{\mu,k}(f_n)l!Z_{\mu,l}(g_n)| \to 0 \quad \text{as } n \to \infty, \tag{B.16}$$

and

$$|\gamma|!E\left|Z_{\mu,|\gamma|}(F_\gamma(f,g)) - Z_{\mu,|\gamma|}(F_\gamma(f_n,g_n))\right| \to 0 \text{ as } n \to \infty$$
$$\text{for all } \gamma \in \Gamma(k,l), \tag{B.17}$$

since then a simple limiting procedure $n \to \infty$, and the already proved part of the theorem for Wiener–Itô integrals of elementary functions imply Theorem 10.2A.

To prove relation (B.16) write with the help of Property c) in Theorem (10.1)

$$\begin{aligned} &E|k!Z,\mu,k(f)l!Z_{\mu,l}(g) - k!Z_{\mu,k}(f_n)l!Z_{\mu,l}(g_n)| \\ &\quad \le k!l!\left(E|Z_{\mu,k}(f)Z_{\mu,l}(g-g_n)| + E|Z_{\mu,k}(f-f_n)Z_{\mu,l}(g_n))\right| \\ &\quad \le k!l!\left(\left(EZ^2_{\mu,k}(f)\right)^{1/2}\left(EZ^2_{\mu,l}(g-g_n)\right)^{1/2} \right. \\ &\qquad \left. + \left(EZ^2_{\mu,k}(f-f_n)\right)^{1/2}\left(EZ^2_{\mu,l}(g_n)\right)^{1/2}\right) \\ &\quad \le (k!l!)^{1/2}\left(\|f\|_2\|g-g_n\|_2 + \|f-f_n\|_2\|g_n\|_2\right). \end{aligned}$$

Relation (B.16) follows from this inequality with a limiting procedure $n \to \infty$.

To prove relation (B.17) write

$$\begin{aligned} &|\gamma|!E\left|Z_{\mu,|\gamma|}(F_\gamma(f,g)) - Z_{\mu,|\gamma|}(F_\gamma(f_n,g_n))\right| \\ &\quad \le |\gamma|!E\left|Z_{\mu,|\gamma|}(F_\gamma(f,g-g_n))\right| + |\gamma|!E\left|Z_{\mu,|\gamma|}(F_\gamma(f-f_n,g_n))\right| \\ &\quad \le |\gamma|!\left(EZ^2_{\mu,|\gamma|}(F_\gamma(f,g-g_n))\right)^{1/2} + |\gamma|!\left(EZ^2_{\mu,|\gamma|}(F_\gamma(f-f_n,g_n))\right)^{1/2} \\ &\quad \le (|\gamma|!)^{1/2}\left(\|F_\gamma(f,g-g_n)\|_2 + \|F_\gamma(f-f_n,g_n)\|_2\right), \end{aligned}$$

and observe that by relation (10.13) $\|F_\gamma(f,g-g_n)\|_2 \le \|f\|_2\|g-g_n\|_2$, and $\|F_\gamma(f-f_n,g_n)\|_2 \le \|f-f_n\|_2\|g_n\|_2$. Hence

$$\begin{aligned} &|\gamma|!E\left|Z_{\mu,|\gamma|}(F_\gamma(f,g)) - Z_{\mu,|\gamma|}(F_\gamma(f_n,g_n))\right| \\ &\quad \le (|\gamma|!)^{1/2}\left(\|f\|_2\|g-g_n\|_2 + \|f-f_n\|_2\|g_n\|_2\right). \end{aligned}$$

The last inequality implies relation (B.17) with a limiting procedure $n \to \infty$. Theorem 10.2A is proved. □

Appendix C
The Proof of Some Results About Wiener–Itô Integrals

First I prove Itô's formula about multiple Wiener–Itô integrals (Theorem 10.3). The proof is based on the diagram formula for Wiener–Itô integrals and a recursive formula about Hermite polynomials proved in Proposition C. In Proposition C2 I present the proof of another important property of Hermite polynomials. This result states that the class of all Hermite polynomials is a *complete* orthogonal system in an appropriate Hilbert space. It is needed in the proof of Theorem 10.5 which provides an isomorphism between a Fock space and the Hilbert space generated by Wiener–Itô integrals with respect to a white noise with an appropriate reference measure. At the end of Appendix C the proof of Theorem 10.4, a limit theorem about degenerate U-statistics is given together with a version of this result about the limit behaviour of multiple integrals with respect to a normalized empirical distribution.

Proposition C About Some Properties of Hermite Polynomials. *The functions*

$$H_k(x) = (-1)^k e^{x^2/2} \frac{d^k}{dx^k} e^{-x^2/2}, \quad k = 0, 1, 2, \dots \tag{C.1}$$

are the Hermite polynomials with leading coefficient 1, i.e. $H_k(x)$ is a polynomial of order k with leading coefficient 1 such that

$$\int_{-\infty}^{\infty} H_k(x) H_l(x) \frac{1}{\sqrt{2\pi}} e^{-x^2/2}\, dx = 0 \quad \text{if } k \neq l. \tag{C.2}$$

Beside this,

$$\int_{-\infty}^{\infty} H_k^2(x) \frac{1}{\sqrt{2\pi}} e^{-x^2/2}\, dx = k! \quad \text{for all } k = 0, 1, 2 \dots. \tag{C.3}$$

The recursive relation

$$H_k(x) = x H_{k-1}(x) - (k-1) H_{k-2}(x) \tag{C.4}$$

P. Major, *On the Estimation of Multiple Random Integrals and U-Statistics*, Lecture Notes in Mathematics 2079, DOI 10.1007/978-3-642-37617-7,

holds for all $k = 1, 2, \dots$.

Remark. It is more convenient to consider relation (C.4) valid also in the case $k=1$. In this case $H_1(x) = x$, $H_0(x) = 1$, and relation holds with an arbitrary function $H_{-1}(x)$.

Proof of Proposition C. It is clear from formula (C.1) that $H_k(x)$ is a polynomial of order k with leading coefficient 1. Take $l \geq k$, and write by means of integration by parts

$$\int_{-\infty}^{\infty} H_k(x)H_l(x)\frac{1}{\sqrt{2\pi}}e^{-x^2/2}\,dx = \int_{-\infty}^{\infty} \frac{1}{\sqrt{2\pi}} H_k(x)(-1)^l \frac{d^l}{dx^l}e^{-x^2/2}\,dx$$
$$= \int_{-\infty}^{\infty} \frac{1}{\sqrt{2\pi}} \frac{d}{dx} H_k(x)(-1)^{l-1} \frac{d^{l-1}}{dx^{l-1}} e^{-x^2/2}\,dx.$$

Successive partial integration together with the identity $\frac{d^k}{dx^k} H_k(x) = k!$ yield that

$$\int_{-\infty}^{\infty} H_k(x)H_l(x)\frac{1}{\sqrt{2\pi}}e^{-x^2/2}\,dx = k! \int_{-\infty}^{\infty} \frac{1}{\sqrt{2\pi}}(-1)^{l-k}\frac{d^{l-k}}{dx^{l-k}}e^{-x^2/2}\,dx.$$

The last relation supplies formulas (C.2) and (C.3).

To prove relation (C.4) observe that $H_k(x) - xH_{k-1}(x)$ is a polynomial of order $k-2$. (The term x^{k-1} is missing from this expression. Indeed, if k is an even number, then the polynomial $H_k(x) - xH_{k-1}(x)$ is an even function, and it does not contain the term x^{k-1} with an odd exponent $k-1$. Similar argument holds if the number k is odd.) Beside this, it is orthogonal (with respect to the standard normal distribution) to all Hermite polynomials $H_l(x)$ with $0 \leq l \leq k-3$. Hence $H_k(x) - xH_{k-1}(x) = CH_{k-2}(x)$ with some constant C to be determined.

Multiply both sides of the last identity with $H_{k-2}(x)$ and integrate them with respect to the standard normal distribution. Apply the orthogonality of the polynomials $H_k(x)$ and $H_{k-2}(x)$, and observe that the identity

$$\int H_{k-1}(x)xH_{k-2}(x)\frac{1}{\sqrt{2\pi}}e^{-x^2/2}\,dx = \int H_{k-1}^2(x)\frac{1}{\sqrt{2\pi}}e^{-x^2/2}\,dx = (k-1)!$$

holds. (In this calculation we have exploited that $H_{k-1}(x)$ is orthogonal to $H_{k-1}(x) - xH_{k-2}(x)$, because the order of the latter polynomial is less than $k-1$.) In such a way we get the identity $-(k-1)! = C(k-2)!$ for the constant C in the last identity, i.e. $C = -(k-1)$, and this implies relation (C.4). □

Proof of Itô's formula for multiple Wiener–Itô integrals. Let $K = \sum\limits_{p=1}^{m} k_p$, the sum of the order of the Hermite polynomials, denote the order of the expression in relation (10.24). Formula (10.24) clearly holds for expressions of order $K = 1$. It will be proved in the general case by means of induction with respect to the order K.

In the proof the functions $f(x_1) = \varphi_1(x_1)$ and

$$g(x_1, \dots, x_{K_m-1}) = \prod_{j=1}^{K_1-1} \varphi_1(x_j) \cdot \prod_{p=2}^{m} \prod_{j=K_{p-1}}^{K_p-1} \varphi_p(x_j),$$

will be introduced and the product $Z_{\mu,1}(f)(K_m-1)!Z_{\mu,K_m-1}(g)$ will be calculated by means of the diagram formula. (The same notation is applied as in Theorem 10.3. In particular, $K = K_m$, and in the case $K_1 = 1$ the convention $\prod_{j=1}^{K_1-1} \varphi_1(x_j)=1$ is applied.) In the application of the diagram formula diagrams with two rows appear. The first row of these diagrams contains the vertex $(1,1)$ and the second row contains the vertices $(2,1), \dots, (2, K_m-1)$. It is useful to divide the diagrams to three disjoint classes. The first class, Γ_0 contains only the diagram γ_0 without any edges. The second class Γ_1 consists of those diagrams which have an edge of the form $((1,1),(2,j))$ with some $1 \le j \le k_1-1$, and the third class Γ_2 is the set of those diagrams which have an edge of the form $((1,1),(2,j))$ with some $k_1 \le j \le K_m-1$. Because of the orthogonality of the functions φ_s for different indices s $F_\gamma \equiv 0$ and $Z_{\mu,K_m-2}(F_\gamma) = 0$ for $\gamma \in \Gamma_2$. The class Γ_1 contains k_1-1 diagrams. Let us consider a diagram γ from this class with an edge $((1,1),(2,j_0))$, $1 \le j \le k_1-1$. We have for such a diagram $F_\gamma = \prod_{j\in\{1,\dots,K_1-1\}\setminus\{j_0\}} \varphi_1(x_{(2,j)}) \prod_{p=2}^{m} \prod_{j=K_{p-1}}^{K_p-1} \varphi_p(x_{(2,j)})$, and by our inductive hypothesis $(K_m-2)!Z_{\mu,K_m-2}(F_\gamma) = H_{k_1-2}(\eta_1) \prod_{p=2}^{m} H_{k_p}(\eta_p)$. Finally

$$K_m!Z_{\mu,K_m}(F_{\gamma_0}) = K_m!Z_{\mu,K_m}\left(\prod_{p=1}^{m}\left(\prod_{j=K_{p-1}+1}^{K_p} \varphi_p(x_j)\right)\right)$$

for the diagram $\gamma_0 \in \Gamma_0$ without any edge.

Our inductive hypothesis also implies the following identity for the expression we wanted to calculate with the help of the diagram formula.

$$Z_{\mu,1}(f)(K_m-1)!Z_{\mu,K_m-1}(g) = \eta_1 H_{k_1-1}(\eta_1) \prod_{p=2}^{m} H_{k_p}(\eta_p).$$

The above calculations together with the observation $|\Gamma_1| = k_1 - 1$ yield the identity

$$
\begin{aligned}
K_m!Z_{\mu,K_m}&\left(\prod_{p=1}^{m}\left(\prod_{j=K_{p-1}+1}^{K_p}\varphi_p(x_j)\right)\right)\\
&= K_m!Z_{\mu,K_m}(F_{\gamma_0})\\
&= Z_{\mu,1}(f)(K_m-1)!Z_{\mu,K_m-1}(g)-\sum_{\gamma\in\Gamma_1}(K_m-2)!Z_{\mu,K_m-2}(F_\gamma)\\
&= \eta_1 H_{k_1-1}(\eta_1)\prod_{p=2}^{m}H_{k_p}(\eta_p)-(k_1-1)H_{k_1-2}(\eta_1)\prod_{p=2}^{m}H_{k_p}(\eta_p)\\
&= [\eta_1 H_{k_1-1}(\eta_1)-(k_1-1)H_{k_1-2}(\eta_1)]\prod_{p=2}^{m}H_{k_p}(\eta_p). \qquad \text{(C.5)}
\end{aligned}
$$

On the other hand, $\eta_1 H_{k_1-1}(\eta_1)-(k_1-1)H_{k_1-2}(\eta_1)=H_{k_1}(\eta_1)$ by formula (C.4). These relations imply formula (10.24), i.e. Itô's formula. □

I present the proof of another important property of the Hermite polynomials in the following Proposition C2.

Proposition C2 on the completeness of the orthogonal system of Hermite polynomials The Hermite polynomials $H_k(x)$, $k=0,1,2,\dots$, defined in formula (C.5) constitute a complete orthonormal system in the L_2-space of the functions square integrable with respect to the Gaussian measure $\frac{1}{\sqrt{2\pi}}e^{-x^2/2}\,dx$ on the real line.

Proof of Proposition C2. Let us consider the orthogonal complement of the subspace generated by the Hermite polynomials in the space of the square integrable functions with respect to the measure $\frac{1}{\sqrt{2\pi}}e^{-x^2/2}\,dx$. It is enough to prove that this orthogonal completion contains only the identically zero function. Since the orthogonality of a function to all polynomials of the form x^k, $k=0,1,2,\dots$ is equivalent to the orthogonality of this function to all Hermite polynomials $H_k(x)$, $k=0,1,2,\dots$, Proposition C2 can be reformulated in the following form:

If a function $g(x)$ on the real line is such that

$$\int_{-\infty}^{\infty} x^k g(x)\frac{1}{\sqrt{2\pi}}e^{-x^2/2}\,dx=0 \quad \text{for all } k=0,1,2,\dots \tag{C.6}$$

and

$$\int_{-\infty}^{\infty} g^2(x)\frac{1}{\sqrt{2\pi}}e^{-x^2/2}\,dx<\infty, \tag{C.7}$$

then $g(x)=0$ for almost all x.

Given a function $g(x)$ on the real line whose absolute value is integrable with respect to the Gaussian measure $\frac{1}{\sqrt{2\pi}}e^{-x^2/2}\,dx$ define the (finite) measure ν_g,

$$\nu_g(A) = \int_A g(x) \frac{1}{\sqrt{2\pi}} e^{-x^2/2}\, dx$$

on the measurable sets of the real line together with its Fourier transform $\tilde{\nu}_g(t) = \int_{-\infty}^{\infty} e^{itx} \nu_g(\,dx)$. (This measure ν_g and its Fourier transform can be defined for all functions g satisfying relation (C.7), because their absolute value is integrable with respect to the Gaussian measure.) First I show that Proposition C2 can be reduced to the following statement: If a function g satisfies both (C.6) and (C.7) then $\tilde{\nu}_g(t) = 0$ for all $-\infty < t < \infty$.

Indeed, if there were a function g satisfying (C.6) and (C.7) which is not identically zero, then the non-negative functions $g^+(x) = \max(0, g(x))$ and $g^-(x) = -\min(0, g(x))$ would be different. Then also their Fourier transform $\tilde{\nu}_{g^+}(t)$ and $\tilde{\nu}_{g^-}(t)$ would be different, since a finite measure is uniquely determined by its Fourier transform. (This statement is equivalent to an important result in probability theory, by which a probability measure on the real line is determined by its characteristic function.) But this would mean that $\tilde{\nu}_g(t) = \tilde{\nu}_{g^+}(t) - \tilde{\nu}_{g^-}(t) \neq 0$ for some t. Hence Proposition C2 can be reduced to the above statement.

Since $\left| e^{itx} - 1 - (itx) - \cdots - \frac{(itx)^k}{k!} \right| \le \frac{|tx|^{(k+1)}}{(k+1)!}$ for all real numbers t, x and integer $k = 1, 2, \dots$ we may write because of relation (C.6)

$$\begin{aligned}
|\tilde{\nu}_g(t)| &= \left| \int_{-\infty}^{\infty} \left(e^{itx} - 1 - (itx) - \cdots - \frac{(itx)^k}{k!} \right) g(x) \frac{1}{\sqrt{2\pi}} e^{-x^2/2}\, dx \right| \\
&\le \int_{-\infty}^{\infty} \frac{|t|^{(k+1)}}{(k+1)!} |x|^{k+1} |g(x)| \frac{1}{\sqrt{2\pi}} e^{-x^2/2}\, dx
\end{aligned}$$

for all $k = 1, 2, \dots$ and real number t if the function g satisfies relation (C.6). If it satisfies both relation (C.6) and (C.7), then from the last relation and the Schwarz inequality

$$\begin{aligned}
|\tilde{\nu}_g(t)|^2 &\le \text{const.} \frac{|t|^{2(k+1)}}{(k+1)!)^2} \int_{-\infty}^{\infty} |x|^{2(k+1)} \frac{1}{\sqrt{2\pi}} e^{-x^2/2}\, dx \\
&= \text{const.} \frac{|t|^{2(k+1)}}{(k+1)!)^2} 1 \cdot 3 \cdot 5 \cdots (2k+1)
\end{aligned}$$

for all real number t and integer $k = 1, 2, \dots$. Simple calculation shows that the right-hand side of the last estimate tends to zero as $k \to \infty$. This implies that $\tilde{\nu}_g(t) = 0$ for all t, and Proposition C2 holds. □

I finish Appendix C with the proof of Theorem 10.4, a limit theorem about a sequence of normalized degenerate U-statistics. It is based on an appropriate representation of the U-statistics by means of multiple random integrals which makes possible to carry out an appropriate limiting procedure.

Proof of Theorem 10.4. For all $n = 1, 2, \dots$, the normalized degenerate U-statistics $n^{-k/2}k!I_{n,k}(f)$ can be written in the form

$$\begin{aligned} n^{-k/2}k!I_{n,k}(f) &= n^{k/2}\int' f(x_1,\dots,x_k)\mu_n(\,dx_1)\dots\mu_n(\,dx_k) \\ &= n^{k/2}\int' f(x_1,\dots,x_k)(\mu_n(\,dx_1)-\mu(\,dx_1)) \\ &\quad \dots(\mu_n(\,dx_k)-\mu(\,dx_k)), \end{aligned} \tag{C.8}$$

where μ_n is the empirical distribution of the sequence $\xi_1,\dots,\xi_n$ defined in (4.5), and the prime in $\int'$ denotes that the diagonals, i.e. the points $x = (x_1,\dots,x_k)$ such that $x_j = x_{j'}$ for some pairs of indices $1 \le j, j' \le k$, $j \ne j'$, are omitted from the domain of integration. The second identity in relation (C.8) can be justified by means of the identity

$$\begin{aligned} &\int' f(x_1,\dots,x_k)(\mu_n(\,dx_1)-\mu(\,dx_1))\dots(\mu_n(\,dx_k)-\mu(\,dx_k)) - I_{n,k}(f) \\ &= \sum_{V\colon V\in\{1,\dots,k\},\,|V|\ge 1} (-1)^{|V|}\int' f(x_1,\dots,x_k) \\ &\quad \prod_{j\in V}\mu(\,dx_j) \prod_{j\in\{1,\dots,k\}\setminus V} \mu_n(\,dx_j)) = 0. \end{aligned} \tag{C.9}$$

This identity holds for a function f canonical with respect to a non-atomic measure μ, because each term in the sum at the right-hand side of (C.9) equals zero. Indeed, the integral of a canonical function f with respect to $\mu(\,dx_j)$ with some index $j \in V$ equals zero for all fixed values $x_1,\dots,x_{j-1},x_{j+1},\dots,x_k$. The non-atomic property of the measure μ was needed to guarantee that this integral equals zero also in the case when the diagonals are omitted from the domain of integration.

We would like to derive Theorem 10.4 from relation (C.8) by means of an appropriate limiting procedure which exploits the convergence of the random fields $n^{1/2}(\mu_n(A)-\mu(A))$, $A \in \mathcal{X}$, to a Gaussian field $\nu(A)$, $A \in \mathcal{X}$, as $n \to \infty$. But some problems arise if we want to carry out such a program, because the fields $n^{1/2}(\mu_n - \mu)$ converge to a non white noise type Gaussian field. The limit we get is similar to a Wiener bridge on the real line. Hence a relation between Wiener processes and Wiener bridges suggests to write the following version of formula (C.8).

Let us take a standard Gaussian random variable η, independent of the random sequence $\xi_1, \xi_2, \dots$. For a canonical function f the following version of (C.8) holds.

$$n^{-k/2}k!I_{n,k}(f) = J'_{n,k}(f) \tag{C.10}$$

with

$$J'_{n,k}(f) = \int' f(x_1,\dots,x_k)\left[\sqrt{n}(\mu_n(dx_1)-\mu(dx_1))+\eta\mu(dx_1)\right]$$
$$\dots\left[\sqrt{n}(\mu_n(dx_k)-\mu(dx_k))+\eta\mu(dx_k)\right]. \tag{C.11}$$

This relation can be seen similarly to (C.8).

The random measures $n^{1/2}(\mu_n-\mu)+\eta\mu$ converge to a white noise with reference measure μ. Hence Theorem 10.4 can be proved by means of formulas (C.10) and (C.11) with the help of an appropriate limiting procedure. More explicitly, I claim that the following slightly more general result holds. The expressions $J'_{n,k}(f)$ introduced in (C.11) converge in distribution to the Wiener–Itô integral $k!Z_{\mu,k}(f)$ as $n\to\infty$ for all functions f square integrable with respect to the product measure μ^k. This result also holds for non-canonical functions f. This limit theorem together with relation (C.10) imply Theorem 10.4.

The convergence of the random variables $J'_{n,k}(f)$ defined in (C.11) to the Wiener–Itô integral $k!Z_{\mu,k}(f)$ can be easily checked for elementary functions $f\in\bar{\mathcal{H}}_{\mu,k}$. Indeed, if $A_1,\dots,A_M$ are disjoint sets with $\mu(A_s)<\infty$, then the multi-dimensional central limit theorem implies that the random vectors $\{\sqrt{n}((\mu_n(A_s)-\mu(A_s))+\eta\mu(A_s),\ 1\le s\le M\}$ converge in distribution to the random vector $\{(\mu_W(A_s),\ 1\le s\le M\}$, i.e. to a set of independent normal random variables ζ_s, $E\zeta_s=0$, $1\le s\le M$, with variance $E\zeta_s^2=\mu(A_s)$ as $n\to\infty$. The definition of the elementary functions given in (10.2) shows that this central limit theorem implies the demanded convergence of the sequence $J'_{n,k}(f)$ to $k!Z_{\mu,k}(f)$ for elementary functions.

To show the convergence of the sequence $J'_{n,k}(f)$ to $k!Z_{\mu,k}(f)$ in the general case take for any function $f\in\mathcal{H}_{\mu,k}$ a sequence of elementary functions $f_N\in\bar{\mathcal{H}}_{\mu,k}$ such that $\|f-f_N\|_2\to0$ as $N\to\infty$. Then $E(Z_{\mu,k}(f)-Z_{\mu,k}(f_N))^2=E(Z_{\mu,k}(f-f_N))^2\to0$ as $N\to\infty$ by Property (c) in Theorem 10.1. Hence the already proved part of the theorem implies that there exists some sequence of positive integers, $N(n)$, $n=1,2,\dots$, in such a way that $N(n)\to\infty$, and the sequence $J'_{n,k}(f_{N(n)})$ converges to $k!Z_{\mu,k}(f)$ in distribution as $n\to\infty$. Thus to complete the proof of Theorem 10.4 it is enough to show that $E(J'_{n,k}(f_{N(n)})-J'_{n,k}(f))^2=E(J'_{n,k}(f_{N(n)}-f))^2\to0$ as $n\to\infty$.

It is enough to show that

$$E(J'_{n,k}(f))^2\le C\|f\|_2^2 \quad \text{for all } f\in\mathcal{H}_{\mu,k} \tag{C.12}$$

with a constant $C=C_k$ depending only on the order k of the function f and to apply inequality (C.12) for the functions $f_{N(n)}-f$. Relation (C.12) is a relatively simple consequence of Corollary 1 of Theorem 9.4.

Indeed,

$$J'_{n,k}(f)=\sum_{V\subset\{1,\dots,k\}}\eta^{k-|V|}|V|!J_{n,|V|}(f_V)$$

with

$$f_V(x_j,\,j\in V)=\int f(x_1,\dots,x_k)\prod_{j'\in\{1,\dots,k\}\setminus V}\mu(dx_{j'})$$

and the random integral $J_{n,k}(\cdot)$ defined in (4.8), hence

$$E(J'_{n,k}(f))^2\le 2^k\sum_{V\subset\{1,\dots,k\}}(|V|!)^2E\eta^{2(k-|V|)}\cdot EJ^2_{n,|V|}(f_V). \tag{C.13}$$

Inequality $\|f_V\|_2\le\|f\|_2$ holds for all sets $V\subset\{1,\dots,k\}$, hence an application of Corollary 1 of Theorem 9.4 to all random integrals $J_{n,|V|}(f)$ supplies (C.12). □

The above proof also yields the following slight generalization of Theorem 10.4. Let us consider a finite sequence of functions $f_j\in\mathcal{H}_{\mu,j}$, $1\le j\le k$, canonical with respect to a non-atomic probability measure μ. The vectors $\{n^{-j/2}I_{n,j}(f_j),1\le j\le k\}$, consisting of normalized degenerate U-statistics defined with the help of a sequence of independent μ-distributed random variables converge to the random vector $\{Z_{\mu,j}(f_j),1\le j\le k\}$ in distribution as $n\to\infty$. This result together with Theorem 9.4 imply the following limit theorem about multiple random integrals $J_{n,k}(f)$.

Theorem 10.4′ (Limit theorem about multiple random integrals with respect to a normalized empirical measure). *Let a sequence of independent and identically distributed random variables $\xi_1,\xi_2,\dots$ be given with some non-atomic distribution μ on a measurable space $(X,\mathcal{X})$ together with a function $f(x_1,\dots,x_k)$ on the k-fold product $(X^k,\mathcal{X}^k)$ of the space $(X,\mathcal{X})$ such that*

$$\int f^2(x_1,\dots,x_k)\mu(dx_1)\dots\mu(dx_k)<\infty.$$

Let us consider for all $n=1,2,\dots$ the random integrals $J_{n,k}(f)$ of order k defined in formulas (4.5) and (4.8) with the help of the empirical distribution μ_n of the sequence $\xi_1,\dots,\xi_n$ and the function f. These random integrals $J_{n,k}(f)$ converge in distribution, as $n\to\infty$, to the following sum $U(f)$ of multiple Wiener–Itô integrals:

$$\begin{aligned}U(f)&=\sum_{V\subset\{1,\dots,k\}}C(k,V)Z_{\mu,|V|}(f_V)\\&=\sum_{V\subset\{1,\dots,k\}}\frac{C(k,V)}{|V|!}\int f_V(x_j,\,j\in V)\prod_{j\in V}\mu_W(dx_j),\end{aligned}$$

where the functions $f_V(x_j,\ j \in V)$, $V \subset \{1,\dots,k\}$, *are those functions defined in formula (9.3) which appear in the Hoeffding decomposition of the function* $f(x_1,\dots,x_k)$, *the constants* $C(k,V)$ *are the limits appearing in the limit relation* $\lim\limits_{n\to\infty} C(n,k,V) = C(k,V)$ *satisfied by the coefficients* $C(n,k,V)$ *in formula (9.16), and* μ_W *is a white noise with reference measure* μ.

An essential step of the proof of Theorem 10.4 was the reduction of the case of general kernel functions to the case of elementary kernel functions. Let me make some comments about it.

It would be simple to make such a reduction if we had a good approximation of a canonical function with such elementary functions which are also canonical. But it is very hard to find such an approximation. To overcome this difficulty we reduced the proof of Theorem 10.4 to a modified version of this result where instead of a limit theorem for degenerate U-statistics a limit theorem for the random variables $J'_{n,k}(f)$ introduced in formula (C.11) has to be proved. In the proof of such a version we could apply the approximation of a general kernel function with not necessarily canonical elementary functions. Theorem 9.4 helped us to work with such an approximation. Another natural way to overcome the above difficulty is to apply a Poissonian approximation of the normalized empirical measure. Such an approach was applied in [16,34], where some generalizations of Theorem 10.4 were proved.

Appendix D
The Proof of Theorem 14.3 About U-Statistics and Decoupled U-Statistics

The proof of Theorem 14.3. It will be simpler to formulate and prove a generalized version of Theorem 14.3 where such generalized U-statistics are considered in which different kernel functions may appear in each term of the sum. More explicitly, let $\ell = \ell(n,k)$ denote the set of all such sequences $l = (l_1,\dots,l_k)$ of integers of length k for which $1 \le l_j \le n$, $1 \le j \le k$. To define generalized U-statistics let us fix a set of functions $\{f_{l_1,\dots,l_k}(x_1,\dots,x_k),\ (l_1,\dots,l_k)\in\ell\}$ which map the space $(X^k,\mathcal{X}^k)$ to a separable Banach space B, and have the property $f_{l_1,\dots,l_k}(x_1,\dots,x_k)\equiv 0$ if $l_j = l_{j'}$ for some indices $j \ne j'$. (The last condition corresponds to that property of U-statistics that the diagonals are omitted from the summation in their definition.) Let us denote this set of functions by $f(\ell)$, and define, similarly to the U-statistics and decoupled U-statistics the generalized U-statistics and generalized decoupled U-statistics by the formulas

$$I_{n,k}(f(\ell)) = \frac{1}{k!}\sum_{(l_1,\dots,l_k)\colon 1\le l_j\le n,\ j=1,\dots,k} f_{l_1,\dots,l_k}(\xi_{l_1},\dots,\xi_{l_k}) \tag{D.1}$$

and

$$\bar I_{n,k}(f(\ell)) = \frac{1}{k!}\sum_{(l_1,\dots,l_k)\colon 1\le l_j\le n,\ j=1,\dots,k} f_{l_1,\dots,l_k}\left(\xi^{(1)}_{l_1},\dots,\xi^{(k)}_{l_k}\right) \tag{D.2}$$

(with the same independent and identically distributed random variables ξ_l and $\xi_l^{(j)}$, $1\le l\le n$, $1\le j\le k$, as in the definition of the original U-statistics and decoupled U-statistics.)

The following generalization of relation (14.13) will be proved.

$$P\left(\|I_{n,k}(f(\ell))\| > u\right) \le A(k)P\left(\|\bar I_{n,k}(f(\ell))\| > \gamma(k)u\right) \tag{D.3}$$

P. Major, *On the Estimation of Multiple Random Integrals and U-Statistics*, Lecture Notes in Mathematics 2079, DOI 10.1007/978-3-642-37617-7,

with some constants $A(k) > 0$ and $\gamma(k) > 0$ depending only on the order k of these generalized U-statistics. The sign $\|\cdot\|$ in (D.3) denotes the norm in the Banach space we are working in.

We concentrate mainly on the proof of the generalization (D.3) of relation (14.13). Formula (14.14) is a relatively simple consequence of it. Formula (D.3) will be proved by means of an inductive procedure which works only in this more general setting. It will be derived from the following statement.

Let us take two independent copies $\xi_1^{(1)}, \dots, \xi_n^{(1)}$ and $\xi_1^{(2)}, \dots, \xi_n^{(2)}$ of our original sequence of random variables $\xi_1, \dots, \xi_n$, and introduce for all sets $V \subset \{1, \dots, k\}$ the function $\alpha_V(j)$, $1 \le j \le k$, defined as $\alpha_V(j) = 1$ if $j \in V$ and $\alpha_V(j) = 2$ if $j \notin V$. Let us define with their help the following version of decoupled U-statistics:

$$I_{n,k,V}(f(\ell)) = \frac{1}{k!} \sum_{(l_1,\dots,l_k)\colon 1 \le l_j \le n,\, j=1,\dots,k} f_{l_1,\dots,l_k}\left(\xi_{l_1}^{(\alpha_V(1))}, \dots, \xi_{l_k}^{(\alpha_V(k))}\right) \quad \text{for all } V \subset \{1, \dots, k\}. \tag{D.4}$$

The following inequality will be proved: There are some constants $C_k > 0$ and $D_k > 0$ depending only on the order k of the generalized U-statistic $I_{n,k}(f(\ell))$ such that for all numbers $u > 0$

$$P\left(\|I_{n,k}(f(\ell))\| > u\right) \le \sum_{V \subset \{1,\dots,k\},\, 1 \le |V| \le k-1} C_k P\left(D_k \|I_{n,k,V}(f(\ell))\| > u\right). \tag{D.5}$$

Here $|V|$ denotes the cardinality of the set V, and the condition $1 \le |V| \le k-1$ in the summation of formula (D.5) means that the sets $V = \emptyset$ and $V = \{1, \dots, k\}$ are omitted from the summation, i.e. the terms where either $\alpha_V(j) = 1$ or $\alpha_V(j) = 2$ for all $1 \le j \le k$ are not considered. Formula (D.3) can be derived from formula (D.5) by means of an inductive argument. The hard part of the problem is to prove formula (D.5). To do this first we prove the following simple lemma.

Lemma D1. *Let ξ and η be two independent and identically distributed random variables taking values in a separable Banach space B. Then*

$$3P\left(|\xi + \eta| > \frac{2}{3}u\right) \ge P(|\xi| > u) \quad \text{for all } u > 0.$$

Proof of Lemma D1. Let ξ, η and ζ be three independent, identically distributed random variables taking values in B. Then

$$3P\left(|\xi+\eta|>\frac{2}{3}u\right)=P\left(|\xi+\eta|>\frac{2}{3}u\right)+P\left(|\xi+\zeta|>\frac{2}{3}u\right)$$
$$+P\left(|-(\eta+\zeta)|>\frac{2}{3}u\right)$$
$$\ge P(|\xi+\eta+\xi+\zeta-\eta-\zeta|>2u)=P(|\xi|>u).$$

□

To prove formula (D.5) we introduce the random variable

$$\begin{aligned}T_{n,k}(f(\ell))&=\frac{1}{k!}\sum_{\substack{(l_1,\dots,l_k),\,(s_1,\dots,s_k):\\ 1\le l_j\le n,\, s_j=1 \text{ or } s_j=2,\ j=1,\dots,k,}} f_{l_1,\dots,l_k}\left(\xi_{l_1}^{(s_1)},\dots,\xi_{l_k}^{(s_k)}\right)\\ &=\sum_{V\subset\{1,\dots,k\}} I_{n,k,V}(f(\ell)).\end{aligned}\tag{D.6}$$

The random variables $I_{n,k}(f(\ell))$, $I_{n,k,\emptyset}(f(\ell))$ and $I_{n,k,\{1,\dots,k\}}(f(\ell))$ are identically distributed, and the last two random variables are independent of each other. Hence Lemma D1 yields that

$$\begin{aligned}P(\|I_{n,k}(f(\ell))\|>u)&\le 3P\left(\|I_{n,k,\emptyset}(f(\ell))+I_{n,k,\{1,\dots,k\}}(f(\ell))\|>\frac{2}{3}u\right)\\ &=3P\left(\left\|T_{n,k}(f(\ell))-\sum_{V:\,V\subset\{1,\dots,k\},\,1\le|V|\le k-1} I_{n,k,|V|}(f(\ell))\right\|>\frac{2}{3}u\right)\\ &\le 3P(3\cdot 2^{k-1}\|T_{n,k}(f(\ell))\|>u)\\ &\quad+\sum_{V:\,V\subset\{1,\dots,k\},\,1\le|V|\le k-1} 3P(3\cdot 2^{k-1}\|I_{n,k,|V|}(f(\ell))\|>u).\end{aligned}\tag{D.7}$$

To derive relation (D.5) from relation (D.7) we need a good upper bound on the probability $P(3\cdot 2^{k-1}\|T_{n,k}(f(\ell))\|>u)$. To get such an estimate we shall compare the tail distribution of $\|T_{n,k}(f(\ell))\|$ with that of $\|I_{n,k,V}(f(\ell))\|$ for an arbitrary set $V\subset\{1,\dots,k\}$. This will be done with the help of Lemmas D2 and D4 formulated below.

In Lemma D2 such a random variable $\|\hat I_{n,k,V}(f(\ell))\|$ will be constructed whose distribution agrees with that of $\|I_{n,k,V}(f(\ell))\|$. The expression $\hat I_{n,k,V}(f(\ell))$, whose norm will be investigated will be defined in formulas (D.8) and (D.9). It is a random polynomial of some Rademacher functions $\varepsilon_1,\dots,\varepsilon_n$. The coefficients of this polynomial are random variables, independent of the Rademacher functions $\varepsilon_1,\dots,\varepsilon_n$. Beside this, the constant term of this polynomial equals $T_{n,k}(f(\ell))$. These properties of the polynomial $\hat I_{n,k,V}(f(\ell))$ together with Lemma D4 formulated below enable us prove such an estimate on the distribution of $\|T_{n,k}(f(\ell))\|$ that together with formula (D.7) imply relation (D.5). Let us formulate these lemmas.

Lemma D2. *Let us consider a sequence of independent random variables $\varepsilon_1,\dots,\varepsilon_n$, $P(\varepsilon_l=1)=P(\varepsilon_l=-1)=\frac12$, $1\le l\le n$, which is also independent of the random variables $\xi_1^{(1)},\dots,\xi_n^{(1)}$ and $\xi_1^{(2)},\dots,\xi_n^{(2)}$ appearing in the definition of the modified decoupled U-statistics $I_{n,k,V}(f(\ell))$ given in formula (D.4). Let us define with their help the sequences of random variables $\eta_1^{(1)},\dots,\eta_n^{(1)}$ and $\eta_1^{(2)},\dots,\eta_n^{(2)}$ whose elements $(\eta_l^{(1)},\eta_l^{(2)})=(\eta_l^{(1)}(\varepsilon_l),\eta_l^{(2)}(\varepsilon_l))$, $1\le l\le n$, are defined by the formula*

$$(\eta_l^{(1)}(\varepsilon_l),\eta_l^{(2)}(\varepsilon_l))=\left(\frac{1+\varepsilon_l}{2}\xi_l^{(1)}+\frac{1-\varepsilon_l}{2}\xi_l^{(2)},\frac{1-\varepsilon_l}{2}\xi_l^{(1)}+\frac{1+\varepsilon_l}{2}\xi_l^{(2)}\right),$$

i.e. let

$$(\eta_l^{(1)}(\varepsilon_l),\eta_l^{(2)}(\varepsilon_l))=(\xi_l^{(1)},\xi_l^{(2)})\quad \text{if } \varepsilon_l=1,$$

and

$$(\eta_l^{(1)}(\varepsilon_l),\eta_l^{(2)}(\varepsilon_l))=(\xi_l^{(2)},\xi_l^{(1)})\quad \text{if } \varepsilon_l=-1,\quad 1\le l\le n.$$

Then the joint distribution of the pair of sequences of random variables $\xi_1^{(1)},\dots,\xi_n^{(1)}$ and $\xi_1^{(2)},\dots,\xi_n^{(2)}$ agrees with that of the pair of sequences $\eta_1^{(1)},\dots,\eta_n^{(1)}$ and $\eta_1^{(2)},\dots,\eta_n^{(2)}$, which is also independent of the sequence $\varepsilon_1,\dots,\varepsilon_n$.

Let us fix some $V\subset\{1,\dots,k\}$, and introduce the random variable

$$\hat I_{n,k,V}(f(\ell))=\frac{1}{k!}\sum_{(l_1,\dots,l_k):\,1\le l_j\le n,\ j=1,\dots,k} f_{l_1,\dots,l_k}\left(\eta_{l_1}^{(\alpha_V(1))},\dots,\eta_{l_k}^{(\alpha_V(k))}\right),\tag{D.8}$$

where similarly to formula (D.4) $\alpha_V(j)=1$ if $j\in V$, and $\alpha_V(j)=2$ if $j\notin V$. Then the identity

$$\begin{aligned}&2^k\hat I_{n,k,V}(f(\ell))\\&=\frac{1}{k!}\sum_{\substack{(l_1,\dots,l_k),\,(s_1,\dots,s_k):\\1\le l_j\le n,\ s_j=1 \text{ or } s_j=2,\\ j=1,\dots,k,}}(1+\kappa_{s_1,V}^{(1)}\varepsilon_{l_1})\cdots(1+\kappa_{s_k,V}^{(k)}\varepsilon_{l_k})f_{l_1,\dots,l_k}\left(\xi_{l_1}^{(s_1)},\dots,\xi_{l_k}^{(s_k)}\right)\end{aligned}\tag{D.9}$$

holds, where $\kappa_{1,V}^{(j)}=1$ and $\kappa_{2,V}^{(j)}=-1$ if $j\in V$, and $\kappa_{1,V}^{(j)}=-1$ and $\kappa_{2,V}^{(j)}=1$ if $j\notin V$, i.e. $\kappa_{1,V}^{(j)}=3-2\alpha_V(j)$ and $\kappa_{2,V}^{(j)}=-\kappa_{1,V}^{(j)}$.

Before the formulation of Lemma D4 another Lemma D3 will be presented which will be applied in its proof.

Lemma D3. *Let Z be a random variable taking values in a separable Banach space B with expectation zero, i.e. let $E\kappa(Z)=0$ for all $\kappa\in B'$, where B' denotes* *the (Banach) space of all (bounded) linear transformations of B to the real line. Then* $P(\|v+Z\|\ge\|v\|)\ge \inf\limits_{\kappa\in B'}\frac{(E|\kappa(Z)|)^2}{4E\kappa(Z)^2}$ *for all* $v\in B$.

Lemma D4. *Let us consider a positive integer n and a sequence of independent random variables $\varepsilon_1,\dots,\varepsilon_n$, $P(\varepsilon_l=1)=P(\varepsilon_l=-1)=\frac12$, $1\le l\le n$. Beside this, fix some positive integer k, take a separable Banach space B and choose some elements $a_s(l_1,\dots,l_s)$ of this Banach space B, $1\le s\le k$, $1\le l_j\le n$, $l_j\ne l_{j'}$ if $j\ne j'$, $1\le j,j'\le s$. With the above notations the inequality*

$$P\left(\left\|v+\sum_{s=1}^k \sum_{\substack{(l_1,\dots,l_s)\colon 1\le l_j\le n,\ j=1,\dots,s,\\ l_j\ne l_{j'} \text{ if } j\ne j'}} a_s(l_1,\dots,l_s)\varepsilon_{l_1}\cdots\varepsilon_{l_s}\right\|\ge\|v\|\right)\ge c_k \tag{D.10}$$

holds for all $v\in B$ with some constant $c_k>0$ which depends only on the parameter k. In particular, it does not depend on the norm in the separable Banach space B.

Proof of Lemma D2. Let us consider the conditional joint distribution of the sequences of random variables $\eta_1^{(1)},\dots,\eta_n^{(1)}$ and $\eta_1^{(2)},\dots,\eta_n^{(2)}$ under the condition that the random vector $\varepsilon_1,\dots,\varepsilon_n$ takes the value of some prescribed ±1 series of length n. Observe that this conditional distribution agrees with the joint distribution of the sequences $\xi_1^{(1)},\dots,\xi_n^{(1)}$ and $\xi_1^{(2)},\dots,\xi_n^{(2)}$ for all possible conditions. This fact implies the statement about the joint distribution of the sequences $(\eta_l^{(1)},\eta_l^{(2)})$, $1\le l\le n$ and their independence of the sequence $\varepsilon_1,\dots,\varepsilon_n$.

To prove identity (D.9) let us fix a set $M\subset\{1,\dots,n\}$, and consider the case when $\varepsilon_l=1$ if $l\in M$ and $\varepsilon_l=-1$ if $l\notin M$. Put $\beta_{V,M}(j,l)=1$ if $j\in V$ and $l\in M$ or $j\notin V$ and $l\notin M$, and let $\beta_{V,M}(j,l)=2$ otherwise. Then we have for all $(l_1,\dots,l_k)$, $1\le l_j\le n$, $1\le j\le k$, and our fixed set V

$$\begin{aligned}\sum_{\substack{(s_1,\dots,s_k)\colon\\ s_j=1 \text{ or } s_j=2,\ j=1,\dots,k}} &(1+\kappa^{(1)}_{s_1,V}\varepsilon_{l_1})\cdots(1+\kappa^{(k)}_{s_k,V}\varepsilon_{l_k}) f_{l_1,\dots,l_k}\left(\xi_{l_1}^{(s_1)},\dots,\xi_{l_k}^{(s_k)}\right)\\ &=2^k f_{l_1,\dots,l_k}\left(\xi_{l_1}^{(\beta_{V,M}(1,l_1))},\dots,\xi_{l_k}^{(\beta_{V,M}(k,l_k))}\right),\end{aligned} \tag{D.11}$$

since the product $(1+\kappa^{(1)}_{s_1,V}\varepsilon_{l_1})\cdots(1+\kappa^{(k)}_{s_k,V}\varepsilon_{l_k})$ equals either zero or 2^k, and it equals 2^k for that sequence $(s_1,\dots,s_k)$ for which $\kappa^{(j)}_{s_j,V}\varepsilon_{l_j}=1$ for all $1\le j\le k$, and the relation $\kappa^{(j)}_{s_j,V}\varepsilon_{l_j}=1$ is equivalent to $\beta_{V,M}(j,l_j)=s_j$ for all $1\le j\le k$. (In relation (D.11) it is sufficient to consider only such products for which $l_j\ne l_{j'}$ if $j\ne j'$ because of the properties of the functions $f_{l_1,\dots,l_k}$.)

Beside this, $\xi_l^{\beta_{V,M}(l,j)}=\eta_l^{\alpha_V(j)}$ for all $1\le l\le n$ and $1\le j\le k$, and as a consequence

$$f_{l_1,\dots,l_k}\left(\xi_{l_1}^{(\beta_{V,M}(1,l_1))},\dots,\xi_{l_k}^{(\beta_{V,M}(k,l_k))}\right)=f_{l_1,\dots,l_k}\left(\eta_{l_1}^{(\alpha_V(1))},\dots,\eta_{l_k}^{(\alpha_V(k))}\right).$$

Summing up the identities (D.11) for all $1\le l_1,\dots,l_k\le n$ and applying the last identity we get relation (D.9), since the identity obtained in such a way holds for all $M\subset\{1,\dots,n\}$. □

Proof of Lemma D3. Let us first observe that if ξ is a real valued random variable with zero expectation, then $P(\xi\ge 0)\ge\frac{(E|\xi|)^2}{4E\xi^2}$ since $(E|\xi|)^2=4(E(\xi I(\{\xi\ge 0\}))^2\le 4P(\xi\ge 0)E\xi^2$ by the Schwarz inequality, where $I(A)$ denotes the indicator function of the set A. (In the above calculation and in the subsequent proofs I apply the convention $\frac{0}{0}=1$. We need this convention if $E\xi^2=0$. In this case we have the identities $P(\xi=0)=1$ and $E|\xi|=0$, hence the above proved inequality holds in this case, too.)

Given some $v\in B$, let us choose a linear operator κ such that $\|\kappa\|=1$, and $\kappa(v)=\|v\|$. Such an operator exists by the Banach–Hahn theorem. Observe that $\{\omega\colon \|v+Z(\omega)\|\ge\|v\|\}\supset\{\omega\colon \kappa(v+Z(\omega))\ge\kappa(v)\}=\{\omega\colon \kappa(Z(\omega))\ge 0\}$. Beside this, $E\kappa(Z)=0$. Hence we can apply the above proved inequality for $\xi=\kappa(Z)$, and it yields that $P(\|v+Z\|\ge\|v\|)\ge P(\kappa(Z)\ge 0)\ge\frac{(E|\kappa(Z)|)^2}{4E\kappa(Z)^2}$. Lemma D3 is proved. □

Proof of Lemma D4. Take the class of random polynomials

$$Y=\sum_{s=1}^{k}\sum_{\substack{(l_1,\dots,l_s)\colon 1\le l_j\le n,\ j=1,\dots,s,\\ l_j\ne l_{j'} \text{ if } j\ne j'}} b_s(l_1,\dots,l_s)\varepsilon_{l_1}\cdots\varepsilon_{l_s},$$

where ε_l, $1\le l\le n$, are independent random variables with $P(\varepsilon_l=1)=P(\varepsilon_l=-1)=\frac{1}{2}$, and the coefficients $b_s(l_1,\dots,l_s)$, $1\le s\le k$, are arbitrary real numbers. The proof of Lemma D4 can be reduced to the statement that there exists a constant $c_k>0$ depending only on the order k of these polynomials such that the inequality

$$(E|Y|)^2\ge 4c_kEY^2. \tag{D.12}$$

holds for all such polynomials Y. Indeed, consider the polynomial

$$Z=\sum_{s=1}^{k}\sum_{\substack{(l_1,\dots,l_s)\colon 1\le l_j\le n,\ j=1,\dots,s,\\ l_j\ne l_{j'} \text{ if } j\ne j'}} a_s(l_1,\dots,l_s)\varepsilon_{l_1}\cdots\varepsilon_{l_s},$$

and observe that $E\kappa(Z)=0$ for all linear functionals κ on the space B. Hence Lemma D3 implies that the left-hand side expression in (D.10) is bounded from below by $\inf\limits_{\kappa\in B'}\frac{(E|\kappa(Z)|)^2}{4E\kappa(Z)^2}$. On the other hand, relation (D.12) implies that $\inf\limits_{\kappa\in B'}\frac{(E|\kappa(Z)|)^2}{4E\kappa(Z)^2}\ge c_k$.

To prove relation (D.12) first we compare the moments EY^2 and EY^4. Let us introduce the random variables

$$Y_s=\sum_{\substack{(l_1,\dots,l_s)\colon 1\le l_j\le n,\ j=1,\dots,s,\\ l_j\ne l_{j'} \text{ if } j\ne j'}} b_s(l_1,\dots,l_s)\varepsilon_{l_1}\cdots\varepsilon_{l_s}\quad 1\le s\le k.$$

We shall show that the estimates of Chap. 13 imply that

$$EY_s^4\le 2^{4s}\left(EY_s^2\right)^2 \tag{D.13}$$

for these random variables Y_s.

Relation (D.13) together with the uncorrelatedness of the random variables Y_s, $1\le s\le k$, imply that

$$\begin{aligned}EY^4&=E\left(\sum_{s=1}^k Y_s\right)^4\le k^3\sum_{s=1}^k EY_s^4\le k^3 2^{4k}\sum_{s=1}^k (EY_s^2)^2\\&\le k^3 2^{4k}\left(\sum_{s=1}^k EY_s^2\right)^2=k^3 2^{4k}(EY^2)^2.\end{aligned}$$

This estimate together with the Hölder inequality with $p=3$ and $q=\frac32$ yield that

$$EY^2=E|Y|^{4/3}\cdot|Y|^{2/3}\le (EY^4)^{1/3}(E|Y|)^{2/3}\le k2^{4k/3}(EY^2)^{2/3}(E|Y|)^{2/3},$$

i.e. $EY^2\le k^3 2^{4k}(E|Y|)^2$, and relation (D.12) holds with $4c_k=k^{-3}2^{-4k}$. Hence to complete the proof of Lemma D4 it is enough to check relation (D.13).

In the proof of relation (D.13) we may assume that the coefficients $b_s(l_1,\dots,l_s)$ of the random variable Y_s are symmetric functions of the arguments $l_1,\dots,l_s$, since a symmetrization of these coefficients does not change the value of Y. Put

$$B_s^2=\sum_{\substack{(l_1,\dots,l_s)\colon 1\le l_j\le n,\ j=1,\dots,s,\\ l_j\ne l_{j'} \text{ if } j\ne j'}} b_s^2(l_1,\dots,l_s),\quad 1\le s\le k.$$

Then

$$EY_s^2=s!B_s^2,$$

and

$$EY_s^4\le 1\cdot3\cdot5\cdots(4s-1)B_s^4=\frac{(4s)!}{2^{2s}(2s)!}B_s^4$$

by Lemmas 13.4 and 13.5 with the choice $M = 2$ and $k = s$. Inequality (D.13) follows from the last two relations. Indeed, to prove formula (D.13) by means of these relations it is enough to check that $\frac{(4s)!}{2^{2s}(2s)!(s!)^2} \le 2^{4s}$. But it is easy to check this inequality with induction with respect to s. (Actually there is a well-known inequality in the literature, known under the name Borell's inequality, which implies inequality (D.13) with a better coefficient at the right hand side of this estimate.) We have proved Lemma D4. □

Let us turn back to the estimation of the probability $P(3 \cdot 2^{k-1}\|T_{n,k}(f)\| > u)$. Let us introduce the σ-algebra $\mathcal{F} = \mathcal{B}(\xi_l^{(1)}, \xi_l^{(2)}, 1 \le l \le n)$ generated by the random variables $\xi_l^{(1)}, \xi_l^{(2)}, 1 \le l \le n$, and fix some set $V \subset \{1, \dots, k\}$. I show with the help of Lemma D4 and formula (D.9) that there exists some constant $c_k > 0$ such that the random variables $T_{n,k} f(\ell))$ defined in formula (D.6) and $\hat{I}_{n,k,V}(f(\ell))$ defined in formula (D.8) satisfy the inequality

$$P\left(\|2^k \hat{I}_{n,k,V}(f(\ell))\| > \|T_{n,k}(f(\ell))\| | \mathcal{F}\right) \ge c_k \quad \text{with probability 1.} \tag{D.14}$$

In the proof of (D.14) we shall exploit that in formula (D.9) $2^k \hat{I}_{n,k,V}(f(\ell))$ is represented by a polynomial of the Rademacher functions $\varepsilon_1, \dots, \varepsilon_n$ whose constant term is $T_{n,k}(f(\ell))$. The coefficients of this polynomial are functions of the random variables $\xi_l^{(1)}$ and $\xi_l^{(2)}$, $1 \le l \le n$. The independence of these random variables from ε_l, $1 \le l \le n$, and the definition of the σ-algebra $\mathcal{F}$ yield that

$$\begin{aligned} &P\left(\|2^k \hat{I}_{n,k,V}(f(\ell))\| > \|T_{n,k}(f(\ell))\| | \mathcal{F}\right) \\ &\quad = P_{\varepsilon V}\left(\left\| \frac{1}{k!} \sum_{\substack{(l_1,\dots,l_k),\, (s_1,\dots,s_k):\\ 1 \le l_j \le n, s_j = 1 \text{ or } s_j = 2,\\ j=1,\dots,k,}} (1 + \kappa_{s_1,V}^{(1)} \varepsilon_{l_1}) \cdots (1 + \kappa_{s_k,V}^{(k)} \varepsilon_{l_k}) \right.\right. \\ &\qquad \left. f_{l_1,\dots,l_k}\left(\xi_{l_1}^{(s_1)}, \dots, \xi_{l_k}^{(s_k)}\right) \right\| \\ &\qquad \left. > \|T_{n,k}(f(\ell))(\xi_l^{(j)}, 1 \le l \le n, \; j = 1, 2)\| \right), \end{aligned} \tag{D.15}$$

where $P_{\varepsilon V}$ means that the values of the random variables $\xi_l^{(1)}, \xi_l^{(2)}, 1 \le l \le n$, are fixed, (their value depend on the atom of the σ-algebra $\mathcal{F}$ we are considering) and the probability is taken with respect to the remaining random variables ε_l, $1 \le l \le n$. At the right-hand side of (D.15) the probability of such an event is considered that the norm of a polynomial of order k of the random variables $\varepsilon_1, \dots, \varepsilon_n$ is larger than $\|T_{n,k}(f(\ell))(\xi_l^{(j)}, 1 \le l \le n, \; j = 1, 2)\|$. Beside this, the constant term of this polynomial equals $T_{n,k}(f(\ell))(\xi_l^{(j)}, 1 \le l \le n, \; j = 1, 2)$. Hence this probability can be bounded by means of Lemma D4, and this result yields relation (D.14).

The distributions of $I_{n,k,V}(f(\ell))$ and $\hat I_{n,k,V}(f(\ell))$ agree by the first statement of Lemma D2 and a comparison of formulas (D.4) and (D.8). Hence relation (D.14) implies that

$$\begin{aligned}
P\left(\|2^k I_{n,k,V}(f(\ell))\|\ge \frac13\cdot 2^{1-k}u\right)&=P\left(\|2^k \hat I_{n,k,V}(f(\ell))\|\ge \frac13\cdot 2^{1-k}u\right)\\
&\ge P\left(\|2^k \hat I_{n,k,V}(f(\ell))\|\ge \|T_{n,k}(f(\ell))\|,\ \|T_{n,k}(f(\ell))\|\ge \frac13\cdot 2^{1-k}u\right)\\
&=\int_{\{\omega:\,\|T_{n,k}(f(\ell))(\omega)\|\ge\frac13\cdot 2^{1-k}u\}} P\left(\|2^k \hat I_{n,k,V}(f(\ell))\|>\|T_{n,k}(f(\ell))\|\,|\mathscr{F}\right)dP\\
&\ge c_k P(3\cdot 2^{k-1}\|T_{n,k}(f(\ell))\|\ge u).
\end{aligned}$$

The last inequality with the choice of any set $V\subset\{1,\dots,k\}$, $1\le|V|\le k-1$, together with relation (D.7) imply formula (D.5).

We shall formulate an inductive hypothesis, and relation (D.3) will be proved together with it by means of an induction procedure with respect to the order k of the U-statistic. In the proof of this inductive procedure we shall apply the already proved relation (D.5). To formulate it some new quantities will be introduced.

Let $\mathscr{W}=\mathscr{W}(k)$ denote the set of all partitions of the set $\{1,\dots,k\}$. Let us fix k independent copies $\xi_1^{(j)},\dots,\xi_n^{(j)}$, $1\le j\le k$, of the sequence of random variables $\xi_1,\dots,\xi_n$. Given a partition $W=(U_1,\dots,U_s)\in\mathscr{W}(k)$ let us introduce the function $s_W(j)$, $1\le j\le k$, which tells for all arguments j the index of that element of the partition W which contains the point j, i.e. the value of the function $s_W(j)$, $1\le j\le k$, in a point j is defined by the relation $j\in V_{s_W(j)}$. Let us introduce the expression

$$I_{n,k,W}(f(\ell))=\frac1{k!}\sum_{(l_1,\dots,l_k):\,1\le l_j\le n,\ j=1,\dots,k} f_{l_1,\dots,l_k}\left(\xi_{l_1}^{(s_W(1))},\dots,\xi_{l_k}^{(s_W(k))}\right)$$

$$\text{for all } W\in\mathscr{W}(k).$$

An expression of the form $I_{n,k,W}(f(\ell))$, $W\in\mathscr{W}_k$, will be called a decoupled U-statistic with generalized decoupling. Given a partition $W=(U_1,\dots,U_s)\in\mathscr{W}_k$ let us call the number s, i.e. the number of the elements of this partition the rank both of the partition W and of the decoupled U-statistic $I_{n,k,W}(f(\ell))$ with generalized decoupling.

Now I formulate the following hypothesis. For all $k\ge2$ and $2\le j\le k$ there exist some constants $C(k,j)>0$ and $\delta(k,j)>0$ such that for all $W\in\mathscr{W}_k$ a decoupled U-statistic $I_{n,k,W}(f(\ell))$ with generalized decoupling satisfies the inequality

$$P(\|I_{n,k,W}(f(\ell))\|>u)\le C(k,j)P\left(\|\bar I_{n,k}(f(\ell))\|>\delta(k,j)u\right)$$

$$\text{for all } 2\le j\le k \text{ if the rank of } W \text{ equals } j. \tag{D.16}$$

It will be proved by induction with respect to k that both relations (D.3) and (D.16) hold for U-statistics of order k. Let us observe that for $k = 2$ relation (D.3) follows from (D.5). Relation (D.16) also holds for $k = 2$, since in this case we have to consider only the case $j = k = 2$. Relation (D.16) also holds in this case with $C(2,2) = 1$ and $\delta(2,2) = 1$. Hence we can start our inductive proof with $k = 3$. First I prove relation (D.16).

In relation (D.16) the tail-distribution of decoupled U-statistics with generalized decoupling is compared with that of the decoupled U-statistic $\bar{I}_{n,k}(f(\ell))$ introduced in (D.2). Given the order k of these U-statistics it will be proved by means of a backward induction with respect to the rank j of the decoupled U-statistics $I_{n,k,W}(f(\ell))$ with generalized decoupling.

Relation (D.16) clearly holds for $j = k$ with $C(k,k) = 1$ and $\delta(k,k) = 1$. If we already know that these relations hold up to $k-1$, then we prove first relation (D.16) for generalized decoupling U-statistics of order k with respect to backward induction for the rank $2 \le j < k$.

For this goal the following observation will be made. If the rank j of a partition $W = (U_1,\dots,U_j)$ satisfies the relation $2 \le j \le k-1$, then it contains an element with cardinality strictly less than k and strictly greater than 1. For the sake of simpler notation let us assume that the element U_j of this partition is such an element, and $U_j = \{t,\dots,k\}$ with some $2 \le t \le k-1$. The investigation of general U-statistics of rank j, $2 \le j \le k-1$, can be reduced to this case by a reindexation of the arguments in the U-statistics if it is necessary. Let us consider the partition $\bar{W} = (U_1,\dots,U_{j-1},\{t\},\dots,\{k\})$ and the decoupled U-statistic $I_{n,k,\bar{W}}(f(\ell))$ with generalized decoupling corresponding to this partition $\bar{W}$. It will be shown that our inductive hypothesis implies the inequality

$$P(\|I_{n,k,W}(f(\ell))\| > u) \le \bar{A}(k) P\left(\|I_{n,k,\bar{W}}(f(\ell))\| > \bar{\gamma}(k)u\right) \tag{D.17}$$

with $\bar{A}(k) = \sup\limits_{2\le p\le k-1} A(p)$, $\bar{\gamma}(k) = \inf\limits_{2\le p\le k-1} \gamma(p)$ if the rank j of W is such that $2 \le j \le k-1$, where the constants $A(p)$ and $\gamma(p)$ agree with the corresponding coefficients in formula (D.3).

To prove relation (D.17) (where $U_j = \{t,\dots,k\}$ is the last element of the partition W) let us define the σ-algebra $\mathscr{F}$ generated by the random variables appearing in the first $t-1$ coordinates of these U-statistics, i.e. by the random variables $\xi_{l_j}^{s_W(j)}$, $1 \le j \le t-1$, and $1 \le l_j \le n$ for all $1 \le j \le t-1$. We have $2 \le t \le k-1$. By our inductive hypothesis relation (D.3) holds for U-statistics of order $p = k-t+1$, since $2 \le p \le k-1$. I claim that this implies that

$$P(\|I_{n,k,W}(f(\ell))\| > u|\mathscr{F}) \le A(k-t+1) P\left(\|I_{n,k,\bar{W}}(f(\ell))\| > \gamma(k-t+1)u|\mathscr{F}\right) \tag{D.18}$$

with probability 1. Indeed, by the independence properties of the random variables $\xi_l^{s_W(j)}$ (and $\xi_l^{s_{\bar{W}}(j)}$), $1 \le j \le k$, $1 \le l \le n$,

$$P(\|I_{n,k,W}(f(\ell))\| > u|\mathscr{F}) = P_{\xi_l^{sW(j)},1\le j\le t-1}(\|I_{n,k,W}(f(\ell)\| > u)$$

and

$$P\left(\|I_{n,k,\bar W}(f(\ell))\| > \gamma(k-t+1)u|\mathscr{F}\right)$$
$$= P_{\xi_l^{sW(j)},1\le j\le t-1}(\|I_{n,k,\bar W} f(\ell)\| > \gamma(k-t+1)u),$$

where $P_{\xi_l^{sW(j)},1\le j\le t-1}$ denotes that the values of the random variables $\xi_l^{sW(j)}(\omega)$, $1\le j\le t-1$, $1\le l\le n$, are fixed, and we consider the probability that the appropriate functions of these fixed values and of the remaining random variables $\xi^{sW(j)}$ and $\xi^{s\bar W(j)}$, $t\le j\le k$, satisfy the desired relation. These identities and the relation between the sets W and $\bar W$ imply that relation (D.18) is equivalent to the identity (D.3) for the generalized U-statistics of order $2\le k-t+1\le k-1$ with kernel functions

$$f_{l_t,\dots,l_k}(x_t,\dots,x_k)$$
$$= \sum_{(l_1,\dots,l_{t-1}):\,1\le l_j\le n,\,1\le j\le t-1} f_{l_1,\dots,l_k}(\xi_{l_1}^{sW(1)}(\omega),\dots,\xi_{l_{t-1}}^{sW(t-1)}(\omega),x_t,\dots,x_k).$$

Relation (D.17) follows from inequality (D.18) if expectation is taken at both sides. As the rank of $\bar W$ is strictly greater than the rank of W, relation (D.17) together with our backward inductive assumption imply relation (D.16) for all $2\le j\le k$.

Relation (D.16) implies in particular (with the applications of partitions of order k and rank 2) that the terms in the sum at the right-hand side of (D.5) satisfy the inequality

$$P\left(D_k\|I_{n,k,V}(f(\ell))\| > u\right) \le \bar C(k,j)P\left(\|\bar I_{n,k}(f(\ell))\| > \bar D_k u\right)$$

with some appropriate $\bar C_k>0$ and $\bar D_k>0$ for all $V\subset\{1,\dots,k\}$, $1\le|V|\le k-1$. This inequality together with relation (D.5) imply that inequality (D.3) also holds for the parameter k.

In such a way we get the proof of relation (D.3) and its special case, relation (14.13). Let us prove formula (14.14) with its help first in the simpler case when the supremum of finitely many functions is taken. If $M<\infty$ functions $f_1,\dots,f_M$ are considered, then relation (14.14) for the supremum of the U-statistics and decoupled U-statistics with these kernel functions can be derived from formula (14.13) if it is applied for the function $f=(f_1,\dots,f_M)$ with values in the separable Banach space B_M which consists of the vectors $(v_1,\dots,v_M)$, $v_j\in B$, $1\le j\le M$, and the norm $\|(v_1,\dots,v_M)\| = \sup\limits_{1\le j\le m}\|v_j\|$ is introduced in it. The application of formula (14.13) with this choice yields formula (14.14) for this supremum. Let us emphasize that the constants appearing in this estimate do not depend on the number M. (We took only $M<\infty$ kernel functions, because with such a choice the Banach space B_M defined above is also separable.) Since the distribution of the random variables $\sup\limits_{1\le s\le M} \|I_{n,k}(f_s)\|$ converge to that of $\sup\limits_{1\le s<\infty} \|I_{n,k}(f_s)\|$, and the distribution of the random variables $\sup\limits_{1\le s\le M} \left\|\bar I_{n,k}(f_s)\right\|$ converge to that of $\sup\limits_{1\le s<\infty} \left\|\bar I_{n,k}(f_s)\right\|$ as $M\to\infty$, relation (14.14) in the general case follows from its already proved special case and a limiting procedure $M\to\infty$. □

Remark. The above proved formula (D.3) can be slightly generalized. It also holds if the expressions $I_{n,k}(f(\ell))$ and $\bar I_{n,k}(f(\ell))$ appearing in this inequality are defined in a more general way. Namely, they are the random functions introduced in formulas (D.1) and (D.2), but the sequences $\xi_1,\dots,\xi_n$ and their independent copies $\xi_1^{(j)},\dots,\xi_n^{(j)}$ in these formulas are independent random variables which may also be non-identically distributed. Such a generalization can be proved without any essential change in the original proof.

References

1. R. Adamczak, Moment inequalities for U-statistics. Ann. Probab. **34**, 2288–2314 (2006)
2. M. Ajtai, J. Komlós, G. Tusnády, On optimal matchings. Combinatorica **4**(4), 259–264 (1984)
3. K. Alexander, The central limit theorem for empirical processes over Vapnik–Červonenkis classes. Ann. Probab. **15**, 178–203 (1987)
4. M.A. Arcones, E. Giné, Limit theorems for U-processes. Ann. Probab. **21**, 1494–1542 (1993)
5. M.A. Arcones, E. Giné, U-processes indexed by Vapnik–Červonenkis classes of functions with application to asymptotics and bootstrap of U-statistics with estimated parameters. Stoch. Process. Appl. **52**, 17–38 (1994)
6. G. Bennett, Probability inequality for the sum of independent random variables. J. Am. Stat. Assoc. **57**, 33–45 (1962)
7. A. Bonami, Étude des coefficients de Fourier des fonctions de $L^p(G)$. Ann. Inst. Fourier (Grenoble) **20**, 335–402 (1970)
8. C. Borell, On the integrability of Banach space valued Walsh polynomials, in *Séminaire de Probabilités XIII*. Lecture Notes in Mathematics, vol 721 (Springer, Berlin, 1979), pp. 1–3
9. V.H. de la Peña, E. Giné, in *Decoupling. From Dependence to Independence.* Springer Series in Statistics. Probability and Its Application (Springer, New York, 1999)
10. V.H. de la Peña, S. Montgomery–Smith, Decoupling inequalities for the tail-probabilities of multivariate U-statistics. Ann. Probab. **23**, 806–816 (1995)
11. R.L. Dobrushin, Gaussian and their subordinated fields. Ann. Probab. **7**, 1–28 (1979)
12. R.M. Dudley, Central limit theorems for empirical measures. Ann. Probab. **6**, 899–929 (1978)
13. R.M. Dudley, A course on empirical processes, in *Lecture Notes in Mathematics* vol 1097 (Springer, New York, 1984), pp. 1–142
14. R.M. Dudley, *Real Analysis and Probability.* (Wadsworth & Brooks, Pacific Grove, 1989)
15. R.M. Dudley, *Uniform Central Limit Theorems* (Cambridge University Press, Cambridge, 1998)
16. E.B. Dynkin, A. Mandelbaum, Symmetric statistics, Poisson processes and multiple Wiener integrals. Ann. Stat. **11**, 739–745 (1983)
17. P. Frankl, J. Pach, On the number of sets in null-t-design. Eur. J. Combin. **4**, 21–23 (1983)
18. E. Giné, A. Guillou, On consistency of kernel density estimators for randomly censored data: Rates holding uniformly over adaptive intervals. Ann. Inst. Henri Poincaré PR **37**, 503–522 (2001)
19. E. Giné, S. Kwapień, R. Latała, J. Zinn, The LIL for canonical U-statistics of order 2. Ann. Probab. **29**, 520–527 (2001)
20. E. Giné, R. Latała, J. Zinn, Exponential and moment inequalities for U-statistics, in *High Dimensional Probability II.* Progress in Probability, vol 47 (Birkhäuser, Boston, 2000), pp. 13–38

P. Major, *On the Estimation of Multiple Random Integrals and U-Statistics*,
Lecture Notes in Mathematics 2079, DOI 10.1007/978-3-642-37617-7,

21. L. Gross, Logarithmic Sobolev inequalities. Am. J. Math. **97**, 1061–1083 (1975)
22. A. Guionnet, B. Zegarlinski, Lectures on Logarithmic Sobolev inequalities, in *Lecture Notes in Mathematics*, vol. 1801 (Springer, New York, 2003), pp. 1–134
23. D.L. Hanson, F.T. Wright, A bound on the tail probabilities for quadratic forms in independent random variables. Ann. Math. Stat. **42**, 52–61 (1971)
24. W. Hoeffding, A class of statistics with asymptotically normal distribution. Ann. Math. Stat. **19**, 293–325 (1948)
25. W. Hoeffding, Probability inequalities for sums of bounded random variables. J. Am. Math. Soc. **58**, 13–30 (1963)
26. K. Itô, Multiple Wiener integral. J. Math. Soc. Jpn. **3**, 157–164 (1951)
27. S. Janson, *Gaussian Hilbert Spaces* (Cambridge University Press, Cambridge, 1997)
28. E.L. Kaplan, P. Meier, Nonparametric estimation from incomplete data. J. Am. Stat. Assoc. **53**, 457–481 (1958)
29. R. Latała, Estimates of moments and tails of Gaussian chaoses. Ann. Probab. **34**, 2315–2331 (2006)
30. M. Ledoux, On Talagrand deviation inequalities for product measures. ESAIM Probab. Stat. **1**, 63–87 (1996). Available at http://www.emath./fr/ps/
31. M. Ledoux, The concentration of measure phenomenon, in *Mathematical Surveys and Monographs*, vol 89 (American Mathematical Society, Providence, 2001)
32. P. Major, Multiple Wiener–Itô integrals, in *Lecture Notes in Mathematics*, vol 849 (Springer, Berlin, 1981)
33. P. Major, On the tail behaviour of the distribution function of multiple stochastic integrals. Probab. Theory Related Fields **78**, 419–435 (1988)
34. P. Major, Asymptotic distributions for weighted U-statistics. Ann. Probab. **22**, 1514–1535 (1994)
35. P. Major, An estimate about multiple stochastic integrals with respect to a normalized empirical measure. Stud. Scientarum Mathematicarum Hung. **42**(3), 295–341 (2005)
36. P. Major, Tail behaviour of multiple random integrals and U-statistics. Probab. Rev. **2**, 448–505 (2005)
37. P. Major, An estimate on the maximum of a nice class of stochastic integrals. Probab. Theory Related Fields **134**, 489–537 (2006)
38. P. Major, A multivariate generalization of Hoeffding's inequality. Electron. Commun. Probab. **2**, (220–229) (2006)
39. P. Major, On a multivariate version of Bernstein's inequality. Electron. J. Probab. **12**, 966–988 (2007)
40. P. Major, L. Rejtő, Strong embedding of the distribution function under random censorship. Ann. Stat. **16**, 1113–1132 (1988)
41. P. Major, L. Rejtő, A note on nonparametric estimations. In the conference volume to the 65. birthday of Miklós Csörgő (1998), pp. 759–774
42. V.A. Malyshev, R.A. Minlos, *Gibbs Random Fields. Method of Cluster Expansion* (Kluwer, Academic, Dordrecht, 1991)
43. P. Massart, About the constants in Talagrand's concentration inequalities for empirical processes. Ann. Probab. **28**, 863–884 (2000)
44. H.P. Mc. Kean, Wiener's theory of non-linear noise, in *Stochastic Differential Equations*. SIAM-AMS Proceedings, vol. 6, (1973), pp. 197–209
45. E. Nelson, The free Markov field. J. Funct. Anal. **12**, 211–227 (1973)
46. I. Nourdin, G. Peccati, in *Normal Approximations with Malliavin Calculus: From Stein's Method to Universality*. Cambridge Tracts in Mathematics, vol 192 (Cambridge University Press, Cambridge, 2012)
47. D. Nualart, *Malliavin Calculus and Related Topics of Probability and Its Applications*, 2nd edn. (Springer, Berlin, 2006)
48. G. Peccati, M.S. Taqqu, *Wiener Chaos: Moments, Cumulants and Diagrams* (Springer, New York, 2010)
49. D. Pollard, *Convergence of Stochastic Processes* (Springer, New York, 1984)

50. G.-C. Rota, C. Wallstrom, Stochastic integrals: a combinatorial approach. Ann. Probab. **25**(3), 1257–1283 (1997)
51. D. Surgailis, On multiple Poisson stochastic integrals and associated Markov semigroups. Probab. Math. Stat. **2**(3), 217–239 (1984)
52. D. Surgailis, Long-range dependence and Appell rank. Ann. Probab. **28**, 478–497 (2000)
53. G. Szegő, *Orthogonal Polynomials*. American Mathematical Society Colloquium Publications, vol. 23 (American Mathematical Society, Providence, RI, 1967)
54. A. Takemura, Tensor Analysis of ANOVA decomposition. J. Am. Stat. Assoc. **78**, 894–900 (1983)
55. M. Talagrand, Sharper bounds for Gaussian and empirical processes. Ann. Probab. **22**, 28–76 (1994)
56. M. Talagrand, New concentration inequalities in product spaces. Invent. Math. **126**, 505–563 (1996)
57. M. Talagrand, *Spin Glasses: A Challenge for Mathematicians* (Springer, Berlin, 2003)
58. M. Talagrand, in *The General Chaining*. Springer Monographs in Mathematics (Springer, Berlin, 2005)
59. V.N. Vapnik, *The Nature of Statistical Learning Theory* (Springer, New York, 1995)
60. N. Wiener, The homogeneous chaos. Am. J. Math. **60**, 879–936 (1838)

Index

B

C

D

E

P. Major, *On the Estimation of Multiple Random Integrals and U-Statistics*, Lecture Notes in Mathematics 2079, DOI 10.1007/978-3-642-37617-7,

LECTURE NOTES IN MATHEMATICS

Springer

Edited by J.-M. Morel, B. Teissier; P.K. Maini

Editorial Policy (for the publication of monographs)

1. Lecture Notes aim to report new developments in all areas of mathematics and their applications - quickly, informally and at a high level. Mathematical texts analysing new developments in modelling and numerical simulation are welcome.
   Monograph manuscripts should be reasonably self-contained and rounded off. Thus they may, and often will, present not only results of the author but also related work by other people. They may be based on specialised lecture courses. Furthermore, the manuscripts should provide sufficient motivation, examples and applications. This clearly distinguishes Lecture Notes from journal articles or technical reports which normally are very concise. Articles intended for a journal but too long to be accepted by most journals, usually do not have this "lecture notes" character. For similar reasons it is unusual for doctoral theses to be accepted for the Lecture Notes series, though habilitation theses may be appropriate.

2. Manuscripts should be submitted either online at www.editorialmanager.com/lnm to Springer's mathematics editorial in Heidelberg, or to one of the series editors. In general, manuscripts will be sent out to 2 external referees for evaluation. If a decision cannot yet be reached on the basis of the first 2 reports, further referees may be contacted: The author will be informed of this. A final decision to publish can be made only on the basis of the complete manuscript, however a refereeing process leading to a preliminary decision can be based on a pre-final or incomplete manuscript. The strict minimum amount of material that will be considered should include a detailed outline describing the planned contents of each chapter, a bibliography and several sample chapters.
   Authors should be aware that incomplete or insufficiently close to final manuscripts almost always result in longer refereeing times and nevertheless unclear referees' recommendations, making further refereeing of a final draft necessary.
   Authors should also be aware that parallel submission of their manuscript to another publisher while under consideration for LNM will in general lead to immediate rejection.

3. Manuscripts should in general be submitted in English. Final manuscripts should contain at least 100 pages of mathematical text and should always include
   - a table of contents;
   - an informative introduction, with adequate motivation and perhaps some historical remarks: it should be accessible to a reader not intimately familiar with the topic treated;
   - a subject index: as a rule this is genuinely helpful for the reader.

   For evaluation purposes, manuscripts may be submitted in print or electronic form (print form is still preferred by most referees), in the latter case preferably as pdf- or zipped psfiles. Lecture Notes volumes are, as a rule, printed digitally from the authors' files. To ensure best results, authors are asked to use the LaTeX2e style files available from Springer's web-server at:

   ftp://ftp.springer.de/pub/tex/latex/svmonot1/ (for monographs) and
   ftp://ftp.springer.de/pub/tex/latex/svmultt1/ (for summer schools/tutorials).

Additional technical instructions, if necessary, are available on request from lnm@springer.com.

4. Careful preparation of the manuscripts will help keep production time short besides ensuring satisfactory appearance of the finished book in print and online. After acceptance of the manuscript authors will be asked to prepare the final LaTeX source files and also the corresponding dvi-, pdf- or zipped ps-file. The LaTeX source files are essential for producing the full-text online version of the book (see http://www.springerlink.com/openurl.asp?genre=journal&issn=0075-8434 for the existing online volumes of LNM). The actual production of a Lecture Notes volume takes approximately 12 weeks.

5. Authors receive a total of 50 free copies of their volume, but no royalties. They are entitled to a discount of 33.3 % on the price of Springer books purchased for their personal use, if ordering directly from Springer.

6. Commitment to publish is made by letter of intent rather than by signing a formal contract. Springer-Verlag secures the copyright for each volume. Authors are free to reuse material contained in their LNM volumes in later publications: a brief written (or e-mail) request for formal permission is sufficient.

Addresses:

Professor J.-M. Morel, CMLA,
École Normale Supérieure de Cachan,
61 Avenue du Président Wilson, 94235 Cachan Cedex, France
E-mail: morel@cmla.ens-cachan.fr

Professor B. Teissier, Institut Mathématique de Jussieu,
UMR 7586 du CNRS, Équipe "Géométrie et Dynamique",
175 rue du Chevaleret
75013 Paris, France
E-mail: teissier@math.jussieu.fr

For the "Mathematical Biosciences Subseries" of LNM:

Professor P. K. Maini, Center for Mathematical Biology,
Mathematical Institute, 24-29 St Giles,
Oxford OX1 3LP, UK
E-mail : maini@maths.ox.ac.uk

Springer, Mathematics Editorial, Tiergartenstr. 17,
69121 Heidelberg, Germany,
Tel.: +49 (6221) 4876-8259

Fax: +49 (6221) 4876-8259
E-mail: lnm@springer.com